Structured Peer-to-Peer Systems

Dmitry Korzun • Andrei Gurtov

Structured Peer-to-Peer Systems

Fundamentals of Hierarchical Organization, Routing, Scaling, and Security

Foreword by Ion Stoica

 Springer

Dmitry Korzun
Helsinki Institute for Information
    Technology
Aalto University
Aalto, Finland

Andrei Gurtov
Centre for Wireless
    Communications
University of Oulu
Oulu, Finland

Department of Computer Science
Petrozavodsk State University
Russia

ISBN 978-1-4899-8694-8   ISBN 978-1-4614-5483-0 (eBook)
DOI 10.1007/978-1-4614-5483-0
Springer New York Heidelberg Dordrecht London

Printed on acid-free paper

Springer is part of Springer Science+Business Media (www.springer.com)

To my parents Antonina and Zhorzh. DK
To my daughters Julia and Sofia, and wife
Anastasia. AG

Foreword

In the short span of a few years, peer-to-peer technologies have irreversibly changed the fields of networking and distributed systems. Today, the traffic delivered by peer-to-peer systems, such as Bittorrent, PPLive, and eDonkey, dominates the traffic of many Internet Service Providers. The structured peer-to-peer systems, in the form of distributed hash tables (DHTs), have been at the core of the modern large-scale key-value storage systems, including Amazon's DynamoDB, Facebook's Cassandra, and LinkedIn's Voldemort.

This book gives a comprehensive treatment of the structured peer-to-peer systems by covering their architecture, protocol design, and security properties. The book provides an excellent overview of the latest developments in the field and includes several recent contributions of the authors themselves. By using extensive and detailed examples, the authors succeed in making some of the most complex protocols accessible to a large audience.

This book is a great resource for both academics who are, or wish to start, doing research in the peer-to-peer field and practitioners who want to design and build large-scale distributed systems and protocols.

Ion Stoica

Preface

The main goal of this book is to cover important issues in optimization of present large-scale P2P systems as well as help in designing future systems. The book contains a comprehensive survey and summary of research results complemented by authors' own contributions to the field in the recent years. The book is extensively illustrated by examples to help the understanding as well as includes an extensive list of references on P2P articles and resources.

Evolution of P2P Systems

The field of structured P2P systems has seen a fast growth and evolution upon introduction of first distributed hash tables (DHTs) in the early 2000s. The first proposals including Chord, CAN, Pastry, and Tapestry were gradually improved to cope with scalability, locality, and security issues. Deployable as an overlay on the application layer without the need to change the network infrastructure, the P2P approach had opened great opportunities for innovation for developers. By utilizing the processing and bandwidth resources of end users, P2P approach enables high performance of data distribution which is hard to achieve with traditional client–server architectures. That enables commercial use of P2P systems such as distributing updates to the World-of-Warcraft virtual world, where patches over 100 MBytes are applied simultaneously to all users using P2P technology. Many popular social networks, such as Facebook, utilize the DHT internally to store tremendous numbers of key-value pairs.

Now P2P computing is a vast research field with multiple conferences and research groups in the area. The P2P computing is being actively utilized in the Internet for software updates, P2PSIP VoIP, video-on-demand, and distributed backups. Recent introduction of identifier–locator split proposal for Future Internet architectures poses another important application for DHTs, namely mapping between host permanent identity and changing IP address. The growing complexity and scale of modern P2P systems requires introduction of hierarchy and intelligence is routing of requests. Additionally, researchers proposed several anti-cheating mechanisms to ensure fair resource distribution and avoiding the "tragedy of commons." Popular P2P systems have been a subject of various attacks thus bringing security and resilience issues to the front.

Perspective on P2P evolution by Prof. Jon Crowcroft

Peer-to-peer systems have been around for a long time. Back in the 1970s, IBM developed the logical unit 6 of the systems network architecture, which incorporated peer-to-peer application communication support. Then things went quiet for a while, until the 1990s. Two different communities emerged. The nascent file sharing world developed largely unstructured peer-to-peer systems which allowed decentralised symmetric communication between uploaders and downloaders. Meanwhile in the academic world, two seminar projects, the Tapestry work in Berkeley and the Consistent hashing and Chord work at MIT, were based on a more structured approach with a view to more scalable operation in the longer term. At first, it seemed that unstructured systems were the way to go, but gradually the manageability of structured systems started to win out, particularly within the emerging large-scale cloud services.

In this book we see that at the very end where examples of commercial applications from Amazon's Dynamo and S3, Facebook's Cassandra system, Linkedin's Voldermort and even now, the trackerless versions of Bittorrent, all make use of DHTs for scalable key-value stores for a wide variety of uses and reasons. The book addresses the underlying tools and techniques of structured peer-to-peer systems ranging from the wide variety of routing approaches that can be taken through neighbour maintenance, localisation and optimisation of the system. The material is descriptive and analytical, affording the reader an understanding of the design principles and performance characteristics of the different approaches. The maturity of structured peer-to-peer is evident, and this book provides a clear guide to the trade-offs in selecting a system for a purpose.

Structure of the Book

The book starts with introductory part providing terminology, main problems in design of P2P systems, and mathematical notation. The description of classic DHTs is rather compact as it can be readily found in the existing books. Nevertheless, it should refresh the reader's knowledge of basic DHT topologies necessary for understanding material in other book parts.

The second part of the book considers P2P systems built on the principle of local knowledge, i.e., interactions with direct neighbors of a node in the system.

It introduces hierarchical structure to classic DHTs, clustering approaches, and ranking of node's neighbors according to their contribution to the system.

The third part of the book looks at the algorithms going beyond the local neighborhood of a node, such as "pure" hierarchical DHTs and their analytic modeling. We also consider the look-ahead routing where the node can utilize information on its neighbor's neighbors and a more general cyclic routing approach. This part also covers advanced techniques for ensuring fairness in resource distribution among the peers. Security aspects of P2P architectures, detection, and exclusion of malicious nodes are discussed as well.

The forth part of the book focuses on P2P applications. We describe CR-Chord, and extension of Chord utilizing Cyclic Routing for routing around malicious or overloaded nodes. Indirection infrastructures are another application which extends $i3$ for use as control plane for host identity protocol (HIP). We conclude this part by a chapter that describes commercial distributed systems used by companies such as Google and Amazon to run their services. Although such systems are not P2P in the traditional sense, many of them have evolved from P2P designs and still use DHT-style algorithms internally.

Throughout this book we use the following conceptual terms: *principle*, *property*, and *theorem*. A principle defines rules or guidelines to be applied in the network design; a solution that follows the principle achieves good properties. A property is a fact, typically empirical and intuitive about the subject; it either clarifies model requirements and assumptions or is very simply derived from them. A theorem is a profound fact derived for the given model; it formally characterizes the efficiency of the model. For majority of theorems we provide either a complete proof or, if the technical details are not very important for the discussion, a sketch with references to appropriate sources.

Intended Audience

The book is meant for advanced students and researchers interested in evolution of P2P systems. The undergraduate students can find an introduction to the basic P2P system and popular DHTs useful, while master and PhD students can find latest information on hierarchy in DHTs, resource ranking, and request routing. The book will be of great value to researchers working on DHTs as it contains novel results and analysis of existing DHT mechanisms, as well as for engineers and programmers implementing DHTs to their software, such as BitTorrent. A course or seminar on advanced P2P systems can be taught to postgraduate students based on this book.

While there are multiple books covering P2P networking and application, there is no good reference book for hierarchical P2P systems that would also cover fair resource allocation and efficient routing of requests. As the scale and complexity of P2P systems is growing, hierarchy becomes a critical factor to efficient operation.

The book is to a large extent based on authors' own scientific contributions as well as diligent survey of existing work in this area. The book fills the missing gap in the current P2P literature for students, researchers, and engineers.

About the Authors

Dmitry Korzun received his B.Sc. (1997) and M.Sc (1999) degrees in applied mathematics and computer science from the Petrozavodsk State University (Russia). He received a Ph.D. degree in physics and mathematics from the St.-Petersburg State University (Russia) in 2002. He is an associate professor at the Department of Computer Science of Petrozavodsk State University PetrSU, Russia (since 2003) and a part-time research scientist at the Helsinki Institute for Information Technology HIIT, Aalto University, Finland (since 2005). His research interests include analysis and evaluation of distributed systems, discrete modeling, ubiquitous computing in smart spaces, Internet of Things, software engineering, algorithm design and complexity, linear Diophantine analysis and its applications, theory of formal languages and parsing. His educational activity started in 1997 at the Faculty of Mathematics of PetrSU. Since that time he has taught more than 20 study courses on hot topics in computer science, applied mathematics, and information and communication technology. He is an author of more than 100 research and educational publications.

   Andrei Gurtov received his M.Sc (2000) and Ph.D. (2004) degrees in computer science from the University of Helsinki, Finland. He was appointed a professor at University of Oulu in the area of Wireless Internet in December 2009. He is also a principal scientist (on leave currently) leading the Networking Research group at the Helsinki Institute for Information Technology focusing on the host identity protocol and next generation Internet architecture. He is co-chairing the IRTF research group on HIP and teaches as an adjunct professor at the Aalto University and University of Helsinki. Previously, his research focused on the performance of transport protocols in heterogeneous wireless networks. In 2000–2004, he served as a senior researcher at Sonera Finland contributing to performance optimization of GPRS/UMTS networks, intersystem mobility, and IETF standardization. In 2003, he spent six months as a visiting researcher in the International Computer Science Institute at Berkeley working with Dr. Sally Floyd on simulation models of transport protocols in wireless networks. In 2004, he was a consultant at the Ericsson NomadicLab. Dr. Gurtov is a coauthor of over 100 publications including a book (Host Identity Protocol (HIP): Towards the Secure Mobile Internet, ISBN 978-0-470-99790-1, Wiley & Sons, June 2008. Hardcover, 332 pp.), book chapters, research papers, patents, and IETF RFCs. His publications received more than 1,000 citations according to Google Scholar. He is a senior member of IEEE.

Acknowledgments

We thank our colleagues and students for help in research on P2P topics that served as a base for several chapters. Our collaborators, for example, RWTH Aachen and Boeing, provided useful feedback on the papers.

We thank our employers, University of Oulu, Helsinki Institute for Information Technology HIIT, and Petrozavodsk State University for supporting creation of this book. Thanks to generous sponsors who funded our research: Academy of Finland, Tekes, TIVIT, NLnet, Google Inc, Ericsson, and other companies in Finland. This work was supported in part by Academy of Finland project SEMOHealth.

We would like to thank personnel of Springer for help in preparing and publishing of manuscript. We thank Prof. Ion Stoica for kindly agreeing to write a foreword for this book and Prof. Jon Crowcroft for introducing his section with perspective on P2P evolution in the preface.

We appreciate comments from anonymous reviewers that aided us in improving the manuscript focus and organization. Feedback from the Internet Research Task Force was helpful in shaping the book's content.

Thanks to our extended families for their support and understanding of the time and effort it took to complete the book.

Helsinki, Finland   Dmitry Korzun
Oulu, Finland   Andrei Gurtov

Contents

Abbreviations

ACID	Atomicity, Consistency, Isolation, Durability
ANLDE system	Associated with CCF-grammar NLDE system
API	Application Programming Interface
AS	Autonomous System
BT	BitTorrent
BGP	Border Gateway Protocol
CAN	Content Addressable Network
CAP	Cluster-Based Architecture
CCF	Commutative Context Free grammar
CR	Cyclic Routing
DDOS	Distributed Denial of Service attack
DHT	Distributed Hash Table
DNS	Domain Name System
DOLR	Decentralized Object Location and Routing
DOS	Denial of Service attack
GRM	Group Membership Rendezvous
GTPP	General Truncated Pyramid P2P architecture
HDHT	Hierarchical Distributed Hash Table
Hi3	Host Identity Indirection Infrastructure
HIIT	Helsinki Institute for Information Technology
HIP	Host Identity Protocol
HIT	Host Identity Tag
homNLDE	homogenous Nonnegative Linear Diophantine Equation
HTML	Hypertext Markup Language
i3	Internet Indirection Infrastructure
ID	Identifier
IETF	Internet Engineering Task Force
IP	Internet Protocol
JXTA	JuXTApose, an acronym for a programming language and platform independent Open Source protocol started by Sun Microsystems for P2P networking

LSI	Latent Semantic Indexing
NAT	Network Address Translation
NLDE	Nonnegative Linear Diophantine Equation
NoN	Neighbor of Neighbor
OSPF	Open Shortest Path First
PDF	Probability Density Function
PNS	Proximity Neighbor Selection
P2P	Peer-to-Peer
P2PSIP	Peer-to-Peer Session Initiation Protocol
PRR	Plaxton, Rajaraman, Richa tree
RDBMS	Relational Database Management System
REST	Representational State Transfer
RIP	Routing Information Protocol
RPC	Remote Procedure Call
S3	Simple Storage Service
SIP	Session Initiation Protocol
SLAC	Selfish Link-based Adaptation for Cooperation
SOAP	Simple Object Access Protocol
SQL	Structured Query Language
TCP	Transmission Control Protocol
TFT	Tit-for-tat
TTL	Time To Live
URL	Uniform Resource Locator
VoIP	Voice over IP
WWW	World Wide Web
XML	Extensible Markup Language

Part I
Introduction

Overview of Part I

Structured P2P systems have been an active research field since the 1990s. The achieved maturity of theoretical P2P development is such that it is possible to affirm the existence of effective technology. It is used today in practice for constructing complex real-life applications and services. For instance, the key technique of structured P2P systems—Distributed Hash Tables (DHT)—are presently used in several large-scale distributed systems in the Internet, such as Amazon Simple Storage Service (Amazon S3), Peer-to-Peer SIP (P2PSIP), and BitTorrent. Furthermore, DHTs are envisaged as a key mechanism to provide identifier-locator separation for mobile hosts in Future Internet.

This part provides a basic introduction to P2P systems and to problems of their design. Tremendously many papers exist on concrete P2P designs and techniques, on their core models and algorithms, and on theoretical properties and experimental study conclusions. Most essential theoretical and practical achievements have been already summarized in various surveys and books. We present our terminology and mathematical notation to be used in this book and provide appropriate references to the existing literature. In parallel, fundamental problems that almost any P2P system design meets are defined, and the discussion relates them with the dedicated topics of the book: hierarchical organization, routing, scaling, and security.

Chapter 1 introduces basic concepts and terminology of structured P2P systems as well as presents our formal notation for this book. We provide common understanding about intuition and reasons behind the term "structured P2P system" and focus the discussion on theoretical properties and practical issues of such systems.

Chapter 2 complements the introduction providing concrete examples of designs for existing classical DHTs. It focuses on network topology structures that DHT designs use. We consider how such structures are related with the corresponding classes of graphs and what algorithms are supported for routing.

Chapter 1
Terminology, Problems, and Design Issues

Abstract The aim of this chapter is to introduce first the basic concepts, terminology, and notation we use throughout this book. Although the term "structure" is central in our discussion on structured P2P networks, its definition is quite fuzzy in recent P2P literature. Secondly, the chapter aims at common understanding about intuition and reasons behind this term. The description is assisted with introduction to a set of challenging problems that the P2P networking faces today.

1.1 Introduction

A P2P network is a distributed system in nature, without any centralized control. Nodes with symmetrical client-server roles form a self-organizing overlay network on top of the Internet Protocol (IP) network. A number of topologies have been proposed for structured P2P networks in the literature, e.g., see [4, 23, 40, 41, 46, 52, 54, 56, 62, 77] and our brief overview in Chap. 2. All they share common "structural" properties. The most widespread properties are low degree, low diameter, and fair load balance. They make the P2P network scalable: when its size increases (e.g., the number of alive nodes or shared resources) the efficiency (e.g., lookup performance or security) is preserved.

Distributed Hash Table (DHT) is known a primary concept for making structure in P2P overlay networks. Since there is no centralized control in a P2P system, its participants have to self-organize by maintaining control data locally and by performing control operations for the collaboration. It leads to the routing problem in the global P2P network. Section 1.3 introduces basics of DHT network, forming the "classical" understanding of the P2P network structure.

The characteristic property of any P2P network is the symmetry of participants and of their roles. Nevertheless, the homogeneity assumption is unrealistic for P2P systems and many P2P designs apply the hierarchical approach to deal with

D. Korzun and A. Gurtov, *Structured Peer-to-Peer Systems: Fundamentals of Hierarchical Organization, Routing, Scaling, and Security*, DOI 10.1007/978-1-4614-5483-0_1, © Springer Science+Business Media New York 2013

the heterogeneity. Section 1.4 shows that advanced P2P network structures appear because of hierarchy-based solutions to the heterogeneity problem.

In contrast to many other decentralized systems, P2P participants have only partial information about the entire network. The lack of knowledge can be compensated if the network has certain structure appropriate to extrapolation of available local knowledge to the scale of the whole system. Section 1.5 explains that the variety of P2P network structures are due to the local knowledge property.

1.2   Mathematical Preliminaries

The sets of real numbers and integer numbers are denoted $\mathbb{R}$ and $\mathbb{Z}$, respectively. Their non-negative parts are $\mathbb{R}_+$ and $\mathbb{Z}_+$. In n-dimensional spaces $\mathbb{R}^n$ and $\mathbb{Z}^n$, the zero vector is $\mathbf{0}$. A standard unit vector is $\mathbf{e}_i$ where all components are zeros abut the i's one that holds a value of 1. The vector with all components equal to 1 is $\mathbf{1}$. The vector comparison is component-wise: we write $x \leq y$ iff $x_i \leq y_i$ for $i = 1, 2, \ldots, n$.

The algorithm complexity analysis uses asymptotic methods. Let $f(x)$ and $g(x)$ be two functions defined on some upper-unbounded subset of $\mathbb{R}$ and take values in $\mathbb{R}_+$. (The case of $\mathbb{Z}$ is similar.) Our asymptotic notation is summarized in the following table.

Notation	Definition	Intuition
$f = O(g)$	There are $C \in \mathbb{R}_+, C \neq 0$, and $x_0 \in \mathbb{R}$ such that $f(x) \leq Cg(x)$ for all $x > x_0$.	f is bounded above by g (up to constant factor).
$f = \Theta(g)$	There are $C_1, C_2 \in \mathbb{R}_+, C_1 \neq 0$ and $C_2 \neq 0$, and $x_0 \in \mathbb{R}$ such that $C_1 g(x) \leq f(x) \leq C_2 g(x)$ for all $x > x_0$.	f grows similarly to g (within constant bounds).
$f = \Omega(g)$	There are $C \in \mathbb{R}_+, C \neq 0$, and $x_0 \in \mathbb{R}$ such that $f(x) \geq Cg(x)$ for all $x > x_0$.	f is bounded below by g (up to constant factor).
$f = o(g)$	For any $\varepsilon > 0$ there is $x_0 \in \mathbb{R}$ such that $f(x) \leq \varepsilon g(x)$ for all $x > x_0$.	f is dominated by g.
$f = \omega(g)$	For any $C > 0$ there is $x_0 \in \mathbb{R}$ such that $f(x) \geq Cg(x)$ for all $x > x_0$.	f dominates g.

We write $f(x) = \text{polylog}\, x$ if f is a polynomial on $\log x$. The fact $f(x) = O(\log x)$ means that $f(x) = \alpha \log x$ in the worst case. Since the constant $\alpha > 0$ can always be reflected in the logarithm base, we use $f(x) = \log x$ when the base is not essential.

The following notation is opportune in model assumptions. Given $x, y \in \mathbb{R}$, we write $x \ll y$ (resp. $x \gg y$) if the values of x and y differ significantly by the meaning of the problem domain. A possible formalization is exponent-based: assuming $x \geq 0$ and $y > 1$, then $x \ll y$ if $x < y^\alpha$ for some $0 < \alpha < 1$. Similarly, we use $x \approx y$ when the problem domain provides meaning for the closeness.

Given a random variable X, we denote the expectation $E[X]$. In the continuous case, it can be computed as $E[X] = \int_{x \in \mathbb{R}} x f(x) dx$, where $f(\cdot)$ is the probability density function of X. In the discrete case, $\mathsf{E}[X] = \sum_i x_i p_i$, where $p_i = \Pr\{X = x_i\}$ is the probability distribution of X, $\sum_i p_i = 1$. Let A be an event in a N-element system where N is large. We say that A happens with high probability if $\Pr\{A\} = 1 - \Theta(N^{-c}) + o(N^{-c})$ for a given constant $c \geq 1$. A widespread particular case is $\Pr\{A\} = 1 - O(1/N)$.

Alphabet Σ is a finite non-empty set of symbols. A string over Σ is $\alpha = a_1 a_2 \cdots a_n$, where $a_i \in \Sigma$, $1 \leq i \leq n$, $n \geq 0$. The string length $n = |\alpha|$ is the number of symbols in the string. The empty string ε has zero length. Denote Σ^n and Σ^+ the set of all n-length strings and the set of all positive-length strings over Σ, respectively. If $\alpha = \beta\gamma$ then β is a prefix of α and γ is a suffix of α.

1.3 Distributed Hash Tables

A P2P system can be broadly defined as a distributed system with no centralized infrastructure. Participants are called nodes (also known as peers) and they are symmetric in their role in the system. Background on P2P systems in general, their application aspects and characteristics, and the benefits they provide can be found in existing books [8, 9, 65]. Key features of P2P systems are decentralization, self-organization, dynamism, load-balancing, and fault-tolerance. It makes P2P systems very attractive to use in many areas, including such prominent paradigms as ubiquitous computing, cloud computing, and Internet of Things.

Nodes of a P2P system form a network. Its topology is the layout pattern of interconnections between the nodes. In networking, topology is typically understood as the structure of a network. The early class of P2P overlays consists of unstructured systems where the topology follows no special structure. Given a search query, a node does not have any information about which its neighbors may best be able to resolve further the query, Thus, blind and random strategies for search probes are applied. Consequently, search queries are flooded widely, and their cost becomes enormous when the number of nodes is large.

The opposite class—structured P2P overlays—aims at more efficient searching that guarantees location of a target within a moderate number of hops. The fundamental point is that the location of data (resources) is related to the network topology, the set of nodes receiving a particular query becomes related to the content of the query. Instead of "blind" search probes, a node can arrange its neighbors and select the best ones to resolve the query effectively. Consequently, the efficiency is preserved even with massive numbers of participants.

Deployment of a P2P system requires an underlying network, which the nodes use as a communication base, so forming an overlay. A popular class for underlying networks is IP-based networks, e.g., P2P nodes are located at hosts distributed over the Internet. One node can contact another node using communication primitives and infrastructure of the IP.

Table 1.1 Symbol notation for a structured P2P network

Notation	Description
D	Resource key space (application-specific). Resource distribution can be non-uniform and with semantic structure.
S	Node ID space. Typically, IDs are numeric (scalar or vector). Let u, v, and w stand for P2P nodes and their IDs.
R	The number of resources (data items or data keys) stored (and available) in the overlay. In some contexts, R also denotes the full set of currently available resources.
N	The number of alive nodes in overlay. In some contexts, N also denotes the full set of alive nodes.
u, v, w	P2P nodes by their IDs.
$\rho(u,v)$	Distance metric in S that satisfies (i) $\rho(u,v) > 0 \; \forall u, v \in S, u \neq v$, (ii) $\rho(u,u) = 0$ $\forall u \in S$. The symmetry property and the triangle inequality are optional.
N_{uv}	The set of all nodes in between u and v (according to the metric ρ).
$S(u)$	The set of keys that a node u is responsible for.
T_u	Routing table (neighbors) of a node $u \in N$. Although it consists of pairs (v, IP_v) writing $v \in T_u$ does not lead to confusion.
$u \rightarrow^+ v$	Multi-hop overlay path $u \rightarrow w_1 \rightarrow w_2 \rightarrow \cdots \rightarrow w_{l-1} \rightarrow v$.
$\tau(u,v)$	Distance metric in network, e.g., the number of overlay hops or the sum latency of $u \rightarrow^+ v$.

Table 1.1 summarizes our basic notation for structured P2P overlay networks. Consider a P2P overlay of N nodes. We also denote N the set of nodes when it does not lead to confusion. In the following, the term "node" refers both the node and its ID. Let node IDs be assigned from a space S with a distance metric ρ. If $\rho(u,d) > \rho(v,d)$ then v is closer to d. In other words, v is in between u and d. Examples of space distance metric are the Euclidean distance adopted in CAN [54], the length of the common prefix adopted in Pastry [56] and Tapestry [77], and the clockwise distance on the integer ring $[0, 2^m)$ mod 2^m adopted in Chord [62].

Let N_{uv} be the number of nodes in between u and v. A widespread assumption for flat P2P designs is that the mapping from N to S distributes nodes uniformly in S and N_{uv} is proportional to $\rho(u,v)$. For instance, cryptographic secure hash functions can be used to map consistently arbitrary string names into m-digit values [28].

Each overlay node u maintains a local routing table T_u of entries (v, IP_v), where v is a neighbor and IP_v is its IP address. In a dynamic environment, nodes collaborate and adopt their routing tables to an up-to-date state. The number of neighbors $|T_u|$ is the node degree. Typically $|T_u| \ll N$, i.e., only local knowledge about the overlay network is available at u.

At the system level, overlay network topology can be modeled as a graph, embedded into S. The graph can be directional, when $v \in T_u$ does not necessarily lead to $u \in T_v$. It means that u processes incoming requests from some $w \in N$ for $w \notin T_u$, and the node in-degree $|\{w \mid u \in T_w, w \in N\}|$ must be used for characterizing the load of u, in addition to the node out-degree $|T_u|$.

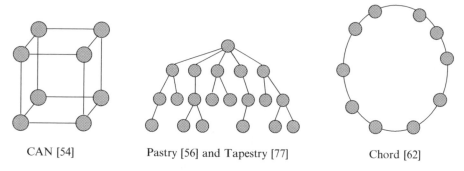

CAN [54]   Pastry [56] and Tapestry [77]   Chord [62]

Fig. 1.1 Examples of network topology classes for some popular DHTs

In a structured P2P system, its topology is tightly controlled via specific selection of neighbors. Typically, it is conditioned on the tradeoff between the node degree and the diameter of the global topology graph. There are two variants of neighbor selection: deterministic and randomized. In deterministic overlays, the only variant of T_u is possible at u for a given N. In randomized overlays, u may consider several candidates when selecting a neighbor. A P2P design implements one of these variants in its maintenance algorithm, which repairs the topology graph in the face of network dynamics.

A P2P system takes care of distributed assigning and locating resources, i.e., the standard operations "put" and "get" implement a mapping from D to N, associating resources with participating nodes. For efficiency, structured P2P systems use the Distributed Hash Table (DHT) as a substrate for the lookup service [5]. A DHT can be thought as a distributed "indexing" mechanism to resolve search queries. It takes (k, data) value as input ($k \in D$), hashes k to establish a mapping from keys to nodes (the lookup service), then assigns the data with the node identified. Such indexing essentially uses the structural network topology properties, and any DHT has to inevitably preserve the predefined structure, see Fig. 1.1 for illustration. Examples of DHT designs include CAN [54] and Symphony [41], Chord [62] and Kademlia [46], Pastry [56] and Tapestry [77], Koorde [27] and Broose [15], Viceroy [40] and Pappilon [1].

Consider a typical flat DHT. The resource key space D and node ID space S coincide. Each resource has a unique key $k \in S$; DHT implements hashing the key to the node ID. Resources are assigned to nodes deterministically; the node d with the closest ID is responsible for a key k (greedy strategy),

$$d = \arg\min_{u \in N} \rho(u, k). \tag{1.1}$$

That is, at any instant, d is responsible for a bucket of keys, $S(d) \subset S$,

$$S(d) = \{k \in S \mid \rho(d, k) < \rho(u, k) \; \forall u \in N, u \neq d\}.$$

Note that grouping the resources into buckets is not hierarchical since all nodes are equal, and their buckets do not form an ordered set.

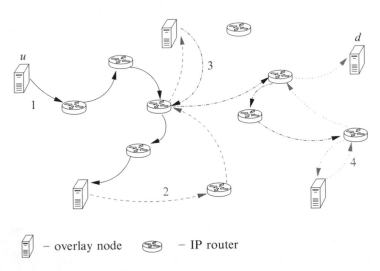

$\boxed{}$ – overlay node $\otimes$ – IP router

Fig. 1.2 Multi-hop routing in a P2P overlay network

Some DHTs additionally replicate the same resource among several nodes (e.g., in $\{u \mid \rho(u,k) < \alpha\rho(d,k)\}$ for some $\alpha > 1$), hence there can be responsible nodes further to the primary responsible node d.

The DHT lookup service searches the node closest to a given ID. Let a node u initiate a lookup for a given key k. The result is an IP address of some responsible node d. Since T_u does not contain all nodes, the lookup needs multi-hop overlay routing from u to d, where d is unknown beforehand. Each intermediate node w finds the next-hop node $v \in T_u$ to forward the lookup, thus forming the one-hop path $w \to v$ in the overlay network. Eventually, an l-hop path $u \to^+ d$ is constructed for $l = |u \to^+ d| \geq 1$:

$$u \to w_1 \to w_2 \to \cdots \to w_{l-1} \to d. \tag{1.2}$$

For the convenience, denote $w_0 = u$ and $w_l = d$. This routing process is depicted in Fig. 1.2. One hop in the overlay leads to several hops in the underlying network.

Depending on who makes forwarding—the source node itself or every intermediate node—routing algorithms are classified onto iterative and recursive. Iterative routing provides more control to the source (e.g., for security). Recursive routing has lower lookup latency (no extra communication between u and w_i) and is more resilient to IP connectivity restrictions (e.g., a firewall can prohibit direct IP communication between u and w_i).

A DHT routing algorithm is essentially based on the properties of the metric space (S, ρ). In progressive routing, lookup path (1.2) consists of nodes whose IDs are progressively closer to d, i.e., $\rho(w_i, d) < \rho(w_{i-1}, d)$. Since d is unknown beforehand, progressive routing applies the key-based criterion

$$\rho(w_i, k) < \rho(w_{i-1}, k), \quad i = 1, 2, \ldots, l-1. \tag{1.3}$$

Greedy routing is the extreme instance of progressive routing when each next-hop node w_i is the neighbor closest to k,

$$w_i = \arg\min_{u \in T_{w_{i-1}}} \rho(u,k), \quad i = 1, 2, \ldots, l-1. \tag{1.4}$$

Greedy criterion (1.4) makes the locally best choice among all possible ones by progressive criterion (1.3). Note that (1.4) is a local variant of (1.1) since the minimization is restricted within local routing tables $T_{w_{i-1}} \subset N$.

Multi-hop routing imposes certain load to intermediate nodes, in addition to the source and destination. An overlay serves many lookups in parallel, and the total load is distributed among nodes and their links to neighbors. A common test case is uniform all-to-all load [68]: for each pair of nodes $u \neq v$, u initiates a lookup to v and there are $N(N-1)$ lookups in total. A network is node(link)-congestion-free if every node (link) has the equal load.

The lookup availability depends on the number of alternative paths between a source and destination [38, 57, 61]. The more paths, the more chances to go around failed or malicious nodes. However, a mechanism for finding paths consisting of good nodes is needed. The straightforward approach exploits multi-path routing when nodes multicast messages to several neighbors [10, 24, 34, 76], either in parallel or sequentially. It has two disadvantages. First, a lot of duplicate messages are generated; many of them are redundant due to the local route selection (non-optimal). Second, paths converge in some DHTs [20, 57, 76], and bypassing malicious nodes becomes impossible.

The DHT lookup service is a base for self-organizing to adapt to the node dynamics. When a node u joins, it contacts any existing overlay node to lookup for u (the node ID acts as a key). According to (1.1), the lookup returns the closest neighbor v, identifying the location of u in the overlay. Then u communicates with v to initiate T_u and u's resource storage as well as to disseminate the update information. When a node u leaves, it notifies its neighbors and passes its resources to new responsible nodes. In open and dynamic environments, nodes generally leave the system far less gracefully due to network failures and selfish or malicious reasons. Temporal leaves are possible, e.g., because of short-term breaks in the IP connectivity.

To cope with dynamics, nodes perform the maintenance control keeping the overlay correctly connected. The strategies are divided into proactive and reactive. In the proactive case, a node regularly checks its neighbors, e.g., by pinging them periodically. In the reactive case, existing overlay communications are used to detect update events, e.g., a neighbor has not responded for lookup by predefined time. When a node u detects that its neighbor v has left, u performs the actions similar as if v notifies about its voluntary leave.

The properties of structured DHT-based overlays lead to fundamental design tradeoffs. The well-known tradeoff problem is routing table size vs. network diameter [68], or $|T_u|$ vs. $|u \to^+ d|$. Clearly, the larger $|T_u|$ the shorter $|u \to^+ d|$. Many popular DHTs, such as Chord [62] and Pastry [56], provide $|T_u| = O(\log N)$

and $|u \rightarrow^+ d| = O(\log N)$, or in other words, $O(\log N)$ state and $O(\log N)$ routing.[1]
Better tradeoffs are possible, e.g., $O(1)$ state and $O(\log N)$ routing in Viceroy [40]
and Ulysses [68], or $O(\sqrt{N})$ state and $O(1)$ routing in Kelips [22] and 2h-Calot [63].
However, they have more complicated network topologies or large routing tables,
hence the maintenance cost becomes an issue [36].

Selecting the best tradeoff becomes a challenging problem in the case of high
dynamics and heterogeneity. The former is the P2P characteristic property that
distinguishes P2P systems from many other decentralized systems; the property
is also known as churn—the system experiences rapid membership changes. The
heterogeneity is a property of modern networking environments, inevitable in
practice, see Sect. 1.4.

The routing distance $\tau(u,d)$ is one of the key performance indexes in the above
tradeoffs. It characterizes the efficiency of (1.2). For example, $\tau(u,d) = |u \rightarrow^+ d|$.
More accurate metrics take into account the latency due to the underlying network,
e.g., $\tau(u,d) = |u \rightarrow^+ d|_\tau = \sum_{i=0}^{l} \tau(w_i, w_{i+1})$ is the total lookup latency for (1.2).

1.4   Heterogeneity and Hierarchy

Flat DHT designs tend to treat nodes equally and offer a flat structure of resource
key space with the uniform partition among participants. That has certain advantages
such as even distribution of workload among nodes.

1.4.1   Consistent Hashing and Uniform Partitioning

A flat DHT uses the concept of consistent hashing to partition the resource items
(keys) R among a distributed set of nodes. The concept is a special kind of hashing
originally devised by Karger et al. [28]. A DHT client uses a consistent hash function
to map a resource item $k \in R$ to one of the nodes in its local view.

The key property of consistent hashing is the "Smoothness" property. It implies
that smooth changes in the set of nodes are matched by a smooth change in the
location of the resources. When a node is added to or removed from the system, the
expected fraction of resource items that must be moved is the minimum needed to
maintain a balanced load across the nodes. Actually, if N changes then only R/N
items need to be remapped on average.

In many traditional hash tables changes of N are very expensive. That is, a
common way of load balancing N nodes is to map item $k \in R$ to node $h(k) \mod N$

---

[1] We do not specify such details like "in the worst case", "with high probability", or "expected"
since they are minor in this high-level discussion.

using some standard hash function h. When nodes are joining or leaving, the hash range N changes, and almost every resource would be hashed to a new location.

In contrast, consistent hashing maps resources to the same node, as far as possible. When a node u joins the system, it takes its share of resources from all the other nodes. When u leaves, its resources are shared among the remaining nodes.

Other properties of consistency hashing are the following.

1. The "Spread" property implies that the total number of different nodes to which a resource item is assigned is small. Distributing the item to this small set will insure access for all clients, without using a lot of storage in the system. Typically, the node set size is assumed small if it is logarithmic on N.
2. The "Load" property implies that any particular node is not assigned an unreasonable number of resources. Typically, the load is assumed reasonable if its close to the average R/N.
3. The "Balance" property implies that items are distributed to nodes randomly. In other words, the probability of $k \in R$ to be assigned to $u \in N$ is $1/N$.
4. The 'Monotonic' property implies that when a node joins the system, only the items assigned to the new node are reassigned.

The main idea behind a rich family of the consistent hashing schemes is to hash both resources and nodes using the same unit interval $D = S = [0, 1)$. A hash function h_r maps resource items randomly, and a hash function h_n does the same for nodes. A node u is assigned an interval $S(u) \subset S$ ($u \in S(u)$) that contains "closest" resource keys to u. If u leaves then the interval $S(u)$ is taken over by a node with an adjacent interval. All the remaining nodes are unchanged.

1.4.2 Inaccuracy of the Homogeneity Assumption

The homogeneity assumption, however, is mostly violated in the P2P world. De jure equal nodes must be de facto of different responsibilities. There are two major issues concerning the heterogeneity in P2P networks. First, the participating nodes are of different capacities. Consequently, a non-flat design can achieve better efficiency by assigning responsibility proportional to the capacity. Second, the environment and domain requirements are heterogeneous, e.g., D is not uniform due to data semantics or the underlying network exhibits non-flat structure with hierarchical domain organization. Consequently, a non-flat design can optimize the performance by introducing appropriate data distribution among the nodes or by adopting the routing algorithm to the underlying network properties.

Empirical studies such as [33] have shown that diversity exists among nodes in their capacity and behavior. Many nodes are not stable and have short session times; they do not contribute much resources to the system. This behavior is typical for regular nodes. A few nodes are more stable and powerful than regular ones. They are supernodes and able to perform more burden such as keeping more resources, serving more requests, performing more maintenance. Flat DHTs do not exploit the heterogeneity and assign equal duties to heterogeneous nodes, which is inefficient.

Internet has several levels of its organization. That is, routing between Autonomous Systems (ASes) uses BGP; OSPF, RIP and some other protocols support intra-AS routing. This global hierarchy allows scalable architectures and administrative autonomy. It is unused in flat DHTs, which intentionally assume that there is no "higher" or "lower" nodes. It is a basic reason of high latency and other types of degradation observable experimentally in flat DHT deployments.

Hierarchy is a well-known concept to deal with the heterogeneity problem in all its various manifestations. We assume the following broad definition.[2] A *hierarchy is an arrangement of items (objects, names, values, categories, etc.) in which the items are represented as being "above", "below", or "at the same level as" one another and with only one "neighbor" above and below each level. These classifications are made with regard to rank, importance, seniority, power status or authority. Abstractly, a hierarchy is simply an ordered set or an acyclic graph.*

In the traditional sense, a P2P hierarchy organizes nodes into a tree-like structure according to their heterogeneous characteristics, i.e., the network topology follows a global hierarchy. It is an instance of decomposition or "divide and conquer" principle—a widespread principle for solving complex problems. Initially, this kind of hierarchy appeared in pure unstructured P2P systems to deal with node heterogeneity in capacity, bandwidth, CPU power, and lifetime [33, 70]. Popular hierarchy-based implementations include Gnutella,[3] KaZaA,[4] EMule,[5] and BitTorrent.[6] Typically, the hierarchy is two-level where each node is either super or regular. Supernodes take more responsibility acting as centralized index servers or network access points. Regular nodes may be of lower availability, capacity and performance; they form networks at the bottom layer. Supernodes connect them forming the top layer.

A similar trend is progressing in structured P2P systems. In the last decade, many *hierarchical DHT* (HDHT) designs have been announced: Brocade [75], Super-peer based lookup [78], Coral [14], HIERAS [69], Hierarchical systems [17] and TOPLUS [18], Kelips [22], Structured superpeers [49], OneHop [13, 21], Canon [16], Diminished Chord [29], GTap [73], Rings of unstructured clouds [60], Multi-ring hierarchy [48], Cyclone [3], Chordella [80–82] and Content-based hierarchy [79], HONet [64], P3ON [51], Chord^k [25], Hierarchy-adaptive topology [72], Hierarchical DHT-based overlays [43–45], Hierarchical small-world networks [19], GTPP [50], and some others.

Basically, an HDHT follows the idea of traditional two-level hierarchy with supernodes and regular nodes. It either approximates the global heterogeneity in

---

[2]http://en.wikipedia.org/wiki/Hierarchy.

[3]http://www.gnutella2.com.

[4]http://www.kazaa.com.

[5]http://www.emule-project.net/.

[6]http://www.bittorrent.com/, http://www.bittorrent.org/.

node capacities or hierarchical Internet organization by appropriate overlay network structuring. As a result, the HDHT design attempts to benefit from available global hierarchies.

There are many other instances of global hierarchies, applicable in P2P designs. In social networks, trust relations between nodes form a global hierarchy. For instance, arrangement of nodes according to their observable reputation or bootstrap process. Hence, nodes are categorized into several levels.

P2P resource semantics leads to an application-aware global hierarchy. Typically, it is due to either partitioning into domains or semantic indexing. For example, resources in multimedia services (e.g., SIP) or name services (e.g., DNS) are classified by the location domain: a resource R of a user U from a group G. Such media content as songs are globally categorized by singer, genre, album, etc. For each document in a data set, a vector of semantic indexes (or keys) are computed and the documents form a hierarchical structure by clustering in the semantic vector space.

Hierarchy, however, has deeper roots in structured P2P designs than global construction of tree-like network topology. Various kinds of arranging items occur very frequently, leading to multiple tree-like structures embedded into the network topology. Due to the decentralized P2P nature, nodes can construct local hierarchies arranging other nodes and P2P resources by own criteria. It is typical for open P2P environments, where selfish nodes with own interests exist. Clearly, generous nodes and free-riders have different strategies in resource contribution and consumption. Each node makes local ranking; the ranks may differ from those that computed at other nodes. For example, even if u and v are friends, w trusts v less, e.g., because of w has detected some incorrectness.

Furthermore, local hierarchies appear because of lack of knowledge about the global hierarchy. In this case, each node estimates the global hierarchy locally, leading to similar (but not exact) local hierarchies at different nodes. For example, nodes u and v may estimate the hierarchy of available paths to the same destination d differently.

1.4.3 Arrangement Models

There are three basic arrangement models that are applicable in P2P systems for taking the heterogeneity of nodes into account [31]. Let $\mathscr{X}$ be a set that represent some knowledge about the system. For example, $\mathscr{X} = N$ is the set of all nodes in the system or $\mathscr{X} = N_u$ consists of all neighbors of some $u \in N$. In general case, u's knowledge of $\mathscr{X}$ may differ from the knowledge of other nodes $v \neq u$.

Ordering: A node u uses a binary relation $\prec$ such that for any $x, y \in \mathscr{X}$ either $x \prec y$, $y \prec x$ or $x = y$. In other words, u can arrange elements of $\mathscr{X}$ in accordance with some "preference". The following two models are extensions (continuous and discrete) of the ordering model.

Ranking: There is a rank function $r : \mathcal{X} \to \mathbb{R}$, and u computes a real numerical value $r(x)$ for each $x \in \mathcal{X}$. Thus, elements of $\mathcal{X}$ are ordered on the real line $\mathbb{R}$. The important additional information is the value $|r(x) - r(y)|$, which is the preference level for u to compare x and y.

Classifying: The elements of $\mathcal{X}$ are categorized into groups or levels $i = 1, 2, \dots, M$ according to the preference. Although this model has less precision than the ranking model the former allows tradeoffs between the complexity and accuracy. For convenience, we assume that $i = 1$ is "most preferable" and $i = M$ is "least preferable".

In the supernode approach, the ordering model arranges nodes from most powerful (or stable, generous, etc.) to weakest, and a node can select a representative for any of these two extremes. The ranking model provides more knowledge for this selection, useful when a set of representatives is needed. However, it requires for a node to compute and maintain $r(v)$ for some nodes v. The classification model defines two classes: supernodes and regular nodes—an easy rule for implementation.

1.5   Local Knowledge and Network Structure

The main reason behind structured P2P networks is the lookup efficiency when the knowledge of a node is limited to its neighbors, $|T_u| \ll N$. Each node u dynamically maintains T_u optimizing $\tau(u, d)$ for potential destinations $d \in N$. The maintenance follows strict rules common for all nodes.

1.5.1   Local Knowledge Size vs. Routing Performance

The performance problem for large networks is closely related with the scalability issue. It requires the routing distance to grow slower than the network size, e.g., $|u \to^+ d| = o(N)$ for all u and d. Similarly, the local knowledge cannot expand fast, e.g., $|T_u| = o(N)$ for any u.

The characteristic property of network topology graph is low node degrees and every pair (u, v) is connected by many short paths [30, 38]. Although u does not know the whole network, u can select the next hop estimating the cost of the rest of the path. The network structure does not provide u with concrete paths; instead, u knows that for a given key good paths are likely available beyond some of the neighbors.

The fundamental limit the local knowledge problem induces is illustrated by the following theoretical fact.

Theorem 1.1. *The diameter of a network with N nodes and equally fixed node degree $m = |T_u| \; \forall u \in N$ has the logarithmic lower bound*

$$D = \max_{u,v \in N} \min_{u \to^+ v} |u \to^+ v| \geq \lceil \log_m (N(m-1)+1) \rceil - 1. \tag{1.5}$$

Proof. It is a direct consequence of the Moore bound

$$N \le 1 + m + m^2 + \cdots + m^D = \frac{m^{D+1} - 1}{m - 1},$$

see more discussion in [38].   □

The bound is provably not achievable for any non-trivial graph. Directed de Bruijn graphs have $D = \log_m N$, coming close to the bound and approaching it asymptotically when $N \to \infty$. Butterfly graphs have $D = 2\log_m N(1 - o(1))$, which is asymptotically twice the diameter of de Bruijn graphs when $N \to \infty$.

Most P2P designs today assume some structure [55]. The characteristic property is an estimator of how well structured a design is. Flat DHT designs with $D = S$ give a wide class of very structured systems. Their neighbor selection rules are semantic-free since S does not capture the resource semantics in D.

In semantic-aware P2P networks, u maintains T_u using additionally some available knowledge of D. For example, nodes with thematically close resources become neighbors. A more complicated mechanism than scalar ID space and routing distance metrics $\rho(u,v)$ and $\tau(u,v)$ is needed to take into account the semantic relation (application domain-aware distance) between resources $D(u)$ and $D(v)$ stored at u and v. In this case, overlay topology should take into account the graph of semantic relations.

1.5.2   System Operation Quality

Maintenance of T_u can involve additional aspects of system operation quality, which do not target directly the minimization of $\tau(u,d)$. Open environments assume diversity and heterogeneity of participants, including unreliable, selfish, and even malicious nodes. In randomized neighbor selection, u may decide a neighbor among several candidates. In particular, u improves routing dependability and security when the neighbor selection takes into account reputation of each candidate, other its trust and reliability metrics, and similar estimates about the paths beyond the candidate [10, 20, 32].

Neighbor maintenance can employ local knowledge on cooperation activity, important in P2P resource sharing systems. A rational node acts as an independent decision maker. Having local knowledge of other nodes, each node makes decisions based on its subjective view. Cooperation activity expands or modifies node's subjective view, providing feedback to the decision making process. Due to the decentralization nature of P2P the decision making process remains local for each node.

Naive approaches to P2P resource sharing lead to imbalances and reduced performance [12, 42, 58], e.g., to free-riding when nodes only consume resources. Cooperation among P2P nodes requires incentives, as used in BitTorrent [11] and its variants [6, 26, 35, 37, 59]. A system incentive mechanism encourages every node to balance rationally when contributing its local resources, consuming external ones, and participating in transit. One of the key point is that low contributing nodes are eventually eliminated from the system.

A rational node ranks its neighbors and resources for deciding its current provision and consumption levels. Intuitively, stingy neighbors should be of low rank, and u reduces its provision to them. Generous neighbors should be of high rank, and u encourages them in the provision process. Decisions are operational since generally behavior of other participants is unpredictable, especially in long-term.

When the global topology is known and nodes behavior can be completely specified as some types of stochastic processes then the methods of queuing networks can be applied, e.g., see [53, 71]. Each node is modeled as a processor with certain arrival and service patterns. For large-scale dynamic P2P systems this approach becomes impractical since a node can track no global and precise view to the dynamic network topology and characteristics of all participants.

As in many dynamic systems, nodes make own decisions based on observations (feedback)—the problem studied in control theory. Nodes consider the entire P2P system as a plant to be controlled [66]. Each node tracks contributions from other nodes and adopts its own contribution accordingly. These observations of others' contributions represent a global state of the system, hence leading to similar difficulties of global knowledge as in queuing models.

Nodes can reduce the observation space to direct observations from neighbors and aggregate indirect observations from the rest of the system. This approach is used in reputation schemes, and distributed learning-propagation algorithms are required [7, 47, 74]. Indirect observations suffer from malicious nodes. The propagation efficiency can be low in large-scale high-dynamic systems.

Game theory focuses on P2P exchange economies [2, 39, 66, 67] with system-wide indexes like the overall node reputation metric or global resource prices. It allows clear economic models with the intuition and properties from classic monetary-based economies. If a node knows the index then the node can rationally select its participation strategy. The system converges to an equilibrium point, e.g., when there is no node that can improve its utility by deviating from the optimal strategy (Nash equilibrium). A distributed algorithm is required for computing such indexes, similarly to reputation learning-propagation schemes.

References

1. Abraham, I., Malkhi, D., Manku, G.S.: Papillon: Greedy routing in rings. In: DISC '05: Proceedings of 19th International Conference on Distributed Computing. Lecture Notes in Computer Science, vol. 3724, pp. 514–515. Springer, Berlin (2005)
2. Aperjis, C., Freedman, M.J., Johari, R.: Bilateral and multilateral exchanges for peer-assisted content distribution. IEEE/ACM Trans. Netw. (2011). doi: http://dx.doi.org/10.1109/TNET.2011.2114898 19(5):1290–1303

3. Artigas, M.S., Lopez, P.G., Ahullo, J.P., Skarmeta, A.F.G.: Cyclone: A novel design schema for hierarchical DHTs. In: IEEE P2P '05: Proceedings of 5th International Conference on Peer-to-Peer Computing, pp. 49–56. IEEE Computer Society, USA (2005). doi: http://dx.doi.org/10.1109/P2P.2005.5

4. Aspnes, J., Shah, G.: Skip graphs. In: SODA '03: Proceedings of 14th Annual ACM-SIAM Symposium on Discrete Algorithms, pp. 384–393. Society for Industrial and Applied Mathematics, USA (2003)

5. Balakrishnan, H., Kaashoek, M.F., Karger, D., Morris, R., Stoica, I.: Looking up data in P2P systems. Commun. ACM, USA 46(2), 43–48 (2003). doi: http://doi.acm.org/10.1145/606272.606299

6. Bharambe, A.R., Herley, C., Padmanabhan, V.N.: Analyzing and improving a BitTorrent network's performance mechanisms. In: Proceedings of IEEE INFOCOM'06, pp. 2884–2895. IEEE (2006)

7. Bickson, D., Malkhi, D.: A unifying framework of rating users and data items in peer-to-peer and social networks. Peer-to-Peer Netw. Appl. 1, 93–103 (2008)

8. Birman, K.P.: Reliable Distributed Systems: Technologies, Web Services, and Applications. Springer New York Inc., Secaucus (2005)

9. Buford, J.F., Yu, H., Lua, E.K.: P2P Networking and Applications. Elsevier, Amsterdam (2009)

10. Castro, M., Drushel, P., Ganesh, A., Rowstron, A., Wallach, D.S.: Secure routing for structured peer-to-peer overlay networks. In: Proceedings of 5th USENIX Symposium on Operating System Design and Implementation (OSDI 2002), pp. 299–314. ACM, Boston (2002)

11. Cohen, B.: Incentives build robustness in BitTorrent. In: Proceedings of 1st Workshop on Economics of Peer-to-Peer Systems (2003)

12. Feldman, M., Chuang, J.: Overcoming free-riding behavior in peer-to-peer systems. ACM SIGecom Exch. 5(4), 41–50 (2005). doi: http://doi.acm.org/10.1145/1120717.1120723

13. Fonseca, P., Rodrigues, R., Gupta, A., Liskov, B.: Full-information lookups for peer-to-peer overlays. IEEE Trans. Parallel Distrib. Syst. 20(9), 1339–1351 (2009)

14. Freedman, M.J., Mazières, D.: Sloppy hashing and self-organizing clusters. In: IPTPS '03: Proceedings of 2nd International Workshop on Peer-to-Peer Systems. Lecture Notes in Computer Science, vol. 2735, pp. 45–55. Springer, Berlin (2003)

15. Gai, A.T., Viennot, L.: Broose: A practical distributed hashtable based on the De-Bruijn topology. In: Proceedings of IEEE 4th International Conference on Peer-to-Peer Computing (P2P'04), pp. 167–164. IEEE Computer Society, USA (2004). doi: http://dx.doi.org/10.1109/P2P.2004.10

16. Ganesan, P., Gummadi, K., Garcia-Molina, H.: Canon in G major: Designing DHTs with hierarchical structure. In: ICDCS '04: Proceedings of 24th International Conference on Distributed Computing Systems, pp. 263–272. IEEE Computer Society (2004)

17. Garcés-Erice, L., Biersack, E., Felber, P.A., Ross, K.W., Urvoy-Keller, G.: Hierarchical peer-to-peer systems. In: Euro-Par 2003: Proceedings of ACM/IFIP International Conference on Parallel and Distributed Computing, ACM/IFIP pp. 643–657 (2003)

18. Garcés-Erice, L., Ross, K.W., Biersack, E.W., Felber, P., Urvoy-Keller, G.: Topology-centric look-up service. In: Proceedings of 5th International Conference on Group Communications and Charges (NGC 2003), Workshop on Networked Group Communication. Lecture Notes in Computer Science, vol. 2816, pp. 58–69. Springer, Berlin (2003)

19. Guisheng, Y., Jie, S., Xianghui, W.: Hierarchical small-world P2P networks. In: ICICSE '08: Proceedings of International Conference on Internet Computing in Science and Engineering, pp. 452–458. IEEE Computer Society (2008). doi: http://dx.doi.org/10.1109/ICICSE.2008.94

20. Gummadi, K., Gummadi, R., Gribble, S., Ratnasamy, S., Shenker, S., Stoica, I.: The impact of DHT routing geometry on resilience and proximity. In: Proceedings of of ACM SIGCOMM'03, pp. 381–394. ACM, New York (2003). doi: http://doi.acm.org/10.1145/863955.863998

21. Gupta, A., Liskov, B., Rodrigues, R.: Efficient routing for peer-to-peer overlays. In: Proceedings of 1st Symposium on Networked Systems Design and Implementation (NSDI'04) (2004). URL: citeseer.ist.psu.edu/gupta04efficient.html

22. Gupta, I., Birman, K., Linga, P., Demers, A., van Renesse, R.: Kelips: Building an efficient and stable P2P DHT through increased memory and background overhead. In: IPTPS '03: Proceedings of 2nd International Workshop on Peer-to-Peer Systems. Lecture Notes in Computer Science, vol. 2735, pp. 160–169. Springer, New York (2003)
23. Harvey, N.J.A., Jones, M.B., Saroiu, S., Theimer, M., Wolman, A.: SkipNet: a scalable overlay network with practical locality properties. In: USITS'03: Proceedings of 4th USENIX Symposium on Internet Technologies and Systems. USENIX Association (2003)
24. Hildrum, K., Kubiatowicz, J.: Asymptotically efficient approaches to fault-tolerance in peer-to-peer networks. In: Proceedings of 17th International Symposium on Distributed Computing (DISC '03), Springer Berlin, Heidelberg, pp. 321–336 (2003)
25. Joung, Y.J., Wang, J.C.: Chord2: A two-layer chord for reducing maintenance overhead via heterogeneity. Comput. Commun. **51**(3), 712–731 (2007)
26. Jun, S., Ahamad, M.: Incentives in BitTorrent induce free riding. In: Proceedings of 2005 ACM SIGCOMM Workshop on Economics of Peer-to-Peer Systems, P2PECON '05, pp. 116–121. ACM, New York (2005). doi: http://doi.acm.org/10.1145/1080192.1080199
27. Kaashoek, M.F., Karger, D.R.: Koorde: A simple degree-optimal distributed hash table. In: IPTPS '03: Proceedings of 2nd International Workshop on Peer-to-Peer Systems. Lecture Notes in Computer Science, vol. 2735, pp. 98–107. Springer, Berlin (2003)
28. Karger, D., Lehman, E., Leighton, T., Panigrahy, R., Levine, M., Lewin, D.: Consistent hashing and random trees: distributed caching protocols for relieving hot spots on the world wide web. In: STOC '97: Proceedings of 29th Annual ACM Symposium on Theory of computing, pp. 654–663. ACM, New York (1997). doi: http://doi.acm.org/10.1145/258533.258660
29. Karger, D.R., Ruhl, M.: Diminished Chord: A protocol for heterogeneous subgroup formation in peer-to-peer networks. In: IPTPS '04: Proceedings of 3rd International Workshop on Peer-to-Peer Systems. Lecture Notes in Computer Science, vol. 3279, pp. 288–297. Springer, Berlin (2004)
30. Kleinberg, J.M.: The small-world phenomenon: an algorithm perspective. In: Proceedings of 32nd Annual ACM Symposium Theory of Computing (STOC '00), pp. 163–170. ACM, New York (2000). doi: http://doi.acm.org/10.1145/335305.335325
31. Korzun, D., Gurtov, A.: Survey on hierarchical routing schemes in "flat" distributed hash tables. Peer-to-Peer Netw. Appl. **4**, 346–375 (2011). doi: http://dx.doi.org/10.1007/s12083-010-0093-z
32. Korzun, D., Nechaev, B., Gurtov, A.: Cyclic routing: Generalizing lookahead in peer-to-peer networks. In: AICCSA2009: Proceedings of 7th IEEE/ACS International Conference on Computer Systems and Applications, pp. 697–704. IEEE Computer Society (2009). doi: http://doi.ieeecomputersociety.org/10.1109/AICCSA.2009.5069403
33. Krishnamurthy, B., Wang, J., Xie, Y.: Early measurements of a cluster-based architecture for P2P systems. In: IMW '01: Proceedings of 1st ACM SIGCOMM Workshop on Internet Measurement, pp. 105–109. ACM, New York (2001). doi: http://doi.acm.org/10.1145/505202.505216
34. Leong, B., Liskov, B., Demaine, E.: Epichord: parallelizing the Chord lookup algorithm with reactive routing state management. In: ICON 2004: Proceedings of 12th International Conference on Networks, IEEE, pp. 270–276 (2004)
35. Levin, D., LaCurts, K., Spring, N., Bhattacharjee, B.: BitTorrent is an auction: analyzing and improving BitTorrent's incentives. SIGCOMM Comput. Commun. Rev. **38**, 243–254 (2008). doi: http://doi.acm.org/10.1145/1402946.1402987
36. Li, J., Stribling, J., Morris, R., Kaashoek, M.F., Gil, T.M.: A performance vs. cost framework for evaluating DHT design tradeoffs under churn. In: Proceedings of IEEE INFOCOM'05, vol. 1, pp. 225–236. IEEE (2005). doi:10.1109/INFCOM.2005.1497894
37. Liao, W.C., Papadopoulos, F., Psounis, K.: Performance analysis of BitTorrent-like systems with heterogeneous users. Perform. Eval. **64**, 876–891 (2007). doi:10.1016/j.peva.2007.06.008
38. Loguinov, D., Kumar, A., Rai, V., Ganesh, S.: Graph-theoretic analysis of structured peer-to-peer systems: Routing distances and fault resilience. IEEE/ACM Trans. Netw. **13**(5), 1107–1120 (2005)

39. Ma, R.T.B., Lee, S.C.M., Lui, J.C.S., Yau, D.K.Y.: Incentive and service differentiation in P2P networks: a game theoretic approach. IEEE/ACM Trans. Netw. **14**(5), 978–991 (2006). doi: http://dx.doi.org/10.1109/TNET.2006.882904

40. Malkhi, D., Naor, M., Ratajczak, D.: Viceroy: a scalable and dynamic emulation of the butterfly. In: PODC '02: Proceedings of 21st Annual Symposium on Principles of Distributed Computing, pp. 183–192. ACM, New York (2002). doi: http://doi.acm.org/10.1145/571825.571857

41. Manku, G.S., Bawa, M., Raghavan, P.: Symphony: distributed hashing in a small world. In: USITS'03: Proceedings of 4th USENIX Symposium on Internet Technologies and Systems, pp. 127–140. USENIX Association (2003)

42. Marti, S., Garcia-Molina, H.: Taxonomy of trust: categorizing P2P reputation systems. Comput. Netw. **50**(4), 472–484 (2006). doi: http://dx.doi.org/10.1016/j.comnet.2005.07.011

43. Martinez-Yelmo, I., Bikfalvi, A., Guerrero, C., Rumin, R.C., Mauthe, A.: Enabling global multimedia distributed services based on hierarchical DHT overlay networks. Int. J. Internet Protoc. Technol. (IJIPT) **3**(4), 234–244 (2008). doi: http://dx.doi.org/10.1504/IJIPT.2008.023772

44. Martinez-Yelmo, I., Cuevas, R., Guerrero, C., Mauthe, A.: Routing performance in a hierarchical DHT-based overlay network. In: PDP 2008: Proceedings of 16th Euromicro Conference on Parallel, Distributed and Network-Based Processing, pp. 508–515. IEEE Computer Society (2008). doi: http://dx.doi.org/10.1109/PDP.2008.79

45. Martinez-Yelmo, I., Guerrero, C., Rumín, R.C., Mauthe, A.: A hierarchical P2PSIP architecture to support skype-like services. In: PDP 2009: Proceedings of 17th Euromicro International Conference on Parallel, Distributed and Network-Based Processing, pp. 316–322. IEEE Computer Society (2009). doi: http://doi.ieeecomputersociety.org/10.1109/PDP.2009.27

46. Maymounkov, P., Mazières, D.: Kademlia: A peer-to-peer information system based on the XOR metric. In: IPTPS '02: Proceedings of 1st International Workshop on Peer-to-Peer Systems. Lecture Notes in Computer Science, vol. 2429, pp. 53–65. Springer, Berlin (2002)

47. Mekouar, L., Iraqi, Y., Boutaba, R.: A contribution-based service differentiation scheme for peer-to-peer systems. Peer-to-Peer Netw. Appl. **2**, 146–163 (2009). doi: http://dx.doi.org/10.1007/s12083-008-0026-2

48. Mislove, A., Druschel, P.: Providing administrative control and autonomy in structured peer-to-peer overlays. In: IPTPS '04: Proceedings of 3rd International Workshop on Peer-to-Peer Systems. Lecture Notes in Computer Science, vol. 3279, pp. 162–172. Springer, New York (2004)

49. Mizrak, A.T., Cheng, Y., Kumar, V., Savage, S.: Structured superpeers: Leveraging heterogeneity to provide constant-time lookup. In: WIAPP 2003: Proceedings of 3rd IEEE Workshop on Internet Applications, IEEE, pp. 104–111 (2003)

50. Ou, Z., Harjula, E., Koskela, T., Ylianttila, M.: GTPP: General truncated pyramid peer-to-peer architecture over structured DHT networks. Mob. Netw. Appl. **15**, 729–749 (2010). doi:10.1007/s11036-009-0193-2

51. Park, K., Pack, S., Kwon, T.: Proximity based peer-to-peer overlay networks (P3ON) with load distribution. In: Proceedings of International Conference on Information Networking (ICOIN 2007). Towards Ubiquitous Networking and Services. Revised Selected Papers, pp. 234–243. Springer, Berlin (2008). doi: http://dx.doi.org/10.1007/978-3-540-89524-4_24

52. Plaxton, C.G., Rajaraman, R., Richa, A.W.: Accessing nearby copies of replicated objects in a distributed environment. In: Proceedings of 9th Annual Symposium on Parallel Algorithms and Architectures (SPAA '97), pp. 311–320, ACM (1997)

53. Ramachandran, K.K., Sikdar, B.: A queuing model for evaluating the transfer latency of peer-to-peer systems. IEEE Trans. Parallel Distrib. Syst. **21**, 367–378 (2010). doi: http://dx.doi.org/10.1109/TPDS.2009.69

54. Ratnasamy, S., Handley, P.F.M., Karp, R., Shenker, S.: A scalable content-addressable network. In: Proceedings of ACM SIGCOMM'01, pp. 161–172. ACM, New York (2001)

55. Risson, J., Moors, T.: Survey of research towards robust peer-to-peer networks: search methods. Comput. Netw. **50**(17), 3485–3521 (2006). doi: http://dx.doi.org/10.1016/j.comnet.2006.02.001

56. Rowstron, A., Druschel, P.: Pastry: Scalable, distributed object location and routing for large-scale peer-to-peer systems. In: Middleware'01: Proceedings of IFIP/ACM International Conference on Distributed Systems Platforms. Lecture Notes in Computer Science, vol. 2218, pp. 329–350. Springer, New York (2001)
57. Seedorf, J., Muus, C.: Availability for structured overlay networks: considerations for simulation and a new bound on lookup success. In: Proceedings of 12th Nordic Workshop on Secure IT Systems, pp. 23–34 (2007)
58. Serbu, S., Bianchi, S., Kropf, P., Felber, P.: Dynamic load sharing in peer-to-peer systems: when some peers are more equal than others. IEEE Internet Comput. **11**(4), 53–61 (2007). doi: http://dx.doi.org/10.1109/MIC.2007.81
59. Sherman, A., Nieh, J., Stein, C.: FairTorrent: bringing fairness to peer-to-peer systems. In: CoNEXT '09: Proceedings of the 5th International Conference on Emerging Networking Experiments and Technologies, pp. 133–144. ACM, New York (2009). doi: http://doi.acm.org/10.1145/1658939.1658955
60. Singh, A., Liu, L.: A hybrid topology architecture for P2P systems. In: ICCCN 2004: Proceedings of 13th International Conference on Computer Communications and Networks, IEEE, pp. 475–480 (2004)
61. Srivatsa, M., Liu, L.: Vulnerabilities and security threats in structured overlay networks: a quantitative analysis. In: ACSAC '04: Proceedings of 20th Annual Computer Security Applications Conference, pp. 252–261. IEEE Computer Society (2004). doi: http://dx.doi.org/10.1109/CSAC.2004.50
62. Stoica, I., Morris, R., Liben-Nowell, D., Karger, D., Kaashoek, M.F., Dabek, F., Balakrishnan, H.: Chord: a scalable peer-to-peer lookup service for Internet applications. IEEE/ACM Trans. Netw. **11**(1), 17–32 (2003)
63. Tang, C., Buco, M.J., Chang, R.N., Dwarkadas, S., Luan, L.Z., So, E., Ward, C.: Low traffic overlay networks with large routing tables. SIGMETRICS Perform. Eval. Rev. **33**(1), 14–25 (2005). doi: http://doi.acm.org/10.1145/1071690.1064216
64. Tian, R., Xiong, Y., Zhang, Q., Li, B., Zhao, B.Y., Li, X.: Hybrid overlay structure based on random walks. In: IPTPS '05: Proceedings of 4th International Workshop on Peer-to-Peer Systems. Lecture Notes in Computer Science, vol. 3640, pp. 152–162. Springer, Berlin (2005)
65. Vu, Q.H., Lupu, M., Ooi, B.C.: Peer-to-Peer Computing: Principles and Applications. Springer, Berlin (2010). doi:10.1007/978-3-642-03514-2
66. Wang, W., Li, B.: To play or to control: a game-based control-theoretic approach to peer-to-peer incentive engineering. In: Proceedings of 11th International Conference on Quality of service (IWQoS'03), pp. 174–192. Springer, Berlin (2003)
67. Wu, F., Zhang, L.: Proportional response dynamics leads to market equilibrium. In: STOC '07: Proceedings of 29th annual ACM Symposium on Theory of Computing, pp. 354–363. ACM, New York (2007). doi: http://doi.acm.org/10.1145/1250790.1250844
68. Xu, J., Kumar, A., Yu, X.: On the fundamental tradeoffs between routing table size and network diameter in peer-to-peer networks. IEEE J. Sel. Areas Commun. **22**(1), 151–163 (2004)
69. Xu, Z., Min, R., Hu, Y.: HIERAS: A DHT based hierarchical P2P routing algorithm. In: ICPP 2003: Proceedings of 32nd International Conference on Parallel Processing, pp. 187–194. IEEE Computer Society (2003)
70. Yang, B., Garcia-Molina, H.: Designing a super-peer network. In: ICDE'03: Proceedings of 19th International Conference on Data Engineering, pp. 49–60 (2003). doi: http://doi.ieeecomputersociety.org/10.1109/ICDE.2003.1260781
71. Yang, X., de Veciana, G.: Performance of peer-to-peer networks: service capacity and role of resource sharing policies. Perform. Eval. **63**, 175–194 (2006). doi:10.1016/j.peva.2005.01.005
72. Zhang, X.M., Wang, Y.J., Li, Z.: Research of routing algorithm in hierarchy-adaptive P2P systems. In: ISPA 2007: Proceedings of 5th International Symposium Parallel and Distributed Processing and Applications. Lecture Notes in Computer Science, vol. 4742, pp. 728–739. Springer, New York (2007)

73. Zhang, Y., Li, D., Chen, L., Lu, X.: Flexible routing in grouped DHTs. In: IEEE P2P '08: Proceedings of 8th International Conference on Peer-to-Peer Computing, pp. 109–118. IEEE Computer Society (2008). doi: http://dx.doi.org/10.1109/P2P.2008.43
74. Zhao, B.Q., Lui, J.C.S., Chiu, D.M.: Analysis of adaptive incentive protocols for P2P networks. In: Proceedings of IEEE INFOCOM'09, pp. 325–333. IEEE (2009)
75. Zhao, B.Y., Duan, Y., Huang, L., Joseph, A.D., Kubiatowicz, J.D.: Brocade: landmark routing on overlay networks. In: IPTPS '02: Proceedings of 1st International Workshop on Peer-to-Peer Systems. Lecture Notes in Computer Science, vol. 2429, pp. 34–44. Springer, Berlin (2002)
76. Zhao, B.Y., Huang, L., Stribling, J., Joseph, A.D., Kubiatowicz, J.D.: Exploiting routing redundancy via structured peer-to-peer overlays. In: ICNP '03: Proceedings of 11th IEEE International Conference on Network Protocols, IEEE, pp. 246–257 (2003)
77. Zhao, B.Y., Huang, L., Stribling, J., Rhea, S.C., Joseph, A.D., Kubiatowicz, J.D.: Tapestry: A resilient global-scale overlay for service deployment. IEEE J. Sel. Areas Commun. 22(1), 41–53 (2004)
78. Zhu, Y., Wang, H., Hu, Y.: A super-peer based lookup in structured peer-to-peer systems. In: Proceedings of ISCA 16th International Conference on Parallel and Distributed Computing Systems (PDCS 2003), International Society for Computers and Their Applications (ISCA), pp. 465–470 (2003)
79. Zoels, S., Eichhorn, M., Tarlano, A., Kellerer, W.: Content-based hierarchies in DHT-based peer-to-peer systems. In: SAINT Workshops 2006: Proceedings of International Symposium Applications and the Internet Workshops, pp. 105–108. IEEE Computer Society (2006). doi: http://dx.doi.org/10.1109/SAINT-W.2006.12
80. Zoels, S., Despotovic, Z., Kellerer, W.: Cost-based analysis of hierarchical DHT design. In: IEEE P2P '06: Proceedings of 6th International Conference on Peer-to-Peer Computing, pp. 233–239. IEEE Computer Society (2006). doi: http://dx.doi.org/10.1109/P2P.2006.13
81. Zoels, S., Despotovic, Z., Kellerer, W.: Load balancing in a hierarchical DHT-based P2P system. In: COLCOM '07: Proceedings of 2007 International Conference on Collaborative Computing: Networking, Applications and Worksharing, pp. 353–361. IEEE Computer Society (2007). doi: http://dx.doi.org/10.1109/COLCOM.2007.4553855
82. Zoels, S., Despotovic, Z., Kellerer, W.: On hierarchical DHT systems — an analytical approach for optimal designs. Comput. Commun. 31(3), 576–590 (2008). doi: http://dx.doi.org/10.1016/j.comcom.2007.08.033

Chapter 2
Flat DHT Routing Topologies

Abstract In this chapter, we survey existing classical DHTs grouped according to their topologies. We describe basic architectures for efficient overlay routing and corresponding classes of graphs. We start with introducing Content Addressable Network (CAN) as an example of Torus topology. It is followed by Chord, Kademlia and Accordion as DHTs using the Ring topology. Pastry, Tapestry and Bamboo DHTs are grouped under the PRR tree topology. Trie and balanced trees are represented by P-Grid, skip graphs. Finally, we present several DHTs that are using De Bruijn and Kautz graphs, Butterfly and O(1)-hop topologies. These topologies differ in how they collect local information in a DHT node about the global network. The next chapter will show how these architectures can be generalized using hierarchical approach.

2.1 Torus

2.1.1 Content Addressable Network

Content Addressable Network (CAN) [27] was one of four DHTs introduced in 2001. In CAN, each node is responsible for an area of key identifier space, called a zone. In addition, it maintains information on nodes responsible for adjacent zones. Request including insert, lookup or delete operations are routed hop-by-hop between zones until the target is reached. The main design goals of CAN where scalability (nodes maintain a limited state), fault tolerance (requests can be routed around failures) and P2P nature (distributed architecture with no single point of failure).

The CAN is organized as a d-dimensional torus with Cartesian coordinates. This is a purely logical construct where the coordinate space is partitioned between participating nodes. Figure 2.1 illustrates a two-dimensional coordinate space which

D. Korzun and A. Gurtov, *Structured Peer-to-Peer Systems: Fundamentals of Hierarchical Organization, Routing, Scaling, and Security*, DOI 10.1007/978-1-4614-5483-0_2, © Springer Science+Business Media New York 2013

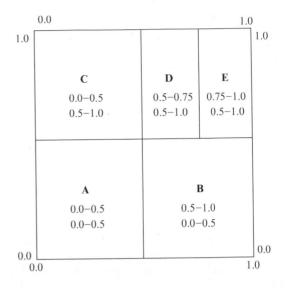

Fig. 2.1 The structure of 2D CAN space

is shared among several nodes. For simplicity, the space is drawn as a plane while in fact it wraps up at the edges, forming a torus. Many paths exist between source and destination and greedy routing is used to reach the destination.

Considering a d-dimensional CAN composed on n nodes, on the average each node would have $2d$ neighbors and the average path length will be $(d/4)(n^{1/d})$. Adding a new node does not affect the node state but increases the average path length as $O(n^{1/d})$.

To reduce the routing delay, CAN maintains several independent coordinate spaces called *realities*. If a node belongs to r realities, it has r independent neighbor sets and r coordinate points in separate zones. Multiple realities improve routing reliability, as a different reality can be used in case routing fails due to node departure within one reality. Furthermore, using different realities routing tables reduces the average routing path length, as distant locations can be often reached with one hop. Both increasing dimensions and number of realities reduces the average path length, but increasing the number of dimensions also increases per-node neighbor state.

Other techniques to improve the performance of CAN include using multiple hash functions to map a single key to several points of the coordinate space and hence increase the availability. Other enhancement is allocating zones taking the node physical location into account. In classic CAN, one logical hop can stretch over many IP hops so that two CAN neighbors can be located in different continents.

The CAN design was evaluated through simulations up to 260,000 nodes. With described enhancements, "knobs on full" CAN can route with less than twice the native IP latency between nodes.

2.2 Ring

The ring is a popular topology for organizing nodes in a DHT. In this section we describe three classic DHTs that use ring topology: Chord, Kademlia and Accordion. Chord represents a ring DHT which is linked in the clockwise direction. Although Kademlia is using a tree for routing, with binary identifiers its XOR metric is similar to the prefix metric and the ring topology. Therefore, Kademlia represents a bi-directional version of Chord.

2.2.1 Chord

Chord [32] is one of the first and most influential DHTs. It was developed by Ian Stoica with colleagues at MIT in 2001. In Chord, node a organized in a circle of up to 2^m nodes, with each node having an ID from that space. IDs of keys and nodes are of m bit length. The IDs are obtained using consistent hashing with SHA-1 algorithm and are uniformly distributed across the identifier space. This ensures even distribution of keys over nodes, and node place within the ring. The ID of a node is a hash of its IP address, and the ID of the key is a hash of its attribute, such as a file name.

Design of Chord aims at achieving following properties. *Load balancing* assumes even distribution of keys over nodes. *Scalability* requires that the lookup costs grow slower than the number of nodes, thus allowing large system construction. *Decentralization* is the main characteristics of P2P systems in general, and in Chord all nodes are equal without introducing single failure points. *Availability* is guaranteed by automatic maintenance of the node organization in response to node joins and leaves.

The ring of nodes is linked together in clockwise direction. The predecessor is the first node in counter-clockwise direction and the successor is a next node in clockwise direction. To ensure operation in the presence of node joins and departures, a node keeps information on r nodes succeeding it in the ring. A key k is assigned to successor of node with ID equal or greater than k.

```
join():
// create a new Chord ring.
n.create()
    predecessor = nil;
    successor = n;

// join a Chord ring containing node n
n.join(n')
    predecessor = nil;
    successor = n'.find_successor(n);
```

Fig. 2.2 The structure of
Chord ring

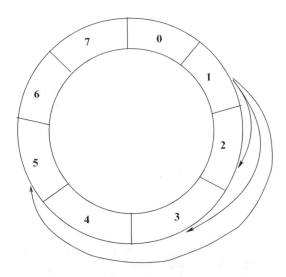

In naive lookup routing, the ring can be traversed sequentially until a node responsible for a key is located. This is quite slow as about half of the ring would need to be traversed on the average. To accelerate the lookup progress, each node keeps a table of *fingers*, each pointing to a successor node of $(n + 2^{i-1}) \mod 2^m$, where m is the number of fingers per node, n is the node ID, and i is the finger table index. Using fingers, the lookup time can be significantly reduced, as only $O(\log N)$ nodes need to be contacted to locate a key in a ring of N nodes. Using each finger halves the remaining distance to the destination node. Figure 2.2 illustrates the position of the fingers. Each entry in the finger table includes the node ID, its IP address and port number.

```
find_successor():
// ask node n to find the successor of id
n.find_successor(id)
    if (id in (n, successor])
        return successor;
    else
        n' = closest_preceding_node(id);
        return n'.find_successor(id);


// search the local table for the highest predecessor
   of id
n.closest_preceding_node(id)
    for i = m downto 1
        if (finger[i] in (n, id))
            return finger[i];
    return n;
```

The lookups can be implemented for recursive or iterative routing. With recursive approach, each new node in the lookup path queries the next hop. With iterative approach, the node that initiated the lookup receives answers from intermediate nodes in the path and itself performs the next hop lookup. The recursive approach reduces the number of required messages to perform a key lookup, while iterative approach is more robust in the presence of nodes joins and leaves. The simulator in the original Chord paper implemented the iterative approach.

```
find_successor():
// ask node n to find the successor of id
n.find_successor(id)
    if (id in (n, successor])
        return successor;
    else
        n' = closest_preceding_node(id);
        return n'.find_successor(id);

// search the local table for the highest predecessor
   of id
n.closest_preceding_node(id)
    for i = m downto 1
        if (finger[i] in (n, id))
            return finger[i];
    return n;
```

Nodes can join the ring and leave dynamically. When a new node joins, it takes over a part of keys that belong to its successor. When a node leaves the ring, all its keys are migrated to its successor. To ensure consistency, each node periodically runs a maintenance protocol to update its successor and finger table.

2.2.2 Kademlia

Kademlia [21] is a DHT introduced in 2002. It reduces the number of administrative messages needed to maintain routing tables, as nodes learn about the peers during lookup operations. The lookups are made in parallel and non-blocking way to minimize the delays. Its performance characteristics are formally proven. Kademlia is widely used for example by BitTorrent clients for distributed tracking of torrents.

Kademlia uses 160-bit IDs for nodes and keys. The keys/value pairs are stored on nodes with sufficiently close ID. XOR metric is used to determine a distance between nodes. Thanks to the metric, parallel queries to any node within a range nearby the target node can be sent and also learn new routes.

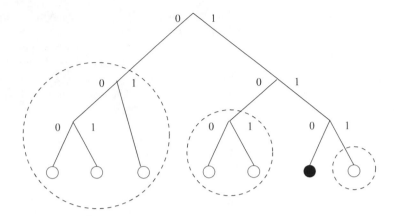

Fig. 2.3 The Kademlia tree for node 110

Kademlia does not separate the routing process into phases but uses the same mechanism until the target node is reached. On each step, the XOR distance to target node is roughly halved. XOR metric $x \oplus y$ is unidirectional that ensures convergence of requests towards the target.

Nodes in Kademlia store information about other nodes in a form of (Node ID, IP address, UDP port) of a distance in the range of 2^i and 2^{i+1}, where i is $0, \ldots, 160$. These lists form *k-buckets* sorted by the time of latest communication. Buckets for smaller i tend to be empty, while for larger i will be of size k (e.g., 20) which is a system parameter. The buckets are filled in whether a new message arrives from another node. Live nodes are never removed from the buckets, only nodes that fail to respond to pings are replaced by new nodes. That ensures that oldest live nodes remain as long as possible in the buckets, as those nodes are least probable to fail also in future.

Figure 2.3 illustrated a Kademlia routing tree for node with ID 110.

2.2.3   Accordion

Accordion [17] fills a gap between DHTs that provide $O(1)$ and $O(\log N)$ lookup times. Achieving $O(1)$ lookups is possible in small and relatively stable DHTs by maintaining a nearly complete state in each DHT node about other nodes ID and addresses. For larger and more dynamic networks, their maintenance traffic can grow prohibitively fast. Traditional DHTs such as Chord providing $O(\log N)$ lookup latency scale better, but require multiple hops to reach the destination that increases the response time. Accordion attempts to tune the DHT parameters to achieve best performance under a constraint of bandwidth budget defined by user. That frees users from the need to manually configure multiple DHT parameters and enables it to operate efficiently under varying network size, churn and lookup rates.

Accordion uses a Chord ring structure with consistent hashing, successor list, and the join protocol. The lookup find the node which ID follows the key. Accordion uses greedy routing approach, each node forwards the lookup request to the node with closest ID preceding the key. The process continues until the predecessor of the key hosting node is reached. The predecessor replies directly to the node that initiated the request.

Accordion explores multiple paths in parallel, sending as many lookups as the bandwidth budget allows. Parallel lookups are preferable compared to explicit probing, as those help to avoid timeouts due to dead nodes and also discover new nodes in a DHT. Each time a node forwards a lookup request to the next hop node, that node returns a list of its neighbors until the key ID. This way, the nodes can expand their routing tables to satisfy the main goal of Accordion, providing the minimum latency given the churn rate and the bandwidth budget.

2.3   PRR Trees

Plaxton, Rajaraman, and Richa proposed a PRR tree in 1997 [26]. PRR trees allow routing in a distributed publication system. They considered static network topology and did not provide a real-world implementation of the system. Later on, PRR trees were adopted by Tapestry and Pastry DHTs that also include management of dynamic membership of nodes in a tree.

Figure 2.4 shows neighbors of a node in PRR tree. Each node has $O(b * \log_b(N))$ neighbors in its routing table on several levels. A neighbor on ith level shares i digits with node ID. Hence, L0 node shares no common digits and L3 node shares three prefix digits. A node with ID closest to the key is the owner of that key.

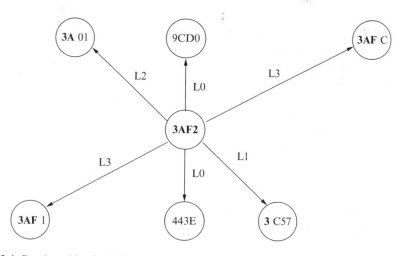

Fig. 2.4 Routing table of a PRR tree

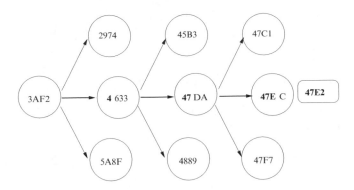

Fig. 2.5 Routing process in a PRR tree

Figure 2.5 shows routing process initiated from node with ID $3AF2$ to a key with ID $47E2$. The routing goal is to find a node with the longest matching prefix of digits. The query proceeds to node 4633 first since it shares the first digit with destination. The second hop is node with ID $47DA$ with two digits and finally node $47EC$ which is numerically closest to the key ID and hence responsible for it.

2.3.1 Pastry

Pastry [29] is one of early DHTs introduced in 2001. Like other DHTs, it distributes keys among participating nodes with IDs, routing a request to the responsible node in $O(\log N)$ hops in a network of N nodes. The destination node is a live node with ID numerically closest to the key ID.

A special feature of Pastry is proximity-aware routing. Each Pastry node keeps information on k nodes closest to it in the identifier space. Such nodes are likely to be spread geographically which could produce very long hops in terms of network latency. Pastry attempts to choose the next hop node which is closest by the routing metric, such as RTT delay.

The basic topology of Pastry is similar to Chord. Node IDs of 128 bits are arranged into a ring. The node ID can be generated as a hash of node's public key or its IP address, which ensures random positioning of the node within the ring.

The Pastry node maintains a routing table, a leaf set and a neighborhood set. Instead of Chord's fingers, a PRR tree is used to route requests to nodes within the ring.

The Pastry authors have evaluated it in an emulated network of 100,000 nodes, confirming its scalability and self-organization capabilities, and ability to take advantage of geographically close nodes.

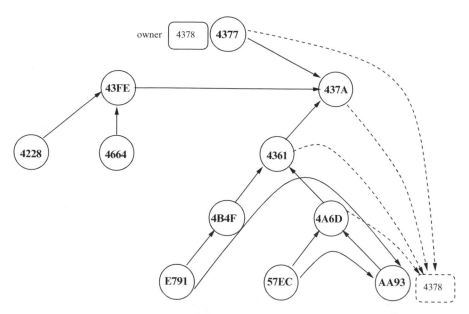

Fig. 2.6 Tapestry PRR topology combined with location mappings

2.3.2   *Tapestry*

Tapestry [36] is one of first DHTs introduced in 2001. It represents a structured peer-to-peer overlay network that guarantees key location in the system when node do not fail badly. Like Pastry, it is based on PRR trees. Tapestry authors use an alternative term to DHT, Decentralized Object Location and Routing (DOLR). A special feature of Tapestry is ability to exploit locality in accessing object replicas.

Figure 2.6 shows Tapestry topology based on PRR trees. Lookup requests are forwarded to the closest copy of an object using location mappings created when an object was published. In the figure, node with ID 4377 is responsible for the key 4378. The key was published by a request to node ID *AA*93 which routed it towards the owner node. At each hop until the destination along the PRR tree, a link to the local copy of the key at node *AA*93 is created. Therefore, a lookup request from a nearby node 57*EC* gets immediately redirected to local copy from a node 4*A6D*, and likewise by node 4361 when request comes from node *E*791.

2.3.3   *Bamboo*

Bamboo [28] is a ring-based DHT that specifically focuses on efficient operation in the presence of high churn. Its authors evaluated the impact of churn on several DHTs and different strategies to cope with churn, including reactive vs periodic failure recovers, timeout calculation during lookups and choice of neighbors.

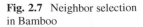

Fig. 2.7 Neighbor selection
in Bamboo

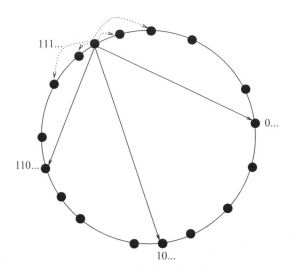

In particular, the churn handling capabilities of Pastry and Chord were investigated. It was found that Pastry recovers poorly even under medium churn, the main reason being that Pastry uses reactive recovery (i.e., repairing failures as soon as they are detected) and thereby is subject to the problem of positive feedback cycles, where network link congestion causes repair packets to be sent, which in turn causes more congestion, mistaken conclusion of whether other neighbors are down, and eventually congestion collapse. On the other hand, the main problem with Chord (which uses periodic recovery) is that under churn, lookup latency increases substantially, which results from inaccurate timeout threshold calculations.

Based on these observations, Bamboo applies three techniques to address churn handling: periodic recovery, timeout calculation algorithms, and proximity neighbor selection (PNS) algorithms. PNS helps to reduce lookup latency by picking up neighbors close in terms of network latency at the expense of additional bandwidth use. Consequently, lower lookup latency leads to better timeout calculation.

The routing algorithm of Bamboo is borrowed from Pastry. It is illustrated in Fig. 2.7. Bamboo maintains the same geometry despite the churn. Two sets of pointers are present in a node, a leaf set to k neighbors before and after the node, as well as pointers to nodes with matching prefixes in PRR tree. Lookups are routed in $O(\log(N))$ hops even in the presence of a large number of broken links.

To cope with node failures, Bamboo uses periodic recovery like Chord. Periodically, a complete routing table is shared with one of a randomly chosen node from the leaf set. A node is considered dead and removed from the routing table after 15 timeouts, although routing to it stops after 5 timeouts. The general conclusion of Bamboo authors is that periodic recovery is well suited for high churn scenarios, while reactive recovery is appropriate when churn is low.

2.4 Tree

2.4.1 Trie

A trie, or prefix tree, is a data structure which name comes from the *retrieval*. In trie, the node position within the tree defines its key value, unlike in the binary search tree. All descendants of a node share a common prefix with a node. Figure 2.8 illustrates the structure of a trie containing a few English words. The character keys are most common for tries, although any ordered lists are suitable.

A trie has a nice property that all basic operations take nearly same time, because lookup, insertion or deletion is largely the same code. That enables tries to be more efficient than binary search trees and hash tables in most important properties. Most importantly, longest-prefix matching or finding a node with a longest prefix with a key is very efficient with a trie.

An example of use of distributed tries in a DHT [9] is described in Sect. 3.5. P-Grid [1, 5] is also using a distributed tree and is described in Sect. 4.6. Another example is ZIGZAG that provides P2P media streaming.

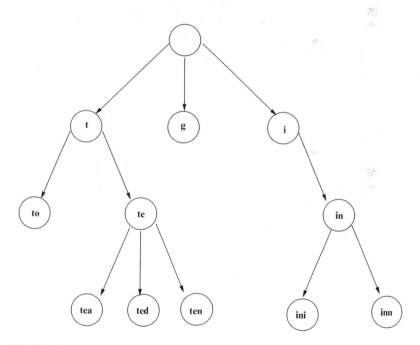

Fig. 2.8 Trie data structure

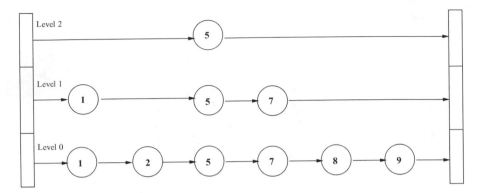

Fig. 2.9 Skip list data structure

2.4.2 Balanced Tree

Skip list is a data structure that consists of several linked lists that connect sorted items, as shown in Fig. 2.9.

Skip list was introduced by W. Pugh in 1990. Each higher list in the hierarchy jumps over multiple list items. The jump span can be selected randomly using a negative binomial distribution. The list item at $i - 1$ level appears at i level with certain probability p, typically 0.5 or 0.25. Then the number of layers in a skip list is $\log(1/p)N$. By varying p it is possible to balance between the storage overhead versus the search time.

The search for a key starts from the top layer list to the right until the key is found or a greater key is located. In latter case, the search returns to the previous node and drops down one layer. The procedure is repeated until the key is found or a greater key is found on the bottom layer, which means the key is not present. Therefore the expected search time is $(\log(1/p)N)/p$.

Skip graphs [3, 4] and SkipNet [13, 14] are examples of DHTs using the skip list data structure. They are described in Sect. 4.5.

2.5 De Bruijn and Kautz Graphs

2.5.1 De Bruijn Graphs

De Bruijn graph of n-dimensions and m symbols is a directed graph of m^n vertices (nodes). Each vertex is a random sequence of any of m symbols of length n, repeating symbols allowed. A directed edge is formed to a neighbor vertex if its ID is formed by shifting one position to the left and adding a new arbitrary symbol.

Fig. 2.10 De Bruijn graph
for three dimensions and two
symbols

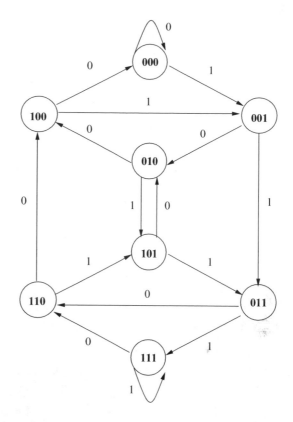

Figure 2.10 illustrates a de Bruijn graph of three dimensions constructed with
two symbols (0,1). It can be seen that each vertex has exactly m incoming and m
outgoing edges. Each de Bruijn graph has an Eulerian path (which visits every edge
exactly once) and Hamiltonian path (that visits each vertex exactly once).

Several DHTs utilize de Bruijn graphs in their topology, including Koorde, D2B,
Distance-halving [24, 25], ODRI [18], and Broose [10].

Koorde [15] is a DHT that modifies Chord to follow the topology of de Bruijn
graphs. Koorde means a chord in Dutch language. Even with two neighbors per
node, Koorde is able to route requests in $O(\log N)$ hops. When users can tune up
the number of neighbor per node up to $O(\log N)$, the number of hop decreases
to $O(\log N/\log \log N)$. Naturally, this comes at the cost of increasing maintenance
overhead. Koorde inherits some algorithms, such as handling concurrent joins, from
Chord.

D2B [8] is another DHT using de Bruijn graphs, but in probabilistic manner.
It ensures with high probability that the nodes have $O(1)$ neighbors. D2B requires
special form of node identifiers to construct a de Bruijn graph.

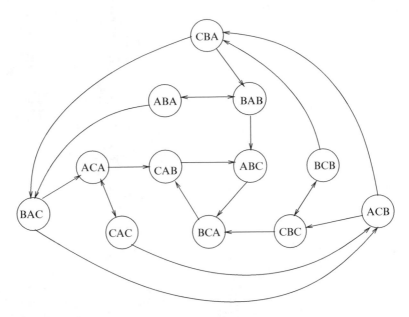

Fig. 2.11 The Kautz graph (2,2)

2.5.2 *Kautz Graph*

Kautz graph is a directed graph where vertices are labeled as all possible combinations of $M + 1$ distinct symbols under condition that adjacent symbols are not the same. M is called the degree of the graph. The dimension of the graph, N defines the length of each vertex label $(N + 1)$. An edge is created to a vertex if its label starts with a different symbol with regard to the source vertex. Figure 2.11 illustrates the Kautz graph for $M = 2$ and $N = 2$.

The Kautz graphs are closely related to De Bruijn graphs. Likewise, they posses Eulerian and Hamiltonian cycles. The Kautz graph has the smallest diameter of any directed graph with degree M and V vertices. The number of vertices in a graph with degree M equals to $(M + 1)M^{N+1}$. The edges of a K_M^{N+1} graph correspond to vertices of K_M^{N+2} graph.

Following DHTs use the Kautz graph as basic topology: FISSIONE [16], Moore [11], BAKE [12], and SKY [35]. They are described in Sect. 4.7.

2.6 Butterfly

Viceroy [19] utilizes the butterfly topology with constant expected number of neighbors. The node degree is less than $O(\log N)$. Figure 2.12 illustrates the multi-layer butterfly topology. The example network contains eight nodes with IDs

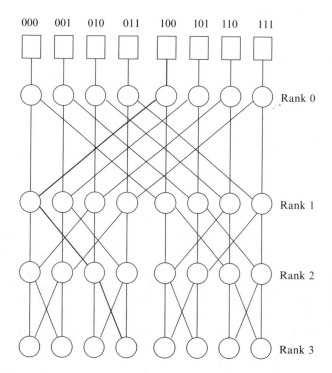

000   001   010   011   100   101   110   111

Rank 0

Rank 1

Rank 2

Rank 3

Fig. 2.12 The butterfly network

000..111. The nodes are linked at four layers, increasingly closer to neighbors. Thick line shows a path from node 100 to node 011. The lookup reaches the destination after three hops, which represents the worst case scenario in a stable network.

Viceroy estimates the size of network to select placement of nodes at several layers. Having a constant node degree helps in managing the cost of joins and leaves, since only a small number of neighbors need to be informed of leave, pinged for keep-alive, or exchanged state information. A small number of neighbors also helps to reduce the number of incoming and outgoing connections for smaller overhead. A possible draw back of this is difficulty in maintaining the topology during high churn.

Other DHTs using butterfly topology include CRN [6, 30], Ulysses [34], Cycloid [31], Pappilon [2], and Mariposa [20]. They are described in Sect. 4.2.

2.7   *O*(1)-Hop

Researchers proposed several DHTs that are able to perform lookups with constant time irrespective of the network size. Such DHTs, known as *O*(1) hop DHTs include OneHop [7], 1h-Calot [33], and D1HT [22]. Their comparison can be found in [23] and [33]. OneHop is described in detail in Sect. 7.3.1.4 among other hierarchical DHTs.

Most of DHTs presented in this chapter use multi-hop routing which requires less maintenance bandwidth and storage costs compared to one-hop DHTs. However, the use of multi-hop DHTs for performance-sensitive applications is problematic due to variable lookup delay. In some cases, one-hop DHTs can consume even less bandwidth compared to multi-hop DHTs when lookup frequency is high because multi-hop DHTs spend more on hop-by-hop forwarding. In general, bandwidth in future networks tends to grow much faster than the latency reduces, therefore technology solutions with small latency have clear potential.

To keep maintenance traffic within acceptable boundaries, D1HT proposes to combine several events, such as peer joins or leaves, within one message. That requires buffering event announcements until a sufficient number is collected to fill a packet. This naturally increases the delay in which other nodes learn about the system changes. To balance between maintenance traffic and event propagation latency, the maximum event holding time is determined from the application requirements. Typically it is assumed that at least 99 % of application requests should be resolved within one hop.

Both D1HT and 1h-Calot use logarithmic tree to distribute updates in a DHT. 1h-Calot uses peer IDs to build the logarithmic tree, while D1HT uses TTL of messages. D1HT is reactive in detecting events through maintenance message but 1h-Calot sends periodic heartbeats with proactive approach. D1HT buffers the events to reduce the traffic, but 1h-Calot requires at least one message and its acknowledgment per event and all DHT nodes.

2.8 Summary

We introduced basic classes of DHT topologies and described popular DHTs representing these classes. The first classic DHTs of CAN, Chord, Pastry and Tapestry are based on d-dimensional torus, a ring, and PRR tree topologies. We also introduced De Bruijn and Kautz graphs and their use in DHTs, skip lists, trie, and butterfly topologies. The chapter is concluded by description of $O(1)$-hop DHTs that aim at resolving most lookups with minimal latency.

References

1. Aberer, K., Cudré-Mauroux, P., Datta, A., Despotovic, Z., Hauswirth, M., Punceva, M., Schmidt, R.: P-Grid: a self-organizing structured P2P system. SIGMOD Rec. **32**(3), 29–33 (2003). doi: http://doi.acm.org/10.1145/945721.945729
2. Abraham, I., Malkhi, D., Manku, G.S.: Papillon: Greedy routing in rings. In: DISC '05: Proceedings of 19th International Conference on Distributed Computing. Lecture Notes in Computer Science, vol. 3724, pp. 514–515. Springer, Berlin (2005)
3. Aspnes, J., Shah, G.: Skip graphs. In: SODA '03: Proceedings of 14th Annual ACM-SIAM Symposium on Discrete Algorithms, pp. 384–393. Society for Industrial and Applied Mathematics (2003)

4. Aspnes, J., Wieder, U.: The expansion and mixing time of skip graphs with applications. In: SPAA '05: Proceedings of 17th Annual ACM Symposium on Parallelism in Algorithms and Architectures, pp. 126–134. ACM, New York (2005).doi: http://doi.acm.org/10.1145/1073970. 1073989

5. Datta, A., Girdzijauskas, S., Aberer, K.: On de bruijn routing in distributed hash tables: There and back again. In: IEEE P2P '04: Proceedings of 4th International Conference on Peer-to-Peer Computing, pp. 159–166. IEEE Computer Society (2004).doi: http://dx.doi.org/10.1109/P2P. 2004.29

6. Fiat, A., Saia, J.: Censorship resistant peer-to-peer content addressable networks. In: SODA '02: Proceedings of 13th Annual ACM-SIAM Symposium on Discrete Algorithms, pp. 94–103. Society for Industrial and Applied Mathematics (2002)

7. Fonseca, P., Rodrigues, R., Gupta, A., Liskov, B.: Full-information lookups for peer-to-peer overlays. IEEE Trans. Parallel Distrib. Syst. **20**(9), 1339–1351 (2009)

8. Fraigniaud, P., Gauron, P.: D2B: a de Bruijn based content-addressable network. Theor. Comput. Sci. **355**(1), 65–79 (2006). doi: http://dx.doi.org/10.1016/j.tcs.2005.12.006

9. Freedman, M.J., Vingralek, R.: Efficient peer-to-peer lookup based on a distributed trie. In: Revised Papers from 1st International Workshop on Peer-to-Peer Systems (IPTPS '01), pp. 66–75. Springer, Berlin (2002)

10. Gai, A.T., Viennot, L.: Broose: A practical distributed hashtable based on the De-Bruijn topology. In: Proceedings of IEEE 4th International Conference on Peer-to-Peer Computing (P2P '04), pp. 167–164. IEEE Computer Society (2004). doi: http://dx.doi.org/10.1109/P2P. 2004.10

11. Guo, D., Wu, J., Chen, H., Luo, X.: Moore: An extendable peer-to-peer network based on incomplete Kautz digraph with constant degree. In: Proceedings of IEEE INFOCOM'07, pp. 821–829. IEEE (2007)

12. Guo, D., Liu, Y., Li, X.Y.: BAKE: A balanced Kautz tree structure for peer-to-peer networks. In: Proceedings of IEEEINFOCOM'08, pp. 2450–2457. IEEE (2008)

13. Harvey, N.J.A., Munro, J.I.: Deterministic SkipNet. Inf. Process. Lett. **90**(4), 205–208 (2004). doi: http://dx.doi.org/10.1016/j.ipl.2004.01.019

14. Harvey, N.J.A., Jones, M.B., Saroiu, S., Theimer, M., Wolman, A.: SkipNet: a scalable overlay network with practical locality properties. In: USITS'03: Proceedings of 4th USENIX Symposium on Internet Technologies and Systems. USENIX Association (2003)

15. Kaashoek, M.F., Karger, D.R.: Koorde: A simple degree-optimal distributed hash table. In: IPTPS '03: Proceedings of 2nd International Workshop on Peer-to-Peer Systems. Lecture Notes in Computer Science, vol. 2735, pp. 98–107. Springer, Berlin (2003)

16. Li, D., Lu, X., Wu, J.: FISSIONE: a scalable constant degree and low congestion DHT scheme based on Kautz graphs. In: Proceedings of IEEE INFOCOM'05, pp. 1677–1688. IEEE (2005)

17. Li, J., Stribling, J., Morris, R., Kaashoek, M.F.: Bandwidth-efficient management of DHT routing tables. In: Proceedings of the 2nd Symposium on Networked Systems Design and Implementation (NSDI '05), pp. 99–114 (2005)

18. Loguinov, D., Kumar, A., Rai, V., Ganesh, S.: Graph-theoretic analysis of structured peer-to-peer systems: Routing distances and fault resilience. IEEE/ACM Trans. Netw. **13**(5), 1107–1120 (2005)

19. Malkhi, D., Naor, M., Ratajczak, D.: Viceroy: a scalable and dynamic emulation of the butterfly. In: PODC '02: Proceedings of 21st Annual Symposium on Principles of Distributed Computing, pp. 183–192. ACM, New York (2002). doi: http://doi.acm.org/10.1145/571825. 571857

20. Manku, G.S.: Routing networks for distributed hash tables. In: PODC '03: Proceedings of 22nd Annual Symposium on Principles of Distributed Computing, pp. 133–142. ACM (2003). doi: http://doi.acm.org/10.1145/872035.872054

21. Maymounkov, P., Mazières, D.: Kademlia: A peer-to-peer information system based on the XOR metric. In: IPTPS '02: Proceedings of 1st International Workshop on Peer-to-Peer Systems. Lecture Notes in Computer Science, vol. 2429, pp. 53–65. Springer, New York (2002)

22. Monnerat, L.R., Amorim, C.L.: D1HT: a distributed one hop hash table. In: Proceedings of 20th IEEE International Symposium on Parallel and Distributed Processing (IPDPS 2006). IEEE Computer Society (2006). doi: http://doi.ieeecomputersociety.org/10.1109/IPDPS.2006. 1639278
23. Monnerat, L.R., Amorim, C.L.: Peer-to-peer single hop distributed hash tables. In: Proceedings of IEEE Globecom'09, pp. 4250–4257, IEEE (2009)
24. Naor, M., Wieder, U.: A simple fault tolerant distributed hash table. In: IPTPS '03: Proceedings of 2nd International Workshop on Peer-to-Peer Systems. Lecture Notes in Computer Science, vol. 2735, pp. 88–97. Springer, New York (2003)
25. Naor, M., Wieder, U.: Novel architectures for P2P applications: the continuous-discrete approach. ACM Trans. Algorithms **3**(3), 37 (2007). doi: http://doi.acm.org/10.1145/1273340. 1273350
26. Plaxton, C.G., Rajaraman, R., Richa, A.W.: Accessing nearby copies of replicated objects in a distributed environment. In: Proceedings of 9th Annual Symposium on Parallel Algorithms and Architectures (SPAA '97), pp. 311–320 (1997)
27. Ratnasamy, S., Handley, P.F.M., Karp, R., Shenker, S.: A scalable content-addressable network. In: Proceedings of ACM SIGCOMM'01, pp. 161–172. ACM, New York (2001)
28. Rhea, S., Geels, D., Roscoe, T., Kubiatowicz, J.: Handling churn in a DHT. In: Proceedings of the USENIX Annual Technical Conference (2004)
29. Rowstron, A., Druschel, P.: Pastry: Scalable, distributed object location and routing for large-scale peer-to-peer systems. In: Middleware'01: Proceedings of IFIP/ACM International Conference on Distributed Systems Platforms. Lecture Notes in Computer Science, vol. 2218, pp. 329–350. Springer, Berlin (2001)
30. Saia, J., Fiat, A., Gribble, S.D., Karlin, A.R., Saroiu, S.: Dynamically fault-tolerant content addressable networks. In: IPTPS '01: Revised Papers from 1st International Workshop on Peer-to-Peer Systems, pp. 270–279. Springer, New York (2002)
31. Shen, H., Xu, C.Z., Chen, G.: Cycloid: a constant-degree and lookup-efficient p2p overlay network. Perform. Eval. **63**(3), 195–216 (2006). doi: http://dx.doi.org/10.1016/j.peva.2005.01.004
32. Stoica, I., Morris, R., Liben-Nowell, D., Karger, D., Kaashoek, M.F., Dabek, F., Balakrishnan, H.: Chord: a scalable peer-to-peer lookup service for Internet applications. IEEE/ACM Trans. Netw. **11**(1), 17–32 (2003)
33. Tang, C., Buco, M.J., Chang, R.N., Dwarkadas, S., Luan, L.Z., So, E., Ward, C.: Low traffic overlay networks with large routing tables. SIGMETRICS Perform. Eval. Rev. **33**(1), 14–25 (2005). doi: http://doi.acm.org/10.1145/1071690.1064216
34. Xu, J., Kumar, A., Yu, X.: On the fundamental tradeoffs between routing table size and network diameter in peer-to-peer networks. IEEE J. Sel. Areas Commun. **22**(1), 151–163 (2004)
35. Zhang, Y., Lu, X., Li, D.: SKY: efficient peer-to-peer networks based on distributed Kautz graphs. Sci. China Ser. F Inf. Sci. **52**(4), 588–601 (2009)
36. Zhao, B.Y., Huang, L., Stribling, J., Rhea, S.C., Joseph, A.D., Kubiatowicz, J.D.: Tapestry: A resilient global-scale overlay for service deployment. IEEE J. Sel. Areas Commun. **22**(1), 41–53 (2004)

Part II
Local Strategies

Overview of Part II

Structured P2P systems achieve their efficiency due to dynamic maintenance of rigorous network topology structure—its topology graph belongs to a class with well-defined invariant properties of connectivity. Resources, functionality, and other types of responsibility are uniformly spread over all the nodes. It ensures low node degree and small network diameter, leading to modest node state and high routing performance. In this case, however, a P2P node knows only a part of the entire network. To approximate the state of the rest network, the node extrapolates its local state based on invariant structural properties of the network topology.

Previously, Part I showed that the first generation of structured P2P networks— flat DHTs—offered a completely flat structure of the key space that is randomly partitioned among participating nodes. Flat DHTs have certain advantages, for example, even distribution of workload among nodes. On the other hand, grouping keys under a single authority or achieving latency guarantees for queries is difficult. The heterogeneity requires differentiation of nodes in a P2P system. Some of them are able to maintain bigger state becoming high degree nodes, which reduce the network diameter. There also can be low-capacity nodes that are not so active and resilient; their state and degree are minimal, which decreases the network connectivity efficiency. A system design adapts the heterogeneity by arranging the nodes such that they are subject to different responsibilities.

To address the node differentiation and adaptation to heterogeneity, various kinds of hierarchy have been proposed over recent years. In this part, we thoroughly go over the evolution of flat DHTs to pre-hierarchical DHTs, where differentiation is done locally, individually by each node. Our focus is on hierarchies that appear due to the the local construction and maintenance of routing tables. We sequentially built a set of design principles; each provides a base for hierarchical topology organization and appropriate routing schemes, improving the system performance. In the extreme case, application of these principles leads to traditional hierarchical DHT (HDHT) designs. These principles are unification points of many existing proposals.

Chapter 3 summarizes the design principles that lead to hierarchical routing schemes in flat DHTs. Although the assumption on node role uniformity is still preserved, careful arrangements in local routing tables allow logarithmic or better routing complexity in the whole system. Therefore, the known routing efficiency of flat DHTs is due to hierarchies that initiated locally by cooperative nodes.

Chapter 4 derives the design principles that lead to a kind of global hierarchies in flat DHTs. In contrast to the previous case of Chap. 3, a hierarchy is not from the point of view of an individual node but differentiates nodes at the global level. A system-wide specialization appears: nodes are specialized in their routing responsibility and resources are specialized by their distribution. As a result, the intrinsic heterogeneity in a P2P system is reflected in the topology to achieve higher performance.

Chapter 5 considers the clustering principle as an evolution step of hierarchical routing schemes in flat DHTs. Differentiation of nodes evolves to clustering when similar nodes or resources form specific groups, which in turn tend to organize a own network. Clustering makes explicit discretization of the intrinsic heterogeneity and reflects the latter in the network topology. We discuss various P2P design problems where the clustering principle appears, including routing, load distribution, and content semantics.

Chapter 6 targets the cooperation in anonymous, dynamic, and autonomous P2P networks and states the local ranking problem a solution to which provides effective means to reward fair cooperating nodes and to punish defect behavior. We consider direct-reciprocity incentive schemes, which are pure local—each node operates with directly observable knowledge about others' activity and performs operational decision-making with no assumption on long-term prediction. BitTorrent-like sharing systems are used as a reference case for single-resource exchange. Then the case is generalized for multi-resource exchange and transit resources.

Chapter 3
Hierarchical Neighbor Maintenance

Abstract In flat DHTs, various kinds of arranging items occur very frequently, leading to tree-like structures and hierarchical routing schemes. Different existing proposals can be unified in the terms of hierarchy, and routing schemes are built into a logical structure. There are certain design principles that define general directions for various hierarchical techniques to improve DHT routing. This chapter shows that the known routing efficiency of flat DHTs is due to hierarchical schemes in the local construction and maintenance of routing tables.

3.1 Introduction

Nodes exploit the small-world phenomena, the lookup paths form hierarchical structures—lookup path hierarchies. They support the existence of short paths and provide effective hierarchical routing schemes with local information. A common case is routing within $O(\log N)$ hops. These schemes are applied in Symphony [31], eCAN [51], Chord [48], Pastry [45], Tapestry [55], Kademlia [33], Koorde [21], Distance-halving [36, 38], D2B [13], ODRI [30], and Broose [15]. Sections 3.2 and 3.3 coin the corresponding design principles of small-world neighborhood and geometrically progressive routing.

Many extensions of flat DHT designs apply appropriate arrangement of neighbors for adapting to a global hierarchy that exists in the network. Nodes locally optimize lookup hops—a key design principle in neighbor and route selection [3,9,17,52,54], multipath routing [18,25,54], sybil-resistance routing [10], load balancing [47], and some others. Section 3.4 elaborates this principle of locally adaptive selection.

The lookup path hierarchy provides every node with an effective way for collecting and using knowledge about the network beyond the neighbors. It is a design principle in neighbor-of-neighbor routing [32,37], distributed trie of popular lookups [14], and cyclic routing [25]. Section 3.5 considers the DHT designs that follows the look-ahead principle.

D. Korzun and A. Gurtov, *Structured Peer-to-Peer Systems: Fundamentals of Hierarchical Organization, Routing, Scaling, and Security*, DOI 10.1007/978-1-4614-5483-0_3, © Springer Science+Business Media New York 2013

This principle evolves to DHTs with large routing tables or with high replication, when more expensive global routing is replaced with faster local routing by the cost of node state and maintenance. The DHT family with large routing tables includes EpiChord [27], Accordion [29], SmartBoa [20], OneHop [12], 1h-Calot [49], and D1HT [34]. The DHT family with high replication is represented by DHash [8], Beehive [41], and Yarqs [53]. Sections 3.6 and 3.7 discuss these DHT families and the problems they are faced with.

Recall from Chap. 1 that DHT routing construct multi-hop paths

$$u \to w_1 \to w_2 \to \cdots \to w_{l-1} \to d \qquad (3.1)$$

from a source node $u = w_0$ to a responsible node $d = w_l$. A widespread strategy is progressive routing when (3.1) satisfies

$$\rho(w_i,k) < \rho(w_{i-1},k), \quad i = 1,2,\ldots,l-1. \qquad (3.2)$$

It leads to greedy routing in the extreme case:

$$w_i = \arg\min_{u \in T_{w_{i-1}}} \rho(u,k), \quad i = 1,2,\ldots,l-1. \qquad (3.3)$$

3.2   Small-World Networks and Progressive Routing

Models for the small-world phenomenon naturally exhibit a certain hierarchy in routing tables. First, a node divides its neighbors onto local and long-range [23], in accordance with the classifying model. Figure 3.1 shows the basic scheme. Local neighbors of a node u are equal. Long-range neighbors are arranged in u's vicinity, in accordance with the ranking model.

3.2.1   Long-Range Neighbors

Kleinberg's small-world construction [22] highlights the idea of selection distant nodes for neighbors. The node ID space is the $n \times n$ grid

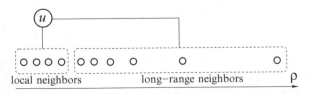

Fig. 3.1 The small-world hierarchy of neighbors: (1) discrete classification onto local and long-range; (2) arrangement of long-range neighbors with vicinity clustering at all distance scales

$$S = \{1, 2, \ldots, n\} \times \{1, 2, \ldots, n\}$$
$$\text{with } \rho\left((x_1, y_1), (x_2, y_2)\right) = |x_2 - x_1| + |y_2 - y_1|,$$

where $N \leq n^2$. Local neighbors of u are all v such that $\rho(u, v) \leq \varepsilon$ for a universal constant $\varepsilon \geq 1$. Additionally, u selects $m \geq 1$ long-range neighbors with probability proportional to $[\rho(u, v)]^{-\alpha}$ for $\alpha \geq 0$ (a kind of Zipf-like or power-law distribution). The parameter α controls the density of long-range neighbors at all distance scales. When $\alpha = 0$ the long-range neighbors are distributed uniformly. As α increases, they become more clustered. Effective greedy routing is possible for $\alpha = 2$; paths are polylogarithmic with the expected length $O(\log^2 N)$.

The small-world hierarchy of node's neighborhood invents the following DHT design principle.

Principle 1 (Small-world neighborhood). *Neighbors are at all distance scales. More distant neighbors are sparser in a local routing table.*

An example DHT that does not follow Principle 1 is CAN [42]. Recall that the CAN ID space is the unit n-dimensional Cartesian coordinate space $[0, 1)^n$ mod 1 (torus) with Euclidean distance, a continuous analog of Kleinberg's small-world grid. The coordinate space is partitioned into hyper-rectangles (zones). Each node is responsible for a zone (for keys from this zone). Two nodes are neighbors (symmetrically) if their zones share an $n - 1$ dimensional hyperplane, hence there are only local neighbors. (Their number is $m' = 2n$.) Greedy routing takes $O(nN^{1/n})$ hops, and paths are longer than polylogarithmic. Only if the dimension satisfies $n = \Theta(\log_b N)$ for some $b > 1$ or a more aggressive asymptotic dependence $n = \Omega(\log_b N)$ then the state and routing complexity can achieve $O(\log N)$ bound.

Consider the following definition. It provides a family of routing strategies, including greedy routing. A hop $u \to v$ for k is called *b-progressive* (geometrically progressive with the ratio $1/b$) if

$$\rho(v, k) \leq \frac{\rho(u, k)}{b} \quad \text{for } b > 1. \tag{3.4}$$

As a result, the current distance is reduced at least by b, so considered as efficient.[1]

As we shall see, many DHT designs with logarithmic-complexity routing achieves geometrically progressive routing by appropriate selection of long-range neighbors.

3.2.2 Case Study: Symphony

Symphony [31] is a particular inspiration of Kleinberg's small-world construction. The ID space is a ring (the unit interval $[0, 1)$ mod 1). The distance $\rho(u, k)$ is the

---

[1] In [52], such a routing strategy is called frugal; the corresponding paths are called geometric.

Fig. 3.2 Ring division for u
in Symphony: long-range
neighbors are from the whole
ring except the short arc
(length $= 1/N$) immediately
clockwise u

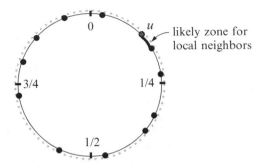

likely zone for
local neighbors

clockwise arc length from u to k. It is asymmetric, and routing is unidirectional
(clockwise). A node u is responsible for the arc $[u, v)$ where v is closest to u.

Assume the uniform distribution of nodes in S. Every node maintains $m' = 2$
local and $m \geq 1$ long-range neighbors. Local neighbors are the closest nodes on
the ring. A node u selects a long-range neighbor v closest to the point $u + \delta$
mod 1 for a random δ. For simplicity, assume $v = u + \delta$ mod 1. The generation
of δ is independent with the probability density function[2] $h(x) = 1/(x \ln N)$ for
$x \in [1/N, 1)$, where $N \geq 2$ is estimated locally. Intuitively, $\delta = 1/N$ is the mean
distance to u's closest node and $\delta = 1$ is supreme for the most distant one. Hence
all candidates for long-range neighbors are in $[u + 1/N, u + 1)$ mod 1, see Fig. 3.2.

The following theorem shows a perfect example of probabilistic application of
geometrically progressive routing (3.4), leading to efficient P2P routing.

Theorem 3.1. *The Symphony routing algorithm has polylogarithmic routing efficiency on average.*

Proof. Let u serve a lookup to key k and $v \in T_u$. The hop $u \to v$ reduces the current
distance at least by half if $v = (u + \delta)$ mod 1 for $\delta \in [\rho(u,k)/2, \rho(u,k))$. The
probability is

$$p = \int_{\rho(u,k)/2}^{\rho(u,k)} h(x)dx = \frac{1}{\log_2 N}, \tag{3.5}$$

which is independent on u and k. Note that $p(N) \to 0$ when $N \to \infty$.

The probability that there is a 2-progressive $v \in T_u$ for a given k is

$$p_m = 1 - (1-p)^m = 1 - \left(1 - \frac{1}{\log_2 N}\right)^m. \tag{3.6}$$

If $m = \Omega(\log_2 N)$, i.e., $m \geq \alpha \log_2 N$ for a constant $\alpha > 0$, then for large N,

$$p_m \geq 1 - \left(1 - \frac{1}{\log_2 N}\right)^{\alpha \log_2 N} \xrightarrow[N \to \infty]{} 1 - e^{-\alpha}. \tag{3.7}$$

---

[2]This PDF is often called harmonic. It is an instance of the power-low family of distribution
functions.

For instance, $p_m \approx 0.63$ for $m \approx \log_2 N$. That is, $O(\log N)$ routing state (typical to many DHTs) leads to the high probability of geometrically progressive hops.

Consider a path $w_i \rightarrow^+ w_{i+1}$ with l_i hops where only the last one is at least 2-progressive. Any forwarding node except the penultima has no 2-progressive hops in its routing table. A sufficient condition for the 2-progress is $\rho(w_i, k) \geq 2\rho(w_{i+1}, k)$. Note that in greedy routing a node always selects a geometrically progressive hop if it is available. The path hop length l_i is a random variable having the geometrical distribution

$$\Pr\left[|w_i \rightarrow^+ w_{i+1}| = l_i\right] = (1 - p_m)^{l_i - 1} p_m$$

with the mean $1/p_m$. In addition, there are cases when the distance $\rho(w_i, k)$ is halved by a sequence of short hops; it can only increase the probability of shorter paths $w_i \rightarrow^+ w_{i+1}$. In total, the expected length is $E[l_i] = O(1/p_m)$, which by (3.7) can be bounded above by a constant if N is large and $m = \Omega(\log_2 N)$.

For the fullness consider a less efficient strategy (in fact, the original proof in [31] used it) is the nodes on $w_i \rightarrow^+ w_{i+1}$ select the next hops randomly. It constructs longer paths before the current distance diminishes at least by half since an intermediate node can have a 2-progressive hop but it was not selected. The path construction also leads to geometrical distribution with the mean $1/p = \log_2 N$:

$$\Pr\left[|w_i \rightarrow^+ w_{i+1}| = l_i\right] = (1 - p)^{l_i - 1} p.$$

The expected length $E[l_i] = O(\log_2 N)$ by (3.5).

Return to greedy routing case (3.6) in Symphony. Since $p(N)$ tends to zero for large N the expansion to Maclaurin series gives

$$p_m = 1 - (1 - p)^m = 1 - \sum_{i=0}^{\infty} C_m^i (-1)^i p^i = mp + o(p),$$

where C_m^i are generalized binomial coefficients. Consequently, the general bound is $E[l_i] = O\left(\frac{\log_2 N}{m}\right)$.

Now consider routing from the source node w_0 to the responsible node $d = w_l$

$$w_0 \rightarrow^+ w_1 \rightarrow^+ \cdots \rightarrow^+ w_{l-1} \rightarrow^+ w_l = d. \tag{3.8}$$

Compared with (3.1), it consists of multi-hop subpaths $w_i \rightarrow^+ w_{i+1}$. Every subpath at least halves the distance. Hence $N_{w_i k} < N/2^i$, and after $l = O(\log_2 N)$ steps $N_{w_l k} = 0$. Therefore, the Symphony greedy routing algorithm constructs paths of the expected length $l E[l_i]$, and two important cases for node state complexity are

$$E\left[|u \rightarrow^+ d|\right] = \begin{cases} O\left(\dfrac{1}{m} \log^2 N\right) & \text{if } m = O(1), \\ O(\log N) & \text{if } m = \Omega(\log N). \end{cases} \tag{3.9}$$

$\square$

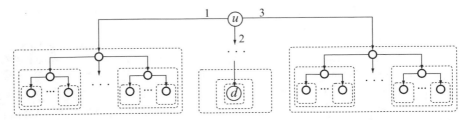

Fig. 3.3 A path construction hierarchy $H[u \rightarrow^+]$. For each key, u selects a b-progressive hop (options 1, 2 and 3 are available). *Dashed rectangles* show the closeness to destinations

Based on Principle 1, a given network can be augmented with long-range neighbors so that $O(\text{polylog} N)$ greedy routing is possible [2, 11]. For instance, eCAN (CAN augmented with "expressways") provides $O(\log N)$ routing [51].

3.2.3 Path-Based Hierarchies

The key property of the b-progress is geometrical distance reduction (3.4). Although nodes select the next hop independently on the previous part of path, the following design principle preserves the relation between hops.

Principle 2 (Geometrically progressive routing). *Given a constant $b > 1$. Any node u has a b-progressive hop $v \in T_u$ for an arbitrary key $k \in S$. In resolving lookups, u always selects b-progressive hops.*

The principle defines *the path construction hierarchy* $H[u \rightarrow^+]$ that describes the path construction starting from a node u, see Fig. 3.3. For any key, u selects a b-progressive hop, and the lookup arrives to a geometrically closer area of the destination d. Every next-hop node does the same. A level in $H[u \rightarrow^+]$ characterizes the distance scale; the lower is the level, the larger part of the overlay is crossed.

In this hierarchy, u does not consider "backward" paths in the overlay since every node keeps neighbors to cover all destinations, including u and its vicinity. Therefore, $H[u \rightarrow^+]$ is a subgraph of the overlay topology graph. The hierarchy is local since it is defined from the point of view of each individual node u.

The following properties clarify the definition of $H[u \rightarrow^+]$.

1. $H[u \rightarrow^+]$ consists of all N nodes. Substantiation: any node d is reachable from u.
2. For any path $w' \rightarrow w \rightarrow w''$ in $H[u \rightarrow^+]$, always w is in between w' and w'' in terms of the space metric ρ, i.e., $\rho(w, w'') < \rho(w', w'')$. Substantiation: there are no backward links; the remaining route $w \rightarrow w''$ is always shorter.
3. For b-progressive routing, the condition from the previous property is reduced to $b\rho(w, w'') \leq \rho(w', w'')$. Substantiation: distance scale decreases geometrically.

Note that $H[u \rightarrow^+]$ is not a tree since the uniqueness in selecting the next hop is not guaranteed. Cycles can exist, see Fig. 3.4 for an illustration.

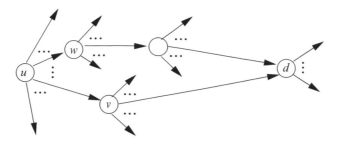

Fig. 3.4 An example of cycle appearance in $H[u \rightarrow^+]$. Although the path $u \rightarrow^+ d$ through w is longer (in hops) than through v, it provides an alternative, e.g., if v is failed

In general, the more cycles are in $H[u \rightarrow^+]$ the lower is the hierarchy factor. In this sense, greedy routing (3.3) is more hierarchical than progressive routing (3.2). The cycle existence is, however, useful for routing, e.g., security and resilience benefit from existence of many alternative paths. Geometrically progressive routing allows tradeoffs preserving the logarithmic bound for path lengths.

Theorem 3.2. *Geometrically progressive routing has logarithmic routing efficiency in the worst case.*

Proof. Given multihop path (3.1). Initially $N_{uk} < N$. Then $\rho(w_1,k) \leq \rho(u,k)/b$, and the proportional reduction is $N_{w_1 k} < N/b$. At the next hop,

$$\rho(w_2,k) \leq \rho(w_1,k)/b \leq \rho(u,k)/b^2 \quad \text{and} \quad N_{w_2 k} < N/b^2.$$

In the worst case the upper bound reduces to $N/b^l \leq 1$ ensuring $N_{dk} = N_{w_l k} = 0$ and yielding $l \leq \log_b(N) = O(\log_b N)$.

Note that only the strict bound $b > 1$, not the concrete value of b, is essential for the logarithmic bound. For any $b > 1$ the majorant sequence $\{N/b^j\}_{j=1}^l$ monotonically decreases, approaching $0 \leq N_{w_l k} < N/b^l \leq 1$ in the worst case. Since $\log_b N = \log_2 N / \log_2 b$, the logarithm base is not important for the O-notation, and $l = O(\log_b N) = O(\log_2 N)$.   □

Note that the progress $b_j > 1$ at each hop j may be different. Then there exists a majorant sequence:

$$\{N/(b_1 b_2 \cdots b_j)\}_{j=1}^l \leq \{N/b^j\}_{j=1}^l \quad \text{with} \quad b = \min\{b_j \mid j = 1, 2, \ldots, l\} > 1.$$

The distance metric is *unidirectional* if there is at most one $y \in S$ such that $\rho(x,y) = \delta$ for any given $x \in S$ and $\delta > 0$. In terms of the ranking model, u can arrange uniquely all points $k \in S$ according to $\rho(u,k)$. Consequently, greedy routing converge all lookups for the same key along the same path, regardless of the source node [18]. It leads to *path destination hierarchy* $H[\rightarrow^+ d]$, a backward structure to the path construction hierarchy $H[u \rightarrow^+]$.

Fig. 3.5 A path destination
hierarchy $H[\to^+ d]$. If
$|u \to^+ d| = l$ then u is on
layer l. Similarly to
Principle 1, the larger l the
more nodes on layer l

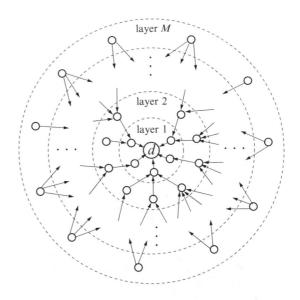

Intuitively, each d is a root of a tree containing all N nodes, a kind of spanning
tree of the network topology graph. As shown schematically in Fig. 3.5, $H[\to^+ d]$ is
constructed by all routing paths from other nodes to d. Different destinations define
different trees. At each step of progressive routing (3.2), the lookup travels to a node
that is closer to the root in (S, ρ). Immediate d's children are those nodes that keep
d in their routing tables; they form layer $l = 1$. Source nodes of two-hop paths to d
form layer $l = 2$. The nodes for which d is most distant in hops form the top layer
$l = M$.

The intersection $H[u \to^+] \cap H[\to^+ d]$ defines a set of possible paths $u \to^+ d$. In
greedy routing (3.3), next hops are chosen deterministically, and there are no cycles
in $H[\to^+ d]$. Consequently, the path $u \to^+ d$ is unique. Progressive routing (3.2)
allows alternatives in selecting next-hop nodes, and the hierarchy factor of $H[\to^+ d]$
becomes lower because of cycles.

3.3   Routing in Greedy DHTs and de Bruijn Graphs

Principle 2 heightens Principle 1 by ensuring that for any key a node knows a
b-progressive hop. In Symphony, the long-range neighbor selection follows the
ranking model for the distance scales. Symphony approximates geometrically
progressive routing since some hops in (3.8) are not 2-progressive. As a result,
constructing $O(\log N)$ paths is not possible when the number of neighbors per node
is constant, see $m = O(1)$ in (3.9) and Theorem 3.1. Now, consider two important
classes of DHT topologies: greedy DHTs, which have become very popular in

Table 3.1 Popular greedy DHTs that employ geometrically progressive routing

DHT design	ID space S	Distance metric $\rho(u,k)$
Chord [48]	n-bit numbers uniformly projected to the ring ($b=2$)	$\rho(u,k) = \begin{cases} k-u, & \text{if } u \leq k, \\ 2^n - (u-k), & \text{otherwise.} \end{cases}$ Clockwise arc length between u and k
Tapestry [55]	n-digit numbers in base $b \geq 2$	$\rho(u,k) = \sum_{j=0}^{n-1} |u_j - k_j| b^j$ where $u = \sum_{j=0}^{n-1} u_j b^j$ and $k = \sum_{j=0}^{n-1} k_j b^j$. Longer the prefix (in digits) shared among u and k, closer these IDs
Pastry [45]	Similarly to Tapestry with $b = 2^c$ for $c \geq 1$	The same as in Tapestry
Kademlia [33]	Similarly to Tapestry with $b = 2$	The same as in Tapestry. Equivalent to the bitwise exclusive OR (XOR), $\rho(u,k) = \sum_{j=0}^{n-1} (u_j \oplus k_j) b^j$

practical implementations, and de Bruijn P2P networks, which approximates de Bruijn graphs in the topology. For these classes, the design can follow Principle 2 accurately.

3.3.1   Conventional Greedy DHTs

Many DHTs with greedy routing use the classifying model (discrete) for neighbors, instead of the ranking model (continuous, ranks reflect the distance). As a result, routing becomes geometrically progressive. Examples include Chord [48], Pastry [45], Tapestry [55], and Kademlia [33].

Recall that the above DHTs use ID space S consisting of all nonnegative integers with $n \geq 1$ digits in base $b \geq 2$, $u = \sum_{j=0}^{n-1} u_j b^j$ for $0 \leq u_j < b$, and $N \leq b^n$ (see Table 3.1). A node u discretely partitions S onto disjoint zones

$$S_i(u) = \left\{ k \in S \mid b^i \leq \rho(u,k) < b^{i+1} \right\}, \quad i = 0, \ldots, n-1. \tag{3.10}$$

According to Principle 1, more distant zones are larger. In terms of the distance ρ, each $S_i(u)$ is an annulus $B_{b^{i+1}}(u) \setminus B_{b^i}(u)$, the area between the two concentric balls, where an r-radius ball with center u is defined

$$B_r(u) = \{ k \in S \mid \rho(u,k) < r \}.$$

In Pastry and Tapestry, u keeps $b - 1$ long-range neighbors from each zone. For each $S_i(u)$, there is a neighbor v satisfying $\rho(v,k) \leq b^{i-1}$ for any $k \in S_i(u)$. Let us make explanation in terms of ID prefixes [40]. If $v \in S_i(u)$ then v, u and any

$k \in S_i(u)$ share the $(n-i-1)$-digit prefix. For each of $b-1$ values for the next digit $j = 0, 1, \ldots, b-1$, $j \neq u_{n-i-1}$ that can follow this prefix in $k \in S_i(u)$, a long-range neighbor $v \in S_i(u)$ is stored in T_u. A lookup for k finds $v \in T_u$ such that v is one digit closer to k, i.e., $\rho(v,k) \leq b^{i-1}$, and the hop is b-progressive:

$$\frac{\rho(u,k)}{\rho(v,k)} \geq \frac{b^i}{b^{i-1}} = b.$$

In Chord and Kademlia, u keeps at least one long-range neighbor from each zone. In Kademlia, the distance ρ is symmetrical, thus $\rho(v,k) < 2^{i+1} - 2^i = 2^i$ for any $k \in S_i(u)$ and ith neighbor v. In Chord, the distance metric is asymmetrical. In deterministic Chord the ith neighbor v is closest to u in $S_i(u)$, thus also $\rho(v,k) < 2^i$ for any $k \in S_i(u)$. In routing a lookup, u uses the ith neighbor v. The hop $u \to v$ is geometrically progressive since $\rho(u,k)/\rho(v,k) > 1$.

The Chord distance metric is transitive; $\rho(u,k) = \rho(u,v) + \rho(v,k)$ if v is in between u and k. Without loss of generality, the latter can always be assumed for deterministic Chord. Hence, having one ith neighbor ensures the 2-progress for deterministic Chord:

$$\frac{\rho(u,k)}{\rho(v,k)} = \frac{\rho(u,v) + \rho(v,k)}{\rho(v,k)} \geq \frac{2^i}{2^{i+1} - 2^i} + 1 = 2.$$

In randomized Chord as well as in Kademlia, v can be any node from $S_i(u)$, thus v is not always in between u and k.

For 2-progress in Kademlia, a node u would follow the same prefix-aware strategy for selecting $v \in S_i(u)$ as in Pastry and Tapestry. Another way is halving $S_i(u)$ and keeping two neighbors $v_1, v_2 \in S_i(u)$ for each half such that v_1 and v_2 are closest to u in the annuli

$$B_{3 \cdot 2^{i-1}}(u) \setminus B_{2^i}(u) \quad \text{and} \quad B_{2^{i+1}}(u) \setminus B_{3 \cdot 2^{i-1}}(u), \quad \text{respectively.}$$

Kademlia requires every u to keep several neighbors for each $S_i(u)$; the recommended value is 20. Hence for a randomly selected $k \in S_i(u)$, the probability of having $v \in T_u$ such that $\rho(v,k) < 2^{i-1}$ is high. This strategy is also suitable for randomized Chord, when some constant increment of the routing table size improves the ratio of geometrically progressive routing.

As in the small-world construction, these greedy DHTs also support local neighbors: a list of successors in Chord,[3] a leaf set[4] in Tapestry, leaf and neighborhood[5] sets in Pastry, nodes from $S_0 \cup S_{n-1}$ in Kademlia.[6]

---

[3] A Chord node also keeps predecessors; they are local neighbors with respect to the absolute symmetrical distance $\min\{\rho(u,v), \rho(v,u)\}$.

[4] A Tapestry or a Pastry leaf set consists of $2m_0$ numerically closest nodes (for the distance $|v-u|$): m_0 clockwise plus m_0 anticlockwise.

[5] Pastry neighborhood set consists of proximity closest nodes.

[6] In Kademlia (as in Chord) the closest successors of u are the first subsequent nodes in $S_0(u)$ and the closest predecessors are the last subsequent nodes in $S_{n-1}(u)$.

Summarizing, the considered greedy DHTs have the following properties.

1. Pastry and Tapestry ensures b-progressive routing by keeping $b-1$ neighbors from each of n zones ($m = O((b-1)\log_b N)$ in total), where b is the base of numerical n-digit IDs.
2. Deterministic Chord ensures 2-progressive routing by keeping the closest neighbor from each power-two zone ($m = n = O(\log_2 N)$ in total).
3. Randomized Chord and Kademlia ensure b-progressive routing for some $b > 1$ by keeping at least one neighbor from each power-two zone. The probability that a hop is 2-progressive becomes higher when keeping more neighbors per zone.

3.3.2 Chord-Like DHTs

A Chord-like DHT is a generalization of the conventional Chord DHT by using arbitrary base $b \in \mathbb{Z}$, $b \geq 2$, instead of the default base $b = 2$. The ID space can be seen as a discrete ring of b^n points from 0 to $b^n - 1$. A node takes its ID $u \in [0, b^n)$ and arithmetic on node IDs is always on modulo b^n. The topology is constructed by links (fingers) that go from u to a node $v \in S_i(u) = [u + b^i, u + b^{i+1})$ for $i = 0, 1, \ldots, n$. In deterministic topology, v is always the first node $S_i(u)$. In randomized topology, v can be taken arbitrary from $S_i(u)$. As a result, local state is $O(\log_b N)$ and lookup complexity is $O(\log_b N)$ hops. Note that a bidirectional variant is also possible when additional links (anti-clockwise fingers) from u to a node $v \in S_i(u) = [u - b^{i+1}, u - b^i)$ are maintained for $i = 0, 1, \ldots, n$.

Intuitively, routing hops are "jumps" of size b^i. The jump sizes $\sigma_i = |S_i(u)| = b^i$ are the same for at all nodes. In the terminology of Xu et al. [50] this property leads to uniform routing algorithms. We will refer to the set $\{\sigma_i\}_{i=0}^{n-1}$ as the jump set. Consider continuous generalization [50] when $x \in \mathbb{R}$, $0 < x < 1$. The jump set is $\{x^{i+1}|S|\}_{i=0}^{n-1}$, where the number of neighbors n is selected such that $x^n|S| \approx 1$. The discrete case is reduced to the continuous one by taking $x = 1/b$.

The problem is to find a jump set that reduces the overlay network diameter. Let the ID space be normalized into a unit interval $[0, 1)$, i.e., we move to the arithmetic on modulo 1. Then the jump sizes are $\sigma_i = x^i$. The goal here is to approximate every real number $y \in [0, 1)$ using these jump sizes in a "greedy" fashion, when allowing a small "remainder". This requirement is achieved by taking $x = \sqrt{2} - 1 \approx 0.414$, which is the root of the equation $1 - 2x = x^2$. Given number $y \in [0, 1)$ to approximate, there are three cases at the very beginning:

(a) If $y \in [0, x)$ then the approximation is made.
(b) If $y \in [x, 2x)$ then subtract x from it (a jump of size x in the normalized space) and the remainder $y - x$ is in $[0, x)$.
(c) If $y \in [2x, 1)$ then subtract x from it two times, and the remainder $y - 2x$ is in $[0, x^2)$.

The above procedure will be repeatedly executed in a recursive and greedy fashion. The intuition of the steps (a)–(c) is the following. If y belongs to case (a), it is already "better-off" in terms of path length. If y belongs to case (b) or (c), then one or two additional jumps of size x are needed to reduce the remainder to case (a). Since case (c) requires one more jump than case (b), we compensate this difference by allowing its remainder to jump to the region $[0, x^2)$ (note that $1 - 2x = x^2$) instead of $[0, x)$ as in case (b). In this way, we "equalize" the cost to approximate numbers in regions $[x, 2x)$ and $[2x, 1)$. Note that such equalization is done in a recursive way, spreading its "equalization" benefit recursively.

The routing protocol is the same as in the original Chord. The number of neighbors is reduced to $n \approx \log_{1/x} N \approx 0.786 \log_2 N$, which is 21.4 % less than in the original Chord with the exactly $\log_2 N$ routing table size. The average path length is approximately $0.614 \log_2 N$, that is 22.7 % greater than in Chord.

Cordasco et al. [7] provided a further advanced scheme, called F-Chord(α). Its jump sizes are calculated based on Fibonacci numbers and $1/2 \le \alpha \le 1$ is a tuning parameter. The model aims at better tradeoffs between overlay path length and routing table size.

Let f_i be Fibonacci numbers for $i = 0, 1, \ldots$, which are defined as $f_0 = 0$, $f_1 = 1$ and $f_i = f_{i-2} + f_{i-1}$ for $i > 1$. Fix n such that $f_{n-1} < N \le f_n$. Note that $m \approx \log_\phi N \approx 1.44 \log_2 N$, where $\phi = (1 + \sqrt{5})/2$ is the golden ratio. The basic idea is that the difference between two consecutive jumps is the preceding jump, i.e., $\sigma_{i+2} - \sigma_{i+1} = \sigma_i$. The maximum number of hops in greedy routing does not exceed half the size of the finger table. This basic case corresponds to $\alpha = 1$. In general, F-Chord(α) uses the set of $[\alpha(n-2)]$ jumps

$$\sigma_i = \begin{cases} f_{2i}, & i = 1, 2, \ldots, [(1-\alpha)(n-2)], \\ f_i, & i = 2[(1-\alpha)(n-2)] + 2, \ldots, n-1, \end{cases}$$

where a certain quantity of the jumps with odd indices are eliminated.

For any value of α, the diameter of an F-Chord(α) overlay network is half the size of the routing table: $[n/2] \approx 0.72 \log_2 N$, i.e., the worst-case path length is lower than in Chord. The average path length is upper bounded with

$$0.398 \log_2 N + (1 - \alpha) 0.248 \log_2 N + 1.$$

The F-Chord(α) scheme can be further extended [6] to base $b \ge 2$, where b is increased in order to reduce the number of hops at the expense of increased routing table size. It remains logarithmic in size, but the constant factor is higher.

3.3.3   De Bruijn Graphs

De Bruijn graph $\mathscr{B}(b, n)$ is a base for a rich family of DHT designs, including Koorde [21], Distance-halving [36, 38], D2B [13], ODRI [30], and Broose [15]. They also follows Principle 2 of geometrically progressive routing.

Fig. 3.6 The de Bruijn graph
$\mathscr{B}(2,3)$; the picture is
from [13]

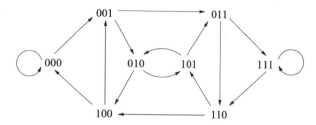

As in greedy DHTs, the ID space S consists of n-digit nonnegative integers in base b. The distance discretely depends on the length $l(u,v)$ of the longest suffix of u that equals the prefix of v,

$$l(u,v) = \max\{j \mid u_{j-1}=v_{n-1}, u_{j-2}=v_{n-2}, \ldots, u_0=v_{n-j}\},$$

where $u = u_{n-1}\cdots u_1 u_0$ and $v = v_{n-1}\cdots v_1 v_0$. If the maximum does not exists then $l(u,v) = 0$. For example, $l(110,101) = 2$, $l(101,110) = 1$, and $l(011,001) = 0$. The larger $l(u,v)$, the closer the nodes. Therefore, $l(u,v)$ follows the classifying model and defines n distance scales.

Clearly, $l(u,v)$ is not symmetric. Furthermore, if $l(u,v) = l(u,w)$ then an additional metric makes further differentiation. For instance, Koorde uses the Chord distance, Distance-halving uses the absolute numerical difference, Broose uses the XOR distance as in Kademlia.

Similarly to the greedy DHTs, a node u partitions S onto several zones and selects a neighbor for each of them. In contrast, the number of zones is equal to b without dependence on the ID length n, hence leading to $O(1)$ state. Each zone $S_i(u)$ is determined by $u_{n-2}u_{n-3}\cdots u_0 i$ for $i = 0, 1, \ldots, b-1$, i.e., the node ID is shifted to the left and appended with digit i on the right. The corresponding de Bruijn graph for $b = 2$ and $n = 3$ is shown in Fig. 2.10 from Chap. 2.

In the greedy DHTs, each zone represents one distance scale. In the de Bruijn graph approach, $S_i(u)$ simultaneously covers all distance scales $j = 0, \ldots, n-1$ and is responsible for all keys k such that $j = l(u,k)$ and the $(j+1)$-digit prefix is $u_{j-1}\cdots u_0 i$. In routing a lookup for $k = k_{n-1}\cdots k_1 k_0$, a node u computes $j = l(u,k)$ and forwards to its neighbor $v \in S_{k_{n-j}}(u)$, close to the point

$$u_{n-2}u_{n-3}\cdots u_j u_{j-1}\cdots u_0 k_{n-j-1}$$
$$= u_{n-2}u_{n-3}\cdots u_j k_{n-1}\cdots k_{n-j}k_{n-j-1}$$

For the example in Fig. 2.10, the lookup for $k = 101$ at node $s = 100$ follows the path $100 \rightarrow 001 \rightarrow 010 \rightarrow 101$.

The same neighbor is applicable at different distance scales; a lookup key determines what scale is needed. Nevertheless, Principle 1 is preserved since the number of nodes at distant scale j is proportional b^{n-j}. Each routing step resolves at least one digit in IDs, reducing the number of nodes in between at least by the factor b in accordance with Principle 2.

The idea of geometrically progressive routing in de Bruijn graphs is further developed in Kautz graphs, which we shall discuss in Sect. 4.7.

3.4   Adaptation to Global Hierarchy

Principle 1 is a base of small-world strategies for selecting neighbors. Its combination with Principle 2 leads to discrete partitions on to distance-scale zones $S_i(u)$ in (3.10) and local hierarchies $H[s \rightarrow^+]$ and $H[\rightarrow^+ d]$ in the network topology. In turn, this structure allows adapting to other hierarchies of the network. For instance, if the topology embeds a global hierarchy then the distance between two nodes depends on the height of their lowest common ancestor in the hierarchy tree.

3.4.1   Kleinberg Tree-Based Model

First, consider an explaining theoretical example from [23]. Let the global hierarchy be represented as a complete b-ary tree $\mathscr{T}$, where b is a constant ($b \in \mathbb{Z}_+$, $b \neq 0$) and the height of $\mathscr{T}$ is $\log_b N$. The leaves are nodes N. For two leaves u and v, the height of their least common ancestor in $\mathscr{T}$ is $h(u,v)$. Clearly, $h(u,u) = 0$.

The network topology is stochastically adapted to the hierarchy as follows. The probability that u selects v a neighbor is equal to

$$p_{uv} = \frac{f(h(u,v))}{\sum\limits_{w \in N, w \neq u} f(h(u,w))}, \quad \text{where } f(h) = b^{-h}.$$

It can be proved that the normalizing constant is bounded

$$\sum\limits_{w \in N, w \neq u} f(h(u,w)) = \sum\limits_{w \in N, w \neq u} b^{-h(u,w)} \leq \log_b N.$$

According with this probability distribution, each u creates m links, choosing the neighbor v independently and with repetition allowed. Assume logarithmic node state $m = \Theta(\log_b^2 N)$. Let us call this topology Kleinberg tree-based.[7]

The following theorem shows that structuring the topology using the tree-based hierarchy allows efficient routing although every node has local knowledge only.

Theorem 3.3. *A P2P network with Kleinberg tree-based topology allows a routing algorithm that achieves $O(\log_b N)$ efficiency with high probability.*

---

[7]For $m = O(1)$ the tree-based topology allows routing in $O(\log_b^4 N)$ hops [23].

Proof. Without loss of generality we assume that lookups are based on node IDs instead of resource keys. Let u serve a lookup for the target node d. Suppose that $l = h(u,d)$ and $t \in \mathscr{T}$ is the least common ancestor. Let $\mathscr{T}'$ be subtree of $\mathscr{T}$ rooted at t and $\mathscr{T}''$ be the subtree of $\mathscr{T}'$ of height $l-1$ that contains d. Obviously, $u \notin \mathscr{T}''$.

If there is $v \in T_u$ such that $v \in \mathscr{T}''$ then u can make a b-progressive hop to d, achieving $h(v,d) \leq h(u,d) - 1$. There are b^{l-1} leaves (network nodes) in $\mathscr{T}''$. For any $v \in \mathscr{T}''$ the probability of being a neighbor of u is

$$p_u = \Pr\{v \in T_u \mid v \in \mathscr{T}''\} \geq \frac{b^{-l}}{\log_b N} \geq \frac{1}{b \log_b N}.$$

Since $T_u = m \geq C \log_b^2 N$ for a some constant C, the probability $(1 - p_u)^m$ that T_u contains no nodes from $\mathscr{T}''$ is bounded for large N

$$\left(1 - \frac{1}{b \log_b N}\right)^m \leq \left(1 - \frac{1}{b \log_b N}\right)^{C \log_b^2 N} \leq \varepsilon^{C_2 \log_b N} = 1/N + o(1/N),$$

where $0 < \varepsilon < 1$ and $C_2 > 0$ are constants. Therefore, T_u contains a node from $\mathscr{T}''$ with high probability.

Then the required algorithm states each node u on the routing path to d to select $v \in T_u$ such that $v \in \mathscr{T}''$, where $\mathscr{T}''$ depends on u and d. From the foregoing it follows that such v is in T_u with high probability.

No explicit construction of $\mathscr{T}''$ is needed since u can select v that satisfies $h(v,d) < h(u,d)$. In the worst case, $h(u,d) = \log_b N$. Each hop reduces this discrete distance at least by 1, so the target is reached in $O(\log_b N)$ hops. $\qquad \square$

3.4.2 *Proximity-Based Selection*

Now consider the crucial practical instance of global hierarchy—the underlying network topology. It reflects the domain-based Internet architecture. The hierarchy defines the proximity of nodes in terms of latency. The height of their lowest common ancestor of u and v positively correlates with the latency of communications between u and v. Let $\tau(u,v)$ be a latency metric estimated at u [39]. The latency of l-hop path (3.1) is

$$|u \to^+ d|_\tau = \sum_{i=0}^{l-1} \tau(w_i, w_{i+1}).$$

An obvious property of geometrically progressive routing is the worst case bound

$$|u \to^+ d|_\tau = O(\Delta \log N) \quad \text{for the latency diameter } \Delta = \max_{u,v \in N} \tau(u,v). \tag{3.11}$$

Neighbor and route selection schemes (PNS and PRS) are known for adapting to the proximity [3,9,17,29,52,54]. In PNS, a node u selects neighbors to T_u based on their proximity. In PRS, u selects next-hop nodes from T_u based on their proximity. The schemes aim at short paths not just in terms of overlay hops but also in terms of network latency. The aim is achieved with local optimization of latency at each hop. As a result, the property $\tau(w_i, w_{i+1}) \ll \Delta$ happens frequently in paths (3.1).

We treat PNS and PRS the concrete instances of the next design principle. It applies the ranking model where τ characterizes "node ranks" based on the global hierarchy.

Principle 3 (Locally adaptive selection). *Each node u locally arranges other nodes according to τ. In neighbor selection, u finds v that minimizes $\tau(u,v)$ at a given distance scale (defined by ρ). In routing, the selection of next hops uses composite criteria with $\tau(u,v)$ and $\rho(u,v)$.*

DHTs with geometrically progressive routing can be modified to follow Principle 3. In geometrically progressive routing, a node u keeps a neighbor for each distance-scale zone (3.10). Randomized strategies allows selection among several candidates in $S_i(u)$, giving the flexibility that PNS requires. Let u probe $m_i > 1$ nodes $\{v_j\}_{j=1}^{m_i}$ from each distance-scale zone $S_i(u)$ and select the long-range neighbor with lowest $\tau(u,v_j)$. Probing all nodes in $S_i(u)$ is enormously expensive, and m_i is a tradeoff parameter.

The PNS scheme introduces a new hierarchy level when u additionally arranges nodes in $S_i(u)$ according to the proximity. In routing a lookup for k, the next-hop node v is proximity closest to u among all geometrically progressive neighbors.

There are networks where the worst case in (3.11) can be approached closely despite the use of PNS. We say that the network has exponential latency expansion if for any $u \in N$

$$N_t(u) = |\{v \in N \mid \tau(u,v) \leq t\}| = \Theta(\alpha^t) \quad \text{for a constant } \alpha > 1.$$

In other words, the number of nodes that are within latency t of u grows exponentially.

Theorem 3.4. *If a network has exponential latency expansion, then the expected latency $\mathsf{E}[|u \to^+ d|_\tau] = \Omega(\Delta \log N)$ for b-progressive paths $u \to^+ d$ with nodes $u,d \in N$ chosen at random.*

Proof (Sketch, other details can be found in [52]). Take a random node $w \in N$. With the assumption on uniform node ID distribution in S, the set $S_i(w)$ itself is chosen uniformly at random from all subsets of N of size $b^{i+1} - b^i = (b-1)b^i$, where $i = 0, 1, \ldots, n-1$. Recall that $N \leq b^n$ and for large N we assume $n = \Theta(\log_b N)$.

From the property of exponential latency expansion one can derive that

$$\tau_i(w) = \mathsf{E}\left[\tau(w,v) \mid v \in S_i(w)\right] \geq C_i \Delta \left(1 - i/n\right) \quad \text{for a constant } C_i > 0,$$

where $\tau_i(w)$ is the expected latency between w and $S_i(w)$.

Given path (3.1) with random $u, d \in N$. Without loss of generality, if the path is b-progressive then assume $l = n$ and $w_{i+1} \in S_{n-i-1}(w_i)$. Consequently,

$$\mathsf{E}\left[|u \to^+ d|_\tau\right] = \sum_{i=0}^{n-1} \tau_i(w_{n-1-i}) \geq C' n \Delta - \frac{C''}{n} \sum_{i=0}^{n-1} i = \Omega(\Delta \log_b N). \qquad \square$$

Theorem 3.4 means that no significant latency improvement is possible due to PNS with probing m_i nodes from each zone $S_i(u)$, especially when m_i or $S_i(u)$ are small. Intuitively, in a network with exponential latency expansion, an overwhelming majority of the nodes will be very far from u, and finding the closest node from a small sample is unlikely to significantly improve the latency.

Now, let a network have power-law latency expansion:

$$N_t(u) = |\{v \in N \mid \tau(u, v) \leq t\}| = \Theta(t^\alpha) \quad \text{for a constant } \alpha \geq 1.$$

In contrast to networks of exponential latency expansion, a network of power-law latency expansion ensures that a node u only needs to sample a small number of nodes from each $S_i(u)$ in order to find a latency-nearby node.

Theorem 3.5. *If a network has α-power-law latency expansion, then applying PNS with m probes per zone S_i leads to*

$$\mathsf{E}[|u \to^+ d|_\tau] = O\left(\frac{\Delta}{m^{1/\alpha}} \log N\right)$$

for b-progressive paths $u \to^+ d$ with nodes $u, d \in N$ chosen at random.

Proof (Sketch, other details can be found in [52]). Let us define the minimal latency from a node $w \in N$ to a set of nodes $V \subset N$:

$$\tau(w, V) = \min_{v \in V} \tau(w, v).$$

Then one can prove that considering all random V of fixed size $p \geq 1$, the expected latency is $\mathsf{E}[\tau(w, V)] = O(\Delta p^{-1/\alpha})$. It asserts that the distance to the "closest" node in V varies as $p^{-1/\alpha}$, a crucial consequence of α-power-law latency expansion.

Any w sampled a set $S_i'(w) \subset S_i(w)$. For simplicity and ease of exposition, we assume[8] that $m_i(w) = |S_i'(w)| = m$ for all w and i. Given path (3.1) with random $u, d \in N$. Without loss of generality, if the path is b-progressive then assume $l = n$ and $w_{i+1} \in S_{n-i-1}'(w_i)$. Consequently,

$$\mathsf{E}\left[|u \to^+ d|_\tau\right] = \sum_{i=0}^{n-1} \tau(w_{n-1-i}, S_i'(w_{n-1-i})) \leq \Delta \sum_{i=0}^{n-1} C_i m_i^{-1/\alpha} = O\left(\frac{\Delta}{m^{1/\alpha}} \log_b N\right).$$

$$\square$$

---

[8]Without this assumption the proof becomes technically more complicated and an insignificantly more precise upper bound $\mathsf{E}[|u \to^+ d|_\tau] = O(\Delta) + O\left(\Delta m^{-1/\alpha} \log N\right)$ is derived, see [52].

Compared with (3.11) Theorem 3.5 provides the reduction by $m^{1/\alpha}$. Numerous experimental studies confirmed the improvement significance in practical settings [3, 9, 17, 29, 52, 54]. Note that since

$$\frac{1}{(m-1)^{1/\alpha}} - \frac{1}{m^{1/\alpha}} \geq \frac{1}{\alpha m^{1+1/\alpha}},$$

the improvement is most significant for the first few samples.

Now consider the PRS scheme. It performs better if u knows many neighbors. DHTs with geometrically progressive routing can support it by allowing u to keep several nodes from each zone $S_i(u)$ if the node capacity is enough. (Note that in Kademlia and Broose it is a principal design rule.) Then u arranges its i-zone neighbors (already in T_u) according to the proximity. In routing a lookup for k, composite criteria for selecting the next hop are applicable, e.g., the additive criterion

$$c_\rho \rho(v,k) + c_\tau \tau(u,v) \rightarrow \min \tag{3.12}$$

for some tradeoff constants $c_\rho, c_\tau > 0$. Minimizing $\rho(v,k)$ preserves geometrically progressive routing within few overlay hops. Minimizing $\tau(u,v)$ makes each hop to be of low latency.

For practical implementations the following simple algorithm can be suggested. It is approximate and separates the minimization on ρ and τ.

1. Given k, u finds m closest neighbors based on ρ; denote the set $T_u(k)$.
2. The next hop is $v = \arg \min_{w \in T_u(k)} \tau(u,w)$.

On one hand, $T_u(k)$ likely provides the best b-progress hops for k, preserving routing efficiency in hops. On the other hand, the proximity is taken into account, reducing the latency. Note that in a network of α-power-low latency expansion, the expected improvement is latency reduction by $m^{1/\alpha}$, similarly to Theorem 3.5.

The hierarchical routing scheme with several neighbors from $S_i(u)$ also supports multi-path routing. A lookup is forwarded in parallel to several next hops. Significantly better routing performance and resilience are possible, see comprehensive experiments in [15, 18, 25, 27, 29, 33, 51, 54]. For example, Kademlia exploits this scheme to tolerate node failures. Every ith set of neighbors is kept sorted by time last seen. In this case, $\tau(u,v)$ is the time elapsed from the last successful contact with v.

3.4.3 Other Criteria for Local Adaptation

In general, Principle 3 supports arbitrary hierarchy-based metrics τ, not only latency-oriented. For instance, [10] adapts in the topology another practical global hierarchy—the bootstrap tree $\mathcal{B}$ from social networking, used for a trust metric. When joining to the network, a node u has to contact an existing node v

(the first contact). The knowledge of v is previous off-line relationship between u and v. The set of all relationships defines $\mathcal{B}$. Each node u for each neighbor v stores additionally (in T_u) the path from u to v in $\mathcal{B}$.

The approach aims at secure routing (sybil-resistance) when an adversary convinces good nodes to allow it to join the network. Then the adversary introduces a large number of sybils via this attachment point. Assuming the complexity of obtaining an attachment point is much higher than introducing sybils, $\mathcal{B}$ includes only a small number of large subtrees of sybils. Consequently, there is a high probability that a path of good nodes exists in $\mathcal{B}$ between any two good nodes.

Routing is iterative. A node u at each hop $u \to v$ receives from v the whole T_v, thus u can construct the path to any $w \in T_v$ in $\mathcal{B}$. Then u selects the appropriate next hop. Similar to the case of proximity, the selection uses a trust metric $\tau(u,v)$ of the path from u to w in $\mathcal{B}$ (diversity routing). The various composite criteria are applicable, convex

$$c\rho(v,k) + (1-c)\tau(u,v) \to \min \quad \text{for } 0 \le c \le 1 \tag{3.13}$$

or zig–zag (alternation of progressive and diversity routing)

$$c\rho(v,k) + (1-c)\tau(u,v) \to \min \tag{3.14}$$

when each step alternates $c = 0, 1, 0, 1, \ldots$. Note that (3.13) is a particular case of (3.12); it needs to measure the distance ρ and trust τ in comparable units.

A global hierarchy can be related to the system load balance. For instance in [47], each node estimates the load of its neighbors and replaces the most loaded neighbors with other nodes. In this scheme, τ is proportional to the load inversion. The goal is minimizing the number of lookups forwarded to the nodes that owns popular resources (hot spots). The idea is evolved further to algorithms that group nodes for efficient load-balancing (see Chap. 4). Neighbor and route selection with τ reflecting node reliability and capacity was introduced in [26].

3.5 Global Routing and Lookup Structure

Principle 3 allows each node to locally adapt to certain characteristics of a global hierarchy defined by a metric τ. The latter is used for extrapolation of the local knowledge. Nevertheless, a node u would prefer additional more precise knowledge about the rest of path $u \to v \to^+ d$ beyond the hop $u \to v$.

In progressive routing, the first hops in $u \to^+ d$ are likely large in distance ρ. For instance, greedy routing tries at each hop to cross as large part of the overlay as possible. Let us consider routing divided logically into global and local parts. *Global routing* delivers a lookup to destination vicinity. In *local routing*, the destination is at a nearby node.

According to Principle 1, a node knows its neighborhood rather well. First, many nearby nodes are local neighbors. Second, the lower distance scale the more neighbors in a routing table. In this sense, local routing has more information than global routing. Therefore, global routing is an important subject for further optimization based on additional knowledge. Principle 2 is mostly for global routing since the geometrical reduction aims at large hops.

Principle 4 (Look ahead). *A node utilizes additional knowledge about paths beyond its neighbors to optimize global routing.*

Various hierarchical schemes can be defined on top of the path construction hierarchies $H[u \rightarrow^+]$ to keep (partially) the current lookup state. The examples include neighbor-of-neighbor (NoN) routing [32, 37], distributed trie of popular lookups [14], and cyclic routing [25].

3.5.1 NoN-routing

Progressive routing is augmented with a lookahead mechanism. For routing decisions, u (in addition to its neighbor v) analyzes l subsequent nodes on paths

$$u \rightarrow v^* \rightarrow w_1 \rightarrow \cdots \rightarrow w_l \rightarrow^+ d. \tag{3.15}$$

NoN-routing is greedy routing with lookahead for $l = 1$ [32, 37]. In a lookup for k, a node u knows nodes $w \in T_v$ for any $v \in T_u$. Note that IP addresses IP_w are unknown to u, thus the maintenance overhead is low (no pinging). Among these w, the node u finds $w_1 \in T_{v^*}$ closest to k and forwards the lookup to the corresponding v^*.

Likely, the lookup then goes to w_1, reducing the distance more effectively than greedy routing without lookahead. Note, however, that it is not guaranteed that v^* forwards the lookup exactly to w_1 since v^* makes own decision on the next hop.

NoN-routing exploits two levels in $H[u \rightarrow^+]$: neighbors $v \in T_u$ (level $i = 1$) $\rightarrow$ neighbors-of-neighbors $w \in T_v$ (level $i = 2$). It improves the routing performance in randomized P2P networks [1, 32]. Symphony with NoN-routing achieves expected path length bound $O(\log^2 N/(m \log m))$ with m long-range neighbors per node [32], i.e., the reduction factor is $\log m$, see (3.9). Pappilon provides $\Theta(\log N/\log m)$ routing in the worst case [1]. For many DHTs, $\Theta(\log N/\log \log N)$ routing is expected [4, 32]. A NoN-routing scheme without additional overhead is introduced in [4]; it "limits" the randomization of the original scheme such that neighborhood information is encoded within the hash-value of the node ID.

3.5.2 Distributed Trie

The lookup structure can be organized locally as a tree-like hierarchy that reflects most popular lookups [14]. Recall from Chap. 2 that a distributed trie utilizes the same idea of prefix-matching routing [40] as in Pastry and Tapestry. In fact, T_u is a tree of tables representing $H[u \rightarrow^+]$. A table consists of 2^m entries, each contains several nodes as triples (v, IP_v, t), arranging nodes by timestamps t (last recent seen).

Each level in T_u sequentially resolves m bits in the prefix of the n-bit key, leading to paths in the tree as the following ($n = 6$, $m = 2$, $k = \texttt{010010}$):

$$\texttt{******} \rightarrow \texttt{01****} \rightarrow \texttt{0100**} \rightarrow \texttt{010010}$$

A non-leaf routing table contains nodes v known to resolve the next m bits at time t; the child routing tables contains the corresponding neighbors of v. An entry of a leaf routing table contains a node known at time t to be responsible for a key with the sum prefix (constructed top-down from the root). Note that if a node v appears in some routing table of T_u, then u keeps an entire path down to v.

In a lookup for k, u searches the tree T_u moving down to the leaf routing tables that keeps an entry for the longest prefix p of k. The entry contains a node v with the latest timestamp, i.e., v is most recently known to u to hold the child routing table, not currently available at u. Then, u requests v for either this table or, if v is responsible for k, the resource.

Each node u exploits of the current lookup structure up to any level based on lookups in which u has participated. Local representation T_u evolves in time. Frequent overlay paths are cached at u; the more popular and stable the resource or node, the more efficient routing. It is expensive when churn is high or there are many nodes with popular resources.

3.5.3   Cyclic Routing

Similarly to the distributed trie approach, Cyclic Routing (CR) provides a systematic way for collecting efficient paths up to any level of the lookup structure [24, 25]. A cycle is a path that starts from the source node to its neighbor, then runs through the overlay and returns to the source:

$$u \rightarrow v \rightarrow^+ w_1 \rightarrow^+ \cdots \rightarrow^+ w_l \rightarrow^+ u.$$

Compared with l-lookahead (3.15), there can be intermediate nodes in $w_i \rightarrow^+ w_{i+1}$ that are not fixed explicitly. Hence, the CR selection of nodes to represent a given cycle is flexible and can be adapted to various routing requirements.

Each node u maintains a collection of cycles additionally to its primary routing table T_u. As in NoN-routing, only node IDs are stored to identify a cycle. In progressive routing, u uses either cycles or the underlying DHT. If there is a cycle with a node close to the key, the lookup is sent along the cycle. Otherwise, the underlying DHT selects the next hop.

This strategy supports l-lookahead for any $l \geq 1$ as well as takes the bidirectional nature of P2P communications into account. The CR focus on global routing is

Fig. 3.7 A node u utilizes
CR for better global routing.
When a lookup appears in d's
neighborhood (a node w_i),
local routing is performed

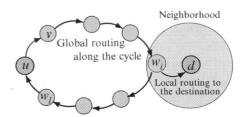

emphasized in Fig. 3.7. Compared with the distributed trie, where all destinations are strictly partitioned among paths in T_u, the same cycle can be used for many destinations and different cycles can be used for the same destination. As a result, CR provides a rather compact data structure for the information about available paths. It can be used in any DHT with progressive routing. When the size of local collection of cycles grows then this local knowledge becomes an interpolation rather than extrapolation for the global network topology, and we defer the detailed analysis of the CR method until Part III.

3.6　Routing with Large Routing Tables

Principle 4 states that u's primary routing table can be augmented with additional information from $H[u \to^+]$ for better global routing. In the extreme case, this idea leads to $O(1)$ routing with large routing tables. The difference between global and local routing disappears, since the majority of nodes becomes in every node's neighborhood. On the other hand, the maintenance cost can become high, as experimental study [28] confirmed.

3.6.1　Designs with Large Routing Tables

For achieving $O(1)$ routing (up to one-hop routing), upper limits for the routing table size are removed, as appeared in EpiChord [27] and Accordion [29]. It leads to systems with large routing tables, up to complete membership information. Extreme examples are OneHop [12], 1h-Calot [49], and D1HT [34] where every node must keep a complete routing table; work [35] makes their comparison. Note that SmartBoa [20] also utilizes large sizes for high-bandwidth nodes (see Sect. 4.4).

EpiChord [27] removes the $O(\log N)$ state upper bound of Chord using a reactive routing state maintenance strategy that amortizes network maintenance costs into existing lookups and uses parallel lookups (multipath routing). The routing table size has no upper limit, and nodes adapt to a wide range of lookup workloads. EpiChord is able to achieve O(1)-hop routing under lookup-intensive workloads and at least $O(\log N)$ routing under churn-intensive workloads.

Accordion [29] adapts to current operating environments and node bandwidth budgets. Nodes proactively search new neighbors. As in Symphony, the density of its long-range neighbors is inversely proportional to the distance. Accordion is not based on a particular data structure, and it has freedom in choosing the size and content of its routing table. The number neighbors m varies from $O(\log N)$ to $O(N)$ states. The bandwidth budget controls the rate of learning about new neighbors. Each node limits its routing table size by evicting failed neighbors. The equilibrium between the learning and eviction processes determines the table size.

3.6.2 Maintenance Traffic Overhead

A comparison of DHTs with variable routing table sizes from $O(1)$ to $O(\log N)$ is given in [49]. In particular, it introduced a model for the tradeoff between the routing table size $m = |T_u|$ and the total traffic generated in a network with logarithmic routing complexity. In addition to N and m, the model parameters are lookup rate λ (workload) and node lifetime μ (churn). Note that $r = 1/\mu$ is the churn rate; in other words, r is the node turnover rate—a fraction of nodes that the churn replaces in the system per time unit.

Theorem 3.6. *Given a P2P network with N nodes, equally fixed routing table size $m = |T_u|$ $\forall u \in N$, and the routing complexity of $\Theta(\log_m N)$ hops. Let λ and μ be node lookup rate and expected node lifetime, respectively. Then the minimum of total traffic is attained for m that is a solution to*

$$m \ln^2 m = C\lambda\mu \ln N, \tag{3.16}$$

where C is a constant determined by the P2P protocol.

Proof. Since λ and μ are characteristics of the expectations, the subsequent results are also for an expected case. At any time instance, the network is preserved to be N nodes in total, i.e., N/μ nodes leave and correspondingly N/μ nodes join per time unit.

When a node joins or leaves, each of its m neighbors should be notified. It requires exchanging a constant portion of traffic per neighbor. Thus, the churn-related maintenance traffic is $B_{\text{churn}} = C_1 mrN$. The constant C_1 characterizes the P2P maintenance protocol.

Each alive node initiates λ lookups per time unit, resulting in λN lookups in total. The traffic for lookups is therefore $B_{\text{lookup}} = C_2\lambda N \log_m N$. The constant C_2 characterizes the P2P routing protocol.

The total traffic is the following function of m:

$$B(m) = B_{\text{churn}} + B_{\text{lookup}} = N(C_1 mr + C_2\lambda \log_m N).$$

Taking the derivative of $B(m)$ and equating it with zero

$$B'(m) = N\left(\frac{C_1}{\mu} - C_2\frac{\lambda\ln N}{m\ln^2 m}\right) = 0,$$

we find that the only minimum is when m satisfies (3.16) for $C = C_2/C_1$.   □

Note that $\lambda\mu$ is the expected number of lookups a node processes during its lifetime. If there are many stable nodes (μ is large) and actively participating nodes (λ is large) then large routing tables are preferable.

Let the network need logarithmic node state, i.e., $m = m(p) = p\log N$ for a parameter $p > 0$, as it happens in many DHT designs. From (3.16) we have

$$p\log N\ln^2(p\log N) = C\lambda\mu\frac{\log N}{\log e}.$$

Let $C' = \sqrt{C/\log e}$. Then

$$m = p\log N = \exp\left(C'\sqrt{\lambda\mu/p}\right). \tag{3.17}$$

The functions $m(p) = p\log N$ and $f(p) = \exp\left(C'\sqrt{\lambda\mu/p}\right)$ are contiguous on $(0,\infty)$. Equation (3.17) has the only solution $p^* \in (0,\infty)$ since $m(p)$ strictly monotonically grows to $+\infty$, $f(p)$ monotonically decays to 1, and there exists a small enough $\varepsilon > 0$ such that $m(\varepsilon) < f(\varepsilon)$. This p^* provides the optimal table size $m^* = p^*\log N$, where $p^* = p^*(N)$, a function of N. The asymptotic $m = \Theta(\log N)$ is not guaranteed since so large routing tables can be non-optimal in terms of the total traffic.

3.6.3 Routing Table Consistency

Large routing tables typically means that the routing state is $m = |T_u| = \Omega(\log N)$, aiming in the high lookup efficiency. Although it supports $O(1)$ routing performance the node join/leave information dissemination becomes an issue for high churn rates. Experimental evaluation [28] showed that the per-node bandwidth consumption in OneHop [12] is proportional rN, where r is the churn rate. Clearly, it supports the intuition that a node routing state of $\Theta(N)$ requires maintenance traffic of $\Theta(N)$.

Consider an analytic model of neighbor turnover in a routing table of a node u. A close variant can be found in EpiChord [27]. Let $x(t)$ and $y(t)$ be the number of alive and stale neighbors in T_u at time t, respectively. For simpler exposition, let us assume that the network consists of u and N other nodes, $N+1$ nodes in total. Consider the proportion of stale neighbors in the steady state,

$$\gamma = \lim_{t \to \infty} \frac{y(t)}{x(t) + y(t)},$$

where $|T_u| = x(t) + y(t) \leq N$.

Form u's point of view the rest network permanently consists of N nodes such that $Nrdt$ of the nodes are renewed uniformly at random due to churn in a time interval dt. At the same time interval, u detects a fraction vdt of stale nodes among its neighbors and u removes them from T_u. Also, u makes λdt lookups uniformly over the ID space, as well as the other N nodes make their own lookups uniformly with the same rate λ. Then we obtain the following differential equation for $x(t)$:

$$\frac{dx}{dt} = \lambda\left(1 - \frac{x}{N}\right) - rx, \tag{3.18}$$

where $\lambda(1 - x/N)$ is the number of incoming lookups to u from newly appeared nodes (u makes them neighbors), rx is the number of the neighbors having become stale in T_u (u still thinks that they are alive). Similarly,

$$\frac{dy}{dt} = rx - vy, \tag{3.19}$$

where vy is the number of stale neighbors that u removes from T_u.

Equations (3.18) and (3.19) forms a system of first-order linear ordinary differential equations. It similar in essence to continuous-time birth-death processes. The solution can be obtained analytically as a steady state and a time-dependent sum of terms with exponents $\exp(-\alpha t)$ for $\alpha > 0$. Hence the solution converges quickly to the steady state solution, for which $dx/dt = dy/dt = 0$. It is a solution to the following linear system:

$$\lambda - (r + \lambda/N)x = 0, \quad rx - vy = 0, \quad x + y \leq N.$$

It leads to

$$x = \frac{\lambda}{r + \lambda/N}, \quad y = \frac{\lambda r}{v(r + \lambda/N)}, \quad \gamma = \frac{r}{r + v}, \quad \text{where } \lambda \leq vN. \tag{3.20}$$

Clearly, $\gamma \to 0$ when the churn is reduced ($r \to 0$) and $\gamma \to 1$ when the churn grows ($r \to \infty$ and $v = o(r)$).

Outgoing lookups provides a way for u to detect stale neighbors in T_u. For instance, a neighbor is treated failed if there is no acknowledgment (in EpiChord). Another way is proactive pinging in background, when periodicity is proportional to the lookup rate (in Accordion). In both cases, the detection rate is $v = c\lambda$ for a constant $c > 0$, leading to $\gamma = r/(r + c\lambda)$. The constant c can be increased with parallel lookups. The condition $\lambda \leq vN$ is satisfied for all N large enough.

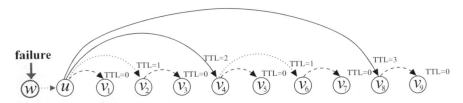

Fig. 3.8 Nodes propagate an event using geometrically progressive hops with corresponding TTLs. If u detects that w is failed then u reports the event to its neighbors v_i setting the event message TTL proportionally to the distance. In turn, each v_i decrements TTL and forwards the message further. The event disseminated over all nodes in the logarithmic time and without duplicates

Theorem 3.7. *Given an N-node network with churn rate $r \geq 0$, lookup rate $\lambda \geq 0$ and stale node detection rate $v \geq 0$ such that $\lambda \leq vN$. Then in the steady state:*

1. *The proportion γ of stale neighbors is independent on the network size N.*
2. *The asymptotic for $N \to \infty$ is $x \to \lambda/r$ and $y \to \lambda/v$.*
3. *If $v \geq r$ then $\gamma \leq 1/2$ (at most half of the neighbors are stale if the stale node detection is at least as fast as the node failure rate).*
4. *If $v = c\lambda$ then $y \to 1/c$ for $N \to \infty$ (the stale node detection that is proportional to the lookup rate leads to the small number of stale neighbors).*
5. *$x = \lambda\mu$ for $N \to \infty$ (the number of alive neighbors is equal to the expected number of lookups that a node processes during its lifetime).*

Proof. All the assertions are obvious consequences of (3.20). Assertion 5 uses the fact that the expected node lifetime is $\mu = 1/r$. $\square$

Assertions 2 and 5 show that large routing tables appear when λ, μ, or v are high (cf. the similar conclusion after Theorem 3.6).

A design can directly apply Principle 2 for efficient (logarithmic) dissemination when each node has a large or even full routing table. For instance in the D1HT design, a node u disseminates an observable failure of a node w to all other nodes in the system such that the number of messages any node processes is $O(\log N)$, see Fig. 3.8 for intuitive explanation. In this case, u assigns $\Theta(\log N)$ long-range neighbors from T_u to be "dissemination gateways". Discussion on other inexpensive dissemination algorithms can be found in [43].

3.7 Local Routing and Replication

Large routing tables reduce the number of levels in path construction hierarchy $H[u \to^+]$ up to one hop $u \to d$ in the extreme case. An opposite approach for achieving $O(1)$ routing exploits replication along lookup paths. Replication is a technique in which data are stored at multiple nodes of the network. For the basic questions we address, it does not matter if the actual data is replicated or if only pointers to the data are replicated.

Replication improves lookup performance and alleviates hot spots since a replica may be found at a nearer node than the primary responsible node. The improvement depends on the number of replicas and the placement strategy. The latter essentially exploits the network structure.

3.7.1   Source-Aware Replication

The extreme case is caching when after a successful lookup the query node stores a replica. For example, such a variant was suggested in CAN [42] where each node can maintain a cache of the resource keys it recently accessed (in addition to its primary store).

The next possible step is replication at neighbors. The generalization follows Principle 4: each node can look ahead in selecting appropriate nodes for replication. Plaxton et al. [40] apply *the source-aware replication*. Let u and k be a random node and key, respectively. One of the basic consequences of Plaxton's model is that if v is the nearest node to u that holds a replica (or original) resource for k then the lookup cost is proportional to $l = |u \rightarrow^+ v|$. Consequently, if interesting resources are replicated in u's vicinity then the lookup performance of u is determined mostly by local routing.

This idea can be implemented based on $H[u \rightarrow^+]$. If u made a lookup for k with path (3.1) to d, then the resource with this key are replicated near u at $r+1$ first nodes $w_0, w_1, \ldots, w_r$. Then subsequent lookups for k are resolved more efficiently. Popular resources are targets of many users, spread over the overlay.

Consider an analytic model, assuming that all resources have the same popularity. Let u make a lookup for k, where u and k are taken at random. Let d be the primary responsible node for k and $|u \rightarrow^+ d| = l$. Assume that each node $w_0 = u, w_1, \ldots, w_r$ on the path $u \rightarrow^+ d$ keeps a replica for k with probability p_k. It characterizes the popularity-based replication fraction of k relatively to u since u has already served lookups for k and some nodes can still keep the replicas.

The probability that there is no node on $u \rightarrow^+ d$ to resolve the lookup earlier than reaching d is $(1 - p_k)^{r+1}$. In this case, the replication mechanism is not involved and the lookup path length is l, e.g., $l = \Theta(\log N)$. It decreases rapidly, especially when p_k is close to 1 (high popularity of k) or r is close to $l-1$ (high replication factor).

With probability $1 - (1 - p_k)^{r+1}$ the lookup path becomes of r hops or shorter. The expectation is

$$\mathsf{E}[L_k] = \sum_{i=0}^{r} i(1 - p_k)^i p_k = p_k \sum_{i=1}^{r} i(1 - p_k)^i,$$

where $(1 - p_k)^i p_k$ is the probability that $u = w_0, w_1, \ldots, w_{i-1}$ have no replica of k and w_i keeps the replica. The sum can be reduced,[9] yielding the following improvement to the expected lookup performance:

$$E[L_k] = \frac{1 - p_k}{p_k} \left(r(1 - p_k)^{r+1} - (r+1)(1 - p_k)^r + 1 \right). \tag{3.21}$$

Theorem 3.8. *Given an N-node network with source-aware replication factor $r \in \mathbb{Z}_+$ and replication fraction $p_k \in [0,1]$ for some $k \in R$. Then the replication provides the following improvement to the expected lookup performance for k*

1. $E[L_k] \leq (1 - p_k)/p_k$.
2. $E[L_k] \to 0$ *for* $p_k \to 1$.
3. $E[L_k] \to (1 - p_k)/p_k$ *for large* r.

Proof. Consider the function $f(q,r) = rq^{r+1} - (r+1)q^r$ for $q = 1 - p \in (0,1]$ and $r \in [0,\infty)$. The derivative is $\partial f/\partial q = r(r+1)q^{r-1}(q - 1) < 0$, thus f decreases monotonically for q and takes values in $[-1,0)$. Applying (3.21) proves the first assertion of the theorem.

The second assertion follows from the first one. Note also that the function $y(x) = (1 - x)/x$ is strictly decreasing on $(0,1)$.

To prove the last assertion consider the derivative

$$\frac{\partial f}{\partial r} = q^r \left((rq - (r+1)) \ln q - (1 - q) \right).$$

It is positive for r large enough since the inequation $(rq - (r+1)) \ln q - (1 - q) > 0$ has solutions $r > -1/\ln q - 1/(1 - q)$. Let $r_0 = \max\{-1/\ln q - 1/(1 - q), 0\}$. Then $f(q,r)$ increases monotonically for $r \in (r_0, \infty)$ and $f(q,r) \approx 0$ for large r. □

Note that the case with non-zero probability for the nodes $w_{r+1}, \ldots, w_{l-1}$ also to replicate k allows further reduction of the lookup path length. Consequently, the above model provides the upper bound. For instance, if k is popular, e.g., $p > 1/2$ then with probability $1 - 1/2^{r+1}$ the resolution of a lookup for k takes 2 or less hops.

As a result of the source-aware replication, popular resources are replicated at many nodes on all layers of $H[\to^+ d]$ (see Fig. 3.5 in Sect. 3.2). If u's neighbors, neighbor of neighbors, and so on are active in some resources then the search of these resources becomes more efficient for u. This replication strategy reduces the need in global routing since the resource can be found by means of local routing with $O(1)$ lookup performance.

---

[9]Denote $f(x) = \sum_{i=1}^r ix^i$ and $g(x) = \sum_{i=1}^r x^{i+1}$. Then $g'(x) = \sum_{i=1}^r (i+1)x^i = f(x) + \frac{1-x^r}{1-x}x$. On the other hand, $g'(x) = \frac{x}{(1-x)^2} \left((r+1)x^{r+1} - (r+2)x^r - x + 2 \right)$ since $g(x) = \frac{1-x^r}{1-x}x^2$. We obtain $f(x) = g'(x) - \frac{1-x^r}{1-x}x = \frac{x}{(1-x)^2} \left(rx^{r+1} - (r+1)x^r + 1 \right)$.

3.7.2 Even Replication

Resource can be replicated along all nodes on the lookup path, similar to the scheme proposed in PAST [46]. Let $p_k > 0$ be the probability that a node $w \in N$ keeps a replica of $k \in R$. Applying $r = l - 1$ in (3.21), we obtain that the expected lookup path length for k is

$$E[|u \xrightarrow{k} {}^+d|] = (1 - (1 - p_k)^l)p_k \sum_{i=1}^{l-1} i(1 - p_k)^i + l(1 - p_k)^l = \frac{1 - p_k}{p_k} + o\left(\frac{1}{N}\right)$$

if $q(N) = (1 - p_k)^l = o(1/N)$ for large N. For instance, the latter happens if $l = \Theta(\log N)$. When u does not keep k (the probability is $1 - p_k$), the expected lookup path length is $E[|u \xrightarrow{k} {}^+d|] \approx 1/p_k$ for large l and $p_k > 0$.

Let k have the same popularity at all nodes in the system, i.e., any $u \in N$, $u \neq d$ initiates lookups for k at equal rate λ_k. This is the case for even replication when the strategy tries to distribute k uniformly among all the nodes.

The resource popularity p_k is the fraction of the nodes that keep a replica of k. If resource popularity is uniform ($p_k = p$ for all $k \in R$) then each resource has the same average number of replicas in the system.

The uniform replication is ineffective when resource popularity is different. It can lead to the bottleneck problem caused by popular resources. If the storage capacity of nodes is limited, there may be a case where popular resources are lack of replicas while there are redundant replicas of unpopular resources. Nevertheless, the path-aware replication in structured DHTs outperforms the uniform replication strategy in unstructured P2P networks. Intuitively, a DHT topology provides more chances for a lookup to reuse a successful lookup path where the resource has been earlier replicated. It is an implicit reflection of the non-uniform resource popularity distribution.

Following [5] consider a model for the optimal number of replicas when resource popularity is uniform. Let q_k denote the probability of a lookup for k in the system, i.e., the popularity of k. We assume the arrangement $1 \geq q_1 \geq q_2 \geq \cdots \geq q_R \geq 0$. Let u be taken at random among nodes that have no replica of k. Assuming the expected lookup path length for k is $1/p_k$ the replication strategy applies for replica allocation such p_k that minimize of the total expectation:

$$\sum_{k=1}^{R} \frac{q_k}{p_k} \to \min$$

$$\sum_{k=1}^{R} p_k = 1, \quad p_k \geq 0. \tag{3.22}$$

Note that this model assumes limited capacity of the system; more replicas for one resource leads to less replicas for another resource.

Obviously, the optimal solution reflects the popularity arrangement as $1 \geq p_1 \geq p_2 \geq \cdots \geq p_R \geq 0$. Uniform replication provides a feasible solution with $p_k = 1/R$ and cost R, making search time equal for any resource k. Interestingly that the same cost has proportional replication with $p_k = q_k$, making search time shorter for popular resources and longer for unpopular ones.

Theorem 3.9. *Given an N-node network where each resource k has popularity $q_k > 0$ and is replicated uniformly at $p_k CN$ random nodes for a capacity parameter $C \geq 1$. Let a lookup for any k find a replica on the lookup path in the $1/p_k$ expected number of hops. Then the replication provides the optimal improvement if*

$$p_k = \frac{1}{\sum_{i \in R} \sqrt{q_i}} \sqrt{q_k}.$$

Proof. Let us apply the Lagrange multipliers method for (3.22). Then

$$\mathscr{L}(p_1, \ldots, p_R, x) = \sum_{k \in R} \frac{q_k}{p_k} + x \left(\sum_{k \in R} p_k - 1 \right) \to \min,$$

where x is a Lagrange multiplier. From $\frac{\partial \mathscr{L}}{\partial p_k} = x - q_k/p_k^2 = 0$ it follows that $p_k = \sqrt{q_k}/\sqrt{x}$. Since $\frac{\partial \mathscr{L}}{\partial x} = \sum_{k \in R} p_k - 1 = 0$, we obtain

$$x = \left(\sum_{k \in R} \sqrt{q_k} \right)^2 \quad \text{and} \quad p_k = \frac{1}{\sum_{i \in R} \sqrt{q_i}} \sqrt{q_k}$$

The expected number of hops before a replica is found is

$$\sum_{k \in R} \frac{q_k}{p_k} = \left(\sum_{k \in R} \sqrt{q_k} \right)^2. \qquad \qquad \square$$

The model assumes random placement strategy for replicas, thus providing the worst case bound. In structured P2P systems, the number of replicas can be reduced further by proper selection of replica placement, e.g., using path-aware replication.

Yarqs [53] is an example of a DHT design with even replication. Nodes on a lookup path are equal in the replication. As in any DHT, resources are located at primary responsible nodes according to their keys via a standard DHT mechanism. Yarqs constructs a cache network on the top of the overlay network. In addition to keys, resources and nodes have attribute values. The cache network supports range queries for semantic attributes. A node analyzes the data that pass through and caches new attribute/key pairs into two buckets. The first bucket stores pairs with attribute values close to the node semantic value. This bucket helps to find the first result for a lookup, a DHT key of semantically close resource. Then DHT can locate a responsible node. In fact, the bucket implements clustering of semantically close resources at nodes (see also Chap. 5). The second bucket collects semantically distant key/attribute pairs. This bucket is used after receiving the first result to find resources in the same range.

3.7.3 Destination-Aware Replication

PAST [46], DHash [8], Beehive [41], and FuzzyNet [16] apply *the destination-aware replication* based on path destination hierarchy $H[\rightarrow^+ d]$. The strategy is an inversion of look ahead, i.e., a lookup path is considered from the destination.

Let d be responsible for resource with a key k. Assume that global routing efficiently resolves lookups from a random node u to a vicinity of the responsible node d. The performance of local routing in $u \rightarrow^+ d$ can be improved by one hop if k is replicated at nodes close to d (a replica is occasionally found at w_{l-1}). This simple variant was suggested in CAN [42] and Chord [48], where each node replicates its resource at several closest neighbors. Iteratively, given lookup path (3.1), each of the last r nodes $w_{l-1}, w_{l-2}, \ldots, w_{l-r}$ also replicates k at nearby nodes.

In this case, each w_i characterizes a replication level indicating the distance scale at which k is replicated. That is, i corresponds to a layer in the hierarchy $H[\rightarrow^+ d]$. Consequently, r is the replication level, which is a parameter in the tradeoff between lookup performance load overhead.

Note that replication only at w_i is passive (caching). In the above scheme, nodes w_i additionally propagate replicas among other (nearby) nodes in the system, the case is called proactive replication. Simulation study in [41] indicated that proactive replication provides better lookup performance than the passive strategy when resource is merely replicated along all nodes on the lookup path.

Considering all nodes as lookup originators, k is proactively replicated at nodes logically preceding d on all lookup paths. Then a node w_i can switch on more efficient local routing to a replica node instead of continuation of global routing along the path $u \rightarrow^+ d$. In contrast to the source-aware replication, popular resources of d are replicated at a few nodes on the topmost layers of $H[\rightarrow^+ d]$ for small r. When r is large then replicas are disseminated widely throughout the system, similarly to the source-aware replication.

If k is popular then many nodes make lookups for k. As a result, d's vicinity becomes full of replicas. In the case of non-uniform resource popularity, resources in replication can be arranged accordingly. Intuitively, more popular resources should be distributed wider in the system. Thus, better lookup performance is achieved if there are more replicas for more popular resources.

As an illustrative case study, consider an analytical model for replication applied in Beehive protocol [41]. The input parameter is a resource popularity metric—the number of lookup queries received for resources. The problem is to minimize the total number of replicas subject to $O(1)$ average lookup performance.

The first basic model assumption is that the lookup rate is a truncated Zipf-like[10] distribution for a finite set of original resources R sorted by decreasing their popularity as $k = 1, 2, \ldots, R$. The lookup rate for the kth most popular resource is proportional to $k^{-\alpha}$, where $0 \leq \alpha \leq 1$ is a system-level parameter—the popularity index. It can be estimated for using as input to the model. The distribution has a heavier tail for smaller values of α. The case $\alpha = 0$ corresponds to a uniform distribution.

In terms of Theorem 3.9, $q_k = ck^{-\alpha}$ for $c = 1/\sum_{i \in R} i^{-\alpha}$. The number of lookups on time interval dt is $q_k dt$. We can assume that the lookup rate for k is equal to $k^{-\alpha}$ by taking appropriate time unit. The total number of lookups to the most popular i resources $k = 1, 2, \ldots, i$ for $i \leq R$ is

$$
\lambda_i = \sum_{k=1}^{i} k^{-\alpha} \approx \int_1^i x^{-\alpha} dx = \begin{cases} \dfrac{i^{1-\alpha} - 1}{1 - \alpha} & \text{if } \alpha \neq 1, \\[2mm] \ln i & \text{if } \alpha = 1. \end{cases}
$$

Then the probability of a lookup for the most popular i resources is

$$
Q(i) = \frac{\lambda_i}{\lambda_R} \approx \begin{cases} \dfrac{i^{1-\alpha} - 1}{R^{1-\alpha} - 1} & \text{if } \alpha \neq 1, \\[2mm] \dfrac{\ln i}{\ln R} & \text{if } \alpha = 1. \end{cases}
$$

Consider a P2P network with b-progressive routing. Denote $l = \log_b N$ the longest lookup path length. The replication levels are $i = 0, 1, \ldots, l$. Let x_i be the fraction of resources replicated at level i or lower. Then

$$
0 \leq x_0 \leq x_1 \leq \cdots \leq x_{l-1} \leq x_l = 1,
$$

where $x_l = 1$ since all resources are replicated at level l (there is a primary responsible node for each resource). There are $x_0 R$ topmost popular resources, which are replicated at all levels in the system.

The second basic model assumption captures the structure of network of topology with b-progressive routing. If d is primary responsible for k then k is replicated at all nodes of level i in $H[\to^+ d]$, i.e., at N/b^i nodes or less. To reach an i-level node for k from an arbitrary node u, routing needs i hops in the worst case.

On average, each node replicates $(x_i - x_{i-1})R/b^i$ resources at level $i = 1, 2, \ldots, l$ and $x_0 R$ at level $i = 0$. In sum, a node is required to keep the following average number of replicas:

---

[10]Sometimes Zipf distribution is referred to as the zeta distribution in the discrete case and the power-law distribution in the continuous case. Originally, Zipf's law appear in linguistics. In the English language, the probability of encountering the kth most common word is given roughly by $p_k = 0.1/k$ for k up to 1,000.

$$x_0 R + \sum_{i=1}^{l} \frac{x_i - x_{i-1}}{b^i} R = \left[\left(1 - \frac{1}{b} \right) \sum_{i=0}^{l-1} \frac{x_i}{b^i} + \frac{1}{b^l} \right] R. \tag{3.23}$$

For simplicity, we limit ourselves with the case $\alpha \neq 1$. The number of lookups that travel $i > 0$ hops is

$$Q(x_i R) - Q(x_{i-1} R) \approx \left(x_i^{1-\alpha} - x_{i-1}^{1-\alpha} \right) \frac{R^{1-\alpha}}{R^{1-\alpha} - 1}$$

The expected lookup performance of the entire system can be given by

$$\sum_{i=1}^{l} i(Q(Rx_i) - Q(Rx_{i-1})) \approx \frac{R^{1-\alpha}}{R^{1-\alpha} - 1} \sum_{i=1}^{l} i \left(x_i^{1-\alpha} - x_{i-1}^{1-\alpha} \right) = \frac{R^{1-\alpha}}{R^{1-\alpha} - 1} \left(l - \sum_{i=0}^{l-1} x_i^{1-\alpha} \right)$$

The lookup performance is constant if the expected number of hops is not exceed a required constant L. Consequently,

$$\sum_{i=0}^{l-1} x_i^{1-\alpha} \geq l - \left(1 - \frac{1}{R^{1-\alpha}} \right) L = h(N, R) \tag{3.24}$$

Based on (3.23) and (3.24) the following optimization problem can be constructed to minimize the expected storage for replicas at nodes and preserving the expected lookup performance within the constant bound.

$$\sum_{i=0}^{l-1} \frac{x_i}{b^i} \to \min, \quad \sum_{i=0}^{l-1} x_i^{1-\alpha} \geq h(N, R), \quad 0 \leq x_{l-1} \leq 1 \tag{3.25}$$

Denote

$$H_j = \frac{1 - b^{\frac{1-\alpha}{\alpha}(j-1)}}{b^{\frac{1-\alpha}{\alpha}(j-1)} \left(1 - b^{\frac{1-\alpha}{\alpha}} \right)} + 1, \quad h_j = j - (1 - R^{\alpha-1})L \text{ for } j = 1, 2, \dots, l. \tag{3.26}$$

Theorem 3.10. *Given an N-node network with b-progressive routing ($b > 1$) and with the lookup rate following a truncated Zipf-like distribution with $\alpha \neq 1$. Let j be the highest value in $\{1, 2, \dots, l\}$ such that $H_j \geq h_j$. Then the replication with*

$$x_i = \begin{cases} b^{-\frac{i}{\alpha}} \left(\frac{1 - b^{\frac{1-\alpha}{\alpha}}}{1 - b^{\frac{1-\alpha}{\alpha}(j-1)}} (j - (1 - R^{\alpha-1})L) \right)^{\frac{1}{1-\alpha}}, & i = 0, 1, \dots, j-1, \\ 1, & i = j, j+1, \dots, l \end{cases} \tag{3.27}$$

minimizes the expected number of replicas per node preserving the expected lookup performance of at most L hops.

Proof. Initially, set $j = l$. Let us apply the Lagrange multipliers method for (3.25).

$$\mathcal{L}_j(x_0,\ldots,x_{j-1},y,z) = \sum_{i=0}^{j-1} \frac{x_i}{b^i} + \left(\sum_{i=0}^{j-1} x_i^{1-\alpha} - h_j\right) y + (x_{j-1} - 1)z \to \min,$$

where $y \in \mathbb{R}_+$ and $z \in \mathbb{R}$ are Lagrange multipliers. Eliminating y from the equations

$$\frac{\partial \mathcal{L}_j}{\partial x_i} = b^{-i} + (1-\alpha)yx_i^{-\alpha} = 0 \text{ for } i \neq j-1 \quad \text{and} \quad \frac{\partial \mathcal{L}_j}{\partial x_{j-1}} = b^{1-j} + (1-\alpha)yx_{j-1}^{-\alpha} = 0$$

yields that

$$x_i = b^{i/\alpha}(z + b^{1-j})^{1/\alpha}x_{j-1} = a_{ij}(z)x_{j-1} \quad \text{for } i = 0,1,\ldots,j-2, \tag{3.28}$$

where $a_{ij}(z) = (b^i(z + 1/b^{j-1}))^{1/\alpha}$.

Applying the lookup performance constraint,

$$\sum_{i=0}^{j-1} x_i^{1-\alpha} = \sum_{i=0}^{j-2} \left(b^i(z + 1/b^{j-1})\right)^{\frac{1-\alpha}{\alpha}} x_{j-1}^{1-\alpha} + x_{j-1}^{1-\alpha}$$

$$= x_{j-1}^{1-\alpha}\left[\left(z + \frac{1}{b^{j-1}}\right)^{\frac{1-\alpha}{\alpha}} \sum_{i=0}^{j-2} b^{\frac{1-\alpha}{\alpha}i} + 1\right]$$

$$= x_{j-1}^{1-\alpha}\left[\left(z + \frac{1}{b^{j-1}}\right)^{\frac{1-\alpha}{\alpha}} \frac{1 - b^{\frac{1-\alpha}{\alpha}(j-1)}}{1 - b^{\frac{1-\alpha}{\alpha}}} + 1\right] \geq h_j.$$

Denote

$$H_j(z) = \left(z + \frac{1}{b^{j-1}}\right)^{\frac{1-\alpha}{\alpha}} \frac{1 - b^{\frac{1-\alpha}{\alpha}(j-1)}}{1 - b^{\frac{1-\alpha}{\alpha}}} + 1,$$

which coincides with (3.26) for $z = 0$. Then

$$\left(\frac{h_j}{H_j(z)}\right)^{\frac{1}{1-\alpha}} \leq x_{j-1} \leq 1.$$

In unconstrained minimization of $\mathcal{L}_j(x_0,\ldots,x_{j-1},y,0)$ with $z = 0$, if $h_j > H_j(0)$ then $x_{j-1} > 1$ in the optimal solution. In this case, we can force $x_{j-1} = 1$ and restate the optimization problem to minimization of the Lagrange function $\mathcal{L}_{j-1}(x_0,\ldots,x_{j-2},y,z)$ of lower dimension. The previous derivations remain the same; the process is repeated $j = l-1, l-2, \ldots$ until j is found such that $h_j \leq H_j$.

Then (3.28) gives the optimal solution for $z = 0$, where $a_{ij}(0) = b^{-\frac{j-i-1}{\alpha}}$ and

$$x_{j-1} = \left(\frac{h_j}{H_j(0)}\right)^{\frac{1}{1-\alpha}} = \left(\frac{j - (1 - R^{\alpha-1})L}{H_j}\right)^{\frac{1}{1-\alpha}}.$$

Substitution of (3.26) yields (3.27). □

A Beehive system uses a replication strategy that follows Theorem 3.10. The parameter α is estimated and replication is performed asynchronously. Each node u has a responsibility to manage replicas stored at u. If the observations show that it is necessary to create more replicas at lower level nodes, then u propagates replicas to nodes preceding u on the lookup path. On the contrary, if the resource popularity is decreased, then u deletes the replica from the local storage.

Data consistency is achieved by propagating changes in data objects from the primary responsible node to other nodes through the replication path. In particular, each resource is associated with a version number. When a resource is modified, a new (higher) version number is assigned for it, and the modification is propagated to nodes keeping replicas. By comparing version numbers, u can determine the latest version of a resource it should keep.

3.7.4 Maintenance Overhead

There is a fundamental tradeoff between replication and resource consumption. More replicas generally improve lookup performance at the cost of space, bandwidth, and aggregate network load. Replication leads to additional load at nodes with replicas; the maintenance of up-to-date copies becomes an issue in systems with very dynamic resources [44]. A challenge in replication is how to determine which resource should be replicated, how many replicas are needed, where replicas are stored, and how to maintain consistency of replicas with their original data.

The destination-aware replication provides the benefit that nodes near to the destination quickly locate the resource without searching in areas that are farther away. Compared with the source-aware replication, the load is lower since a few nodes keeps replicas. It correlates with the locality principle, see Chap. 4. On the other hand, the source-aware replication achieves better global routing performance since the performance improvement is even for nodes that faraway from the primary responsible node.

A simple solution to guarantee consistency of replicas is keeping them only for a period of time. Depending on the resource update frequency, an expiration time can be adjusted. When a replica has expired, it can either be discarded or be refreshed by querying the primary responsible node. The problem is the determination of a suitable expiration time.

A lazy strategy can be employed; a replica is kept at a node as long as possible. When the original resource object is changed, the primary responsible node sends an update request. The problem is that the request must reach all replica nodes, an instance of the information dissemination problem [19,43].

A solution can be obtained at the cost of storage; a primary responsible node keeps links to replica nodes. The straightforward way is to keep links to all replica nodes. Another variant is construction of a disseminating tree of nodes keeping replicas and the primary responsible node as a root. Then updates requests propagate along the tree starting from the root and ending at leaf nodes.

The dissemination tree maintenance is proactive. On a regular basis "heartbeat" messages are sent from the root node downwards to all other nodes in the tree. If a node has no message within a time interval, then the node removes the replica (or rejoins the tree). Each replica node also periodically sends "refresh" messages upward. If the parent node has no message within a time interval, then the child is assumed died and the link to it is removed.

A more advanced maintenance mechanism was implemented in CUP [44]—a protocol for maintaining replicas in P2P networks. Rather than imposing a global propagation policy, nodes receive and propagate updates only when they have personal economic incentive to do so.

When the node capacity is limited, a node can replace less efficient replicas by new ones. Examples of popular metrics of the replica efficiency are least/most recently used, least/most frequently used, minimum/maximum replica size. Various ranking methods can be applied to construct a good replacement strategy. The basic idea is that a node only keeps a replica if it brings benefits to the node. This class of P2P resource and node ranking problem is considered in Chaps. 6 and 10.

3.8 Summary

This chapter started to sequentially build a set of design principles; each provides a base for applying hierarchical schemes. We introduced fundamental principles that applied by nodes locally in pure flat DHT designs.

Principle 1 states the small-world property in construction of local routing tables: closer neighbors are more preferable while all distant scales must be presented. Therefore, each node clusters its nearest nodes to its routing table for efficient local routing as well as defines long-range links for better global connectivity.

Principles 2 and 3 provide more systematic construction of routing tables and routes applying ordering, ranking and classifying models. Geometrically progressive routing aims at efficient global routing. Local adaptation ensures consistency of the overlay performance with the underlying network performance. The principles immediately lead to various tree-like structures involved into DHT routing. In particular, the well-known $O(\log N)$ routing efficiency is a result of local neighborhood arrangements.

Principle 4 extends the power of long-range links with additional knowledge for efficient global routing. A node can look ahead its neighbors, achieving more effective extrapolation of locally available knowledge in lookup path construction.

References

1. Abraham, I., Malkhi, D., Manku, G.S.: Papillon: greedy routing in rings. In: DISC '05: Proceedings of 19th International Conference on Distributed Computing. Lecture Notes in Computer Science, vol. 3724, pp. 514–515. Springer, Berlin (2005)
2. Barrière, L., Fraigniaud, P., Kranakis, E., Krizanc, D.: Efficient routing in networks with long range contacts. In: Proceedings of 15th International Conference on Distributed Computing (DISC '01), pp. 270–284. Springer, Berlin (2001)
3. Castro, M., Drushel, P., Hu, Y., Rowstron, A.: Proximity neighbor selection in tree-based structured peer-to-peer overlays. Technical Report MSR-TR-2002-52, Microsoft Research (2003)
4. Chiola, G., Cordasco, G., Gargano, L., Hammar, M., Negro, A., Scarano, V.: Degree-optimal routing for P2P systems. Theory Comput. Syst. **45**(1), 43–63 (2009). doi: http://dx.doi.org/10.1007/s00224-007-9074-x
5. Cohen, E., Shenker, S.: Replication strategies in unstructured peer-to-peer networks. SIGCOMM Comput. Commun. Rev. **32**, 177–190 (2002). doi: http://doi.acm.org/10.1145/964725.633043
6. Cordasco, G.: Degree-optimal deterministic routing for P2P systems. In: ISCC '05: Proceedings of 10th IEEE Symposium on Computers and Communications, pp. 158–163. IEEE Computer Society (2005). doi: http://dx.doi.org/10.1109/ISCC.2005.45
7. Cordasco, G., Gargano, L., Negro, A., Scarano, V., Hammar, M.: F-Chord: improved uniform routing on Chord. Networks **52**, 325–332 (2008). doi:10.1002/net.v52:4
8. Dabek, F., Kaashoek, M.F., Karger, D., Morris, R., Stoica, I.: Wide-area cooperative storage with CFS. In: Proceedings of 18th ACM Symposium Operating Systems Principles (SOSP '01), pp. 202–215. ACM Press (2001). doi: http://doi.acm.org/10.1145/502034.502054
9. Dabek, F., Li, J., Sit, E., Robertson, J., Kaashoek, M.F., Morris, R.: Designing a DHT for low latency and high throughput. In: Proceedings of 1st Symposium on Networked Systems Design and Implementation (NSDI '04), pp. 85–98 (2004)
10. Danezis, G., Lesniewski-Laas, C., Kaashoek, M.F., Anderson, R.: Sybil-resistant DHT routing. In: Proceedings of 10th European Symposium on Research in Computer Security, pp. 305–318 (2005)
11. Duchon, P., Hanusse, N., Lebhar, E., Schabanel, N.: Towards small world emergence. In: Proceedings of 18th Annual ACM Symposium on Parallelism in Algorithms and Architectures (SPAA '06), pp. 225–232. ACM, New York (2006). doi: http://doi.acm.org/10.1145/1148109.1148145
12. Fonseca, P., Rodrigues, R., Gupta, A., Liskov, B.: Full-information lookups for peer-to-peer overlays. IEEE Trans. Parallel Distrib. Syst. **20**(9), 1339–1351 (2009)
13. Fraigniaud, P., Gauron, P.: D2B: a de Bruijn based content-addressable network. Theor. Comput. Sci. **355**(1), 65–79 (2006). doi: http://dx.doi.org/10.1016/j.tcs.2005.12.006
14. Freedman, M.J., Vingralek, R.: Efficient peer-to-peer lookup based on a distributed trie. In: Revised Papers from 1st International Workshop on Peer-to-Peer Systems (IPTPS '01), pp. 66–75. Springer, Berlin (2002)
15. Gai, A.T., Viennot, L.: Broose: A practical distributed hashtable based on the De-Bruijn topology. In: Proceedings of IEEE 4th International Conference on Peer-to-Peer Computing (P2P '04), pp. 167–164. IEEE Computer Society (2004). doi: http://dx.doi.org/10.1109/P2P.2004.10
16. Girdzijauskas, S., Galuba, W., Darlagiannis, V., Datta, A., Aberer, K.: Fuzzynet: Ringless routing in a ring-like structured overlay. Peer-to-Peer Netw. Appl. **4**(3), 259–273 (2011). doi: http://dx.doi.org/10.1007/s12083-010-0081-3
17. Gummadi, K., Gummadi, R., Gribble, S., Ratnasamy, S., Shenker, S., Stoica, I.: The impact of DHT routing geometry on resilience and proximity. In: Proceedings of ACM SIGCOMM'03, pp. 381–394. ACM, New York (2003). doi: http://doi.acm.org/10.1145/863955.863998

18. Hildrum, K., Kubiatowicz, J.: Asymptotically efficient approaches to fault-tolerance in peer-to-peer networks. In: Proceedings of 17th International Symposium on Distributed Computing (DISC '03), pp. 321–336 (2003)
19. Hromkovic, J., Klasing, R., Pelc, A., Ruzicka, P., Unger, W.: Dissemination of information in communication networks: broadcasting, gossiping, leader election, and fault-tolerance. In: Texts in Theoretical Computer Science. An EATCS Series. Springer, New York (2005). URL: http://www.springer.com/978-3-540-00846-0
20. Hu, J., Li, M., Zheng, W., Wang, D., Ning, N., Dong, H.: Smartboa: constructing P2P overlay network in the heterogeneous Internet using irregular routing tables. In: IPTPS '04: Proceedings of 3rd International Workshop on Peer-to-Peer Systems. Lecture Notes in Computer Science, vol. 3279, pp. 278–287. Springer, Berlin (2004)
21. Kaashoek, M.F., Karger, D.R.: Koorde: a simple degree-optimal distributed hash table. In: IPTPS '03: Proceedings of 2nd International Workshop on Peer-to-Peer Systems. Lecture Notes in Computer Science, vol. 2735, pp. 98–107. Springer, Berlin (2003)
22. Kleinberg, J.M.: The small-world phenomenon: an algorithm perspective. In: Proceedings of 32nd Annual ACM Symposium Theory of Computing (STOC '00), pp. 163–170. ACM, New York (2000). doi: http://doi.acm.org/10.1145/335305.335325
23. Kleinberg, J.M.: Complex networks and decentralized search algorithms. In: Proceedings of International Congress of Mathematicians (ICM 2006). European Mathematical Society (2006)
24. Korzun, D., Gurtov, A.: A Diophantine model of routes in structured P2P overlays. ACM SIGMETRICS Perform. Eval. Rev. 35(4), 52–61 (2008)
25. Korzun, D., Nechaev, B., Gurtov, A.: Cyclic routing: Generalizing lookahead in peer-to-peer networks. In: AICCSA2009: Proceedings of 7th IEEE/ACS International Conference Computer Systems and Applications, pp. 697–704. IEEE Computer Society (2009). doi: http://doi.ieeecomputersociety.org/10.1109/AICCSA.2009.5069403
26. Ledlie, J., Shneidman, J., Amis, M., Mitzenmacher, M., Seltzer, M.: Reliability- and capacity-based selection in distributed hash tables. Computer science technical report, Harvard University (2003)
27. Leong, B., Liskov, B., Demaine, E.: Epichord: parallelizing the Chord lookup algorithm with reactive routing state management. In: ICON 2004: Proceedings of 12th International Conference on Networks, pp. 270–276 (2004)
28. Li, J., Stribling, J., Morris, R., Kaashoek, M.F., Gil, T.M.: A performance vs. cost framework for evaluating DHT design tradeoffs under churn. In: Proceedings of IEEE INFOCOM'05, vol. 1, pp. 225–236. IEEE (2005). doi:10.1109/INFCOM.2005.1497894
29. Li, J., Stribling, J., Morris, R., Kaashoek, M.F.: Bandwidth-efficient management of DHT routing tables. In: Proceedings of the 2nd Symposium on Networked Systems Design and Implementation (NSDI '05), pp. 99–114 (2005)
30. Loguinov, D., Kumar, A., Rai, V., Ganesh, S.: Graph-theoretic analysis of structured peer-to-peer systems: routing distances and fault resilience. IEEE/ACM Trans. Netw. 13(5), 1107–1120 (2005)
31. Manku, G.S., Bawa, M., Raghavan, P.: Symphony: distributed hashing in a small world. In: USITS'03: Proceedings of 4th USENIX Symposium on Internet Technologies and Systems, pp. 127–140. USENIX Association (2003)
32. Manku, G.S., Naor, M., Wieder, U.: Know thy neighbor's neighbor: the power of lookahead in randomized P2P networks. In: STOC '04: Proceedings of 36th Annual ACM Symposium on Theory of Computing, pp. 54–63. ACM, New York (2004). doi: http://doi.acm.org/10.1145/1007352.1007368
33. Maymounkov, P., Mazières, D.: Kademlia: A peer-to-peer information system based on the XOR metric. In: IPTPS '02: Proceedings of 1st International Workshop on Peer-to-Peer Systems. Lecture Notes in Computer Science, vol. 2429, pp. 53–65. Springer, Berlin (2002)
34. Monnerat, L.R., Amorim, C.L.: D1HT: a distributed one hop hash table. In: Proceedings of 20th IEEE International Symposium on Parallel and Distributed Processing (IPDPS 2006). IEEE Computer Society (2006). doi: http://doi.ieeecomputersociety.org/10.1109/IPDPS.2006.1639278

35. Monnerat, L.R., Amorim, C.L.: Peer-to-peer single hop distributed hash tables. In: Proceedings of IEEE Globecom'09, pp. 4250–4257, IEEE (2009)
36. Naor, M., Wieder, U.: A simple fault tolerant distributed hash table. In: IPTPS '03: Proceedings of 2nd International Workshop on Peer-to-Peer Systems. Lecture Notes in Computer Science, vol. 2735, pp. 88–97. Springer, Berlin (2003)
37. Naor, M., Wieder, U.: Know thy neighbor's neighbor: better routing for skip-graphs and small worlds. In: IPTPS '04: Proceedings of 3rd International Workshop on Peer-to-Peer Systems. Lecture Notes in Computer Science, vol. 3279. Springer, New York (2004)
38. Naor, M., Wieder, U.: Novel architectures for P2P applications: the continuous-discrete approach. ACM Trans. Algorithms **3**(3), 37 (2007). doi: http://doi.acm.org/10.1145/1273340. 1273350
39. Ng, T.S.E., Chu, Y.H., Rao, S.G., Sripanidkulchai, K., Zhang, H.: Measurement-based optimization techniques for bandwidth-demanding peer-to-peer systems. In: Proceedings of IEEE INFOCOM'03, vol. 3, pp. 2199–2209 (2003)
40. Plaxton, C.G., Rajaraman, R., Richa, A.W.: Accessing nearby copies of replicated objects in a distributed environment. In: Proceedings of 9th Annual Symposium on Parallel Algorithms and Architectures (SPAA '97), pp. 311–320 (1997)
41. Ramasubramanian, V., Sirer, E.G.: Beehive: $O(1)$ lookup performance for power-law query distributions in peer-to-peer overlays. In: Proceedings of 1st Symposium on Networked Systems Design and Implementation (NSDI '04), pp. 99–112 (2004)
42. Ratnasamy, S., Handley, P.F.M., Karp, R., Shenker, S.: A scalable content-addressable network. In: Proceedings of ACM SIGCOMM'01, pp. 161–172. ACM, New York (2001)
43. Risson, J., Harwood, A., Moors, T.: Topology dissemination for reliable one-hop distributed hash tables. IEEE Trans. Parallel Distrib. Syst. **20**(5), 680–694 (2009)
44. Roussopoulos, M., Baker, M.: CUP: Controlled update propagation in peer-to-peer networks. In: Proceedings of the USENIX Annual Technical Conference, pp. 167–180 (2003)
45. Rowstron, A., Druschel, P.: Pastry: Scalable, distributed object location and routing for large-scale peer-to-peer systems. In: Middleware'01: Proceedings of IFIP/ACM International Conference on Distributed Systems Platforms. Lecture Notes in Computer Science. vol. 2218, pp. 329–350. Springer, Berlin (2001)
46. Rowstron, A., Druschel, P.: Storage management and caching in past, a large-scale, persistent peer-to-peer storage utility. SIGOPS Oper. Syst. Rev. **35**(5), 188–201 (2001). doi: http://doi. acm.org/10.1145/502059.502053
47. Serbu, S., Bianchi, S., Kropf, P., Felber, P.: Dynamic load sharing in peer-to-peer systems: when some peers are more equal than others. IEEE Internet Comput. **11**(4), 53–61 (2007). doi: http://dx.doi.org/10.1109/MIC.2007.81
48. Stoica, I., Morris, R., Liben-Nowell, D., Karger, D., Kaashoek, M.F., Dabek, F., Balakrishnan, H.: Chord: a scalable peer-to-peer lookup service for Internet applications. IEEE/ACM Trans. Netw. **11**(1), 17–32 (2003)
49. Tang, C., Buco, M.J., Chang, R.N., Dwarkadas, S., Luan, L.Z., So, E., Ward, C.: Low traffic overlay networks with large routing tables. SIGMETRICS Perform. Eval. Rev. **33**(1), 14–25 (2005). doi: http://doi.acm.org/10.1145/1071690.1064216
50. Xu, J., Kumar, A., Yu, X.: On the fundamental tradeoffs between routing table size and network diameter in peer-to-peer networks. IEEE J. Sel. Areas Commun. **22**(1), 151–163 (2004)
51. Xu, Z., Zhang, Z.: Building low-maintenance Expressways for P2P systems. Techical Report HPL-2002-41, HP Labs, Palo Alto (2002)
52. Zhang, H., Goel, A., Govindan, R.: Incrementally improving lookup latency in distributed hash table systems. In: Proceedings of 2003 ACM SIGMETRICS International Conference Measurement and Modeling of Computer Systems, pp. 114–125. ACM, New York (2003). doi: http://doi.acm.org/10.1145/781027.781042
53. Zhang, H., Jin, H., Zhang, Q.: Yarqs: Yet another range queries schema in DHT based P2P network. In: Proceedings of 9th IEEE International Conference Computer and Information Technology (CIT '09), pp. 51–56. IEEE Computer Society (2009). doi: http://dx.doi.org/10. 1109/CIT.2009.78

54. Zhao, B.Y., Huang, L., Stribling, J., Joseph, A.D., Kubiatowicz, J.D.: Exploiting routing redundancy via structured peer-to-peer overlays. In: ICNP '03: Proceedings of 11th IEEE International Conference on Network Protocols, pp. 246–257 (2003)
55. Zhao, B.Y., Huang, L., Stribling, J., Rhea, S.C., Joseph, A.D., Kubiatowicz, J.D.: Tapestry: A resilient global-scale overlay for service deployment. IEEE J. Sel. Areas Commun. **22**(1), 41–53 (2004)

Chapter 4
Adaptable Overlay Network Topology

Abstract The previous chapter overviewed hierarchical schemes that a node uses locally to arrange its neighbors, achieving effective routing in a flat network. This chapter focuses on further advanced routing schemes where nodes are intentionally specialized in routing. Such schemes lead to a kind of global hierarchy in the network; nodes become globally differentiated, specializing their roles in routing. According to the participants heterogeneity, the schemes differentiate (1) nodes in their routing responsibility (node specialization) and (2) resources in their distribution in the whole system (resource distribution).

4.1 Introduction

A flat ID space assumes that any node has equal responsibilities in routing. The hierarchy approach, however, applies various kinds of decomposition to differentiate some elements. A P2P design can improve routing complexity if the specialization in routing is distributed non-uniformly across the nodes.

The static node specialization is a design principle in butterfly-graph DHTs like Viceroy [36], Ulysses [46], Cycloid [42], Pappilon [3], Censorship Resistant Network (CRN) [12, 41], and Mariposa [37]. Nodes IDs belong to different levels of routing responsibilities. These levels are defined in advance—the DHT design partitions the ID space. A node is assigned randomly with its ID, appearing on some level. Section 4.2 states the principle of node ID specialization and considers its application in DHT designs.

Node specialization leads to the load balancing problem. The latter is important in any P2P system: computational, storage or communication load should be balanced fairly, even if the node population is heterogeneous. Several DHT design extensions allow nodes to control the ID assignment for better performance and load balancing [29, 30, 33, 39, 43]. A node controls its specialization by selecting appropriate ID and even changes its ID dynamically to preserve the specialization. Section 4.3 discusses local strategies for ID assignment control.

D. Korzun and A. Gurtov, *Structured Peer-to-Peer Systems: Fundamentals of Hierarchical Organization, Routing, Scaling, and Security*, DOI 10.1007/978-1-4614-5483-0_4,
© Springer Science+Business Media New York 2013

Similarly, the differentiation concerns the resource distribution. The P2P system distributes the resources among nodes according to the node specialization—a principle of such DHTs as SmartBoa [23], Diminished Chord [28], 1HS [40], and GTap [50]. Section 4.4 introduces the principle of resource distribution.

A special case for resource distribution is resource semantics. Resources are placed close to their users. It allows keeping routing paths within their domain of a global network hierarchy, as in Skip graphs [5] and SkipNet [20]. Section 4.5 elaborates the corresponding principle of content and path locality.

Non-uniform distribution of resources and their semantics require careful dynamic maintenance of load balance and topology [15, 32]. The corresponding hierarchical schemes appear in P-Grid [1]. Section 4.6 discusses the problem of balanced resources distribution among P2P nodes.

The dynamic maintenance meets the network size change problem when the number of participating nodes can vary significantly during the system evolution. Nodes should adapt their IDs and positions in the network. The adaptation also follows hierarchical schemes. Such DHT designs can be based on Kautz-graphs: FISSIONE [34], Moore [18], BAKE [19], and SKY [51]. Section 4.7 provides an overview of these designs.

4.2   Node ID Specialization

DHT overlay topologies from Chap. 3 are *uniform* [46] meaning that the construction of zones $S_i(u)$ is identical for every node $u \in N$. Recall that for b-progressive routing, zones offsets are the same at any node:

$$S_i(u) = \{k \in S \mid b^i \leq \rho(u,k) < b^{i+1}\}, \quad i = 0,\ldots,n-1.$$

Nodes can be specialized by constructing different offsets of zones for different nodes, leading to improvements in the routing performance.

Principle 5 (Node ID specialization). *Node IDs are partitioned into levels. Levels define different roles in routing.*

The intuition behind comes directly from greedy routing in uniform DHTs. Nodes of level i are defined

$$L_i = \{v \mid \exists u \text{ s.t. } v \in T_u \cap S_i(u)\}, \quad i = 0,1,\ldots,n-1.$$

Levels $L_{n-1}, \ldots, L_0$ define longest to shortest hops in the overlay. The efficiency is due to the use of highest levels first for distant contacts (global routing), then the next levels are for shorter hops, up to L_0 for shortest ones (local routing).

In uniform DHTs, this node specialization is not globally fixed, and on average nodes are equal in their routing function and load. There are DHT designs that do not follow this property, and the node ID space is divided a priori into fixed levels.

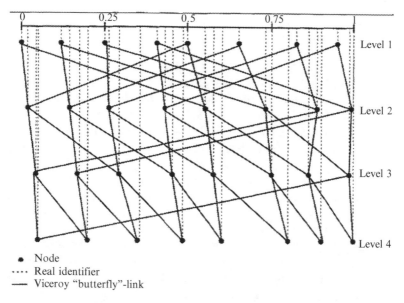

0   0.25   0.5   0.75

Level 1

Level 2

Level 3

Level 4

• Node
···· Real identifier
—— Viceroy "butterfly"-link

Fig. 4.1 An example of the ideal Viceroy topology; up and ring links are not shown. The picture is from [36]

Examples include the DHT topologies that apply a butterfly graph: Viceroy [36], CRN [12, 41], Ulysses [46], Cycloid [42], Pappilon [3], and Mariposa [37]. These designs are mostly for abstract networks; their implementation requires a rigorous maintenance protocol. It is not an easy problem because of a complicated static structure of butterfly topologies. The idea behind the multi-level butterfly topology was illustrated in Fig. 2.12 (on p. 39, Chap. 2).

4.2.1 Viceroy

The design exploits a simple butterfly with several levels. Given a circular $[0, 1)$ space (S, ρ) and levels $i = 1, \dots, n$ for $n = \lceil \log N \rceil$, a full node identity is $u = (ID_u, i)$ for $ID_u \in S$. Local neighbors are u's immediate successor and predecessor in the global ring. Long-range neighbors are either for intra-level contacts (medium hops) or inter-level contacts (long hops). Each level i defines i-ring of nodes at level i, and intra-level contacts are to immediate successor and predecessor in the i-ring. Inter-level contacts are "down-left", "down-right" and "up" links to nodes at levels $i + 1$ (if $i < n$) and $i - 1$ (if $i > 1$). The down-left link is to the closest node at level $i + 1$. The down-right link is to a node v at level $i + 1$ with the distance about $\rho(u, v) = 1/2^i$. The up link is to a close-by node at level $i - 1$. The ideal Viceroy topology is shown in Fig. 4.1.

In fact, the Viceroy topology is the Chord topology decomposed onto n subrings, i.e., the logarithmic number of global groups C_i appears in the network. Since any node selects its level ID at random the groups are approximately equal in size, $|C_i| \approx N/\log N$. The dependence $n = \lceil \log N \rceil$ requires local estimations of N, and this can bias the population of C_i with higher i if N varies rapidly.

This ring hierarchy categorizes roles for nodes at different levels. The higher level, the longer hops the node makes with down-right links. Routing proceeds in three phases. First, the lookup follows up links to reach a level-1 node; the phase is closer to local routing since the hops are short. Second, progressive routing uses i-ring intra-level contacts and down links to reach the destination neighborhood, an instance of global routing. Third, local routing uses successors and predecessors of the global ring, i.e., hops are very short at this phase.

The first two phases are very efficient: $O(\log N)$ hops with high probability. The last phase takes $O(\log^2 N)$ hops with high probability, while the expectation is $O(\log N)$. In total, routing takes $O(\log N)$ hops. There is also a modification of the third routing phase such that routing takes $O(\log N)$ hops with high probability.

The Viceroy design provides the constant number of outgoing links per node. Taking into account the maintenance of incoming links leads to $\Theta(1)$ expected number of records and the high probability bound is $O(\log N)$. For a few nodes this bound can be exceeded, and they have to maintain $\Omega(\log N)$ records. As a result, the design ensures that the majority of nodes has fair load and some nodes can occasionally become "hubs" with higher load.

4.2.2 Ulysses

This DHT design optimizes the tradeoff between the state maintenance and routing costs. It improves upon the Viceroy topology by embedding an (n, m)-butterfly structure, where $n \leq \log_2 N / \log_2 \log_2 N$ is the number of levels and $m = \log_2 N$ is the node out-degree. The diameter of the butterfly is n, the lower bound for routing hops. Recall that in Viceroy the node out-degree is 7.

The ID space S is the n-level n-dimensional m-cuboid with $m^n n$ points in total. Each point is

$$k = (k_0, k_1, \ldots, k_{n-1}; i), \ \ 0 \leq k_j \leq m - 1, 0 \leq i \leq n - 1.$$

A node $u = (u_0, \ldots, u_{n-1}; i)$ is responsible for a zone that is a subcuboid of a level-i cuboid. A zone splitting and merging mechanism is needed in response to node joins and leaves.

Neighbors at the closest levels are nodes for points

$$k = \begin{cases} (k_0, \ldots, k_i, x, k_{i+2} \ldots, k_{n-1}; i+1), & \text{if } i < n - 1 \\ (x, k_1, \ldots, k_{n-1}; 0), & \text{if } i = n - 1 \end{cases}$$

for $0 \leq x < m$ (cf. down links in Viceroy, $m = 2$). Two neighbors at other levels $j \neq i - 1, i + 1$ are simply for points $k = (u_0, \ldots, u_{n-1}; j)$.

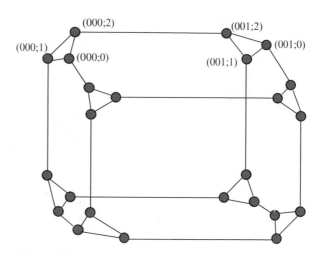

Fig. 4.2 The three-dimensional CCC graph

Routing a lookup for a key $k = (k_0, \ldots, k_{n-1}; j)$ from $u = (u_0, \ldots, u_{n-1}; i)$ first sequentially changes u_{i+1} to k_{i+1}, u_{i+2} to k_{i+2}, ..., u_{j-1} to k_{j-1}, u_j to k_j, i.e., geometrically progressive global routing is performed. Then the lookup goes to a level-j node, i.e., local routing occurs.

Every Ulysses node maintains $O(m + n) = O(\log N)$ state with high probability. The scheme provides $O(n) = O(\log N / \log \log N)$ routing, in spite of node joins and leaves. The static (n, m)-butterfly that Ulysses embeds is node-congestion-free and is not link-congestion-free.

4.2.3 Cycloid

The design is close to the Ulysses scheme and emulates a Cube-Connected-Cycles (CCC) graph, a subgraph of the $(n, 2)$-butterfly. The position of a node in the graph defines the routing role of the node.

Each vertex of n-dimensional cube is a cycle of n points (similarly to level $0 \leq i \leq n - 1$ in the $(n, 2)$-butterfly), see Fig. 4.2. In $u = (u_{n-1} \cdots u_0; i)$, the cubic coordinate is $u_{\text{cub}} = u_{n-1} \cdots u_0$ for binary u_j and the cyclic coordinate is $u_{\text{cyc}} = i$ for $0 \leq i < n$. A key $k = (k_{\text{cub}}; k_{\text{cyc}})$ is assigned to the node whose ID is first closest to k_{cub} and then closest to k_{cyc}.

The neighbor establishment is similar to Pastry with the metric ρ for cubic coordinates. A node with $u_{\text{cyc}} = 0$ has no cubic neighbor. If $i = u_{\text{cyc}} > 0$, a cubical neighbor shares the $(n - i - 1)$-bit prefix with u_{cub} and differs in bit i:

$$v = (u_{n-1} \cdots u_{i+1} \bar{u}_i v_{i-1} \cdots v_0; i - 1).$$

Hence, $2^{i-1} \leq \rho(u_{\text{cub}}, v_{\text{cub}}) < 2^i$. A cubic neighbor allows resolving the ith bit in lookup keys if the node is at level i.

The cyclic neighbors are the first larger and smaller nodes v and w such that (1) $v_{\text{cyc}} = w_{\text{cyc}} = i - 1 \mod n$; (2) u_{cub} shares at least the $(n-i)$-bit prefix with v_{cyc} and w_{cyc}; (3) v_{cub} and w_{cub} are closest to u_{cub} among numerically larger and smaller nodes, respectively. Cyclic neighbors change the current cyclic coordinate without the prefix change in node's cubic coordinate.

Similarly to Pastry, u's leaf set contains nearby nodes. It is divided into inside and outside leaf sets, each consisting of two neighbors. Nodes with the same cubic coordinate organize a local cycle by their cyclic coordinates modulo n (clockwise). The inside leaf set of u includes the immediate predecessor v and the immediate successor w in the local cycle. All local cycles organize the global ring modulo 2^n (clockwise) by their cubic coordinates. In any local cycle, the node with the largest cyclic coordinate is called the primary node of the cycle. The neighbors in u's outside leaf set are the primary nodes of preceding and succeeding cycles (remote cycles).

Routing proceeds in three phases. In the first phase, the lookup is sequentially forwarded using outside leaf set. The phase is a kind of local routing and continues until the cyclic coordinate becomes appropriate to start prefix-matching routing on the second phase. The latter is similar to global routing in Pastry, each step matches at least one next bit using long-range cubic and cyclic neighbors. Outside leaf sets are applicable as well, especially at final steps of the second phase. The third phase finishes the routing process with finding the node closest in the cyclic coordinate using the leaf sets.

4.2.4 Pappilon

The design inherits the Ulysses scheme. The ID space S consists of non-negative integers up to $m^n n$ for $m,n \geq 2$. The Chord asymmetric (clockwise) distance ρ can be used as well as its symmetric version. Let us consider the asymmetric case only.

The network has N nodes arbitrary positioned on a ring. A level is defined $i(u) = n - 1 - (u \mod n)$ for $u \in S$. Each node has m neighbors defined by the links to nodes at the next level $i = i(u) - 1 \mod n$,

$$u + xm^{i(u)}n + 1 \mod N, \quad 0 \leq x \leq m - 1.$$

For $x = 0$, the neighbor is immediate successor. Hence, a power $m^{i(u)}$ in Pappilon plays the same role as a coordinate position in Ulysses.

For a lookup for k, greedy routing proceeds in three phases. First (at most $n - 1$ hops), the lookup sequentially moves from one level to the next one and arrives to a node u such that $i(u) = n - 1$. Note that $\rho(u,k) < m^n n$ at level $n - 1$. Second (at most n hops), if $\rho(u,k) \geq n$, the lookup sequentially goes through levels to a node v at level $i(v) = 0$. Then v forwards to a node u at level $i(u) = n - 1$ such that $\rho(u,k) < n$. Third (at most $n - 1$ hops), the lookup decreases the distance by immediate successors.

This greedy scheme provides $O(\log N / \log n)$ routing in the worst case. Pappilon admits further improvement for shorter paths (reducing the constants inside big-O bounds). As Ulysses, Pappilon guarantees node-congestion-free routing. Furthermore, the improved routing strategies of Pappilon also guarantee link-congestion-free routing.

4.2.5 Censorship Resistant Network

In the previous butterfly-based designs the routing role of a node depends on the level to which the node belongs in the butterfly graph. In CRN a node is presented at many levels.

The design constructs a backbone butterfly-based network of supernodes, where the number of levels is $n = \lceil \log N - \log \log N \rceil$. Supernodes are categorized into top, middle and bottom, depending on their level in the butterfly.

The backbone network defines the structure of the overlay network topology. Each supernode is associated with a set of nodes. Each node u belongs to several supernodes: when joining the system u associates itself with C top supernodes, C bottom supernodes, and $C \log N$ middle supernodes, where C is a system-level parameter. Nodes that belong to the same supernode form a clique in the overlay network. If two supernodes are connected in the backbone network then the sets of their nodes are "high" connected in the overlay (up to the complete bipartite graph).

Resources are stored at bottom supernodes. For each resource several supernodes are selected randomly. Then the resource is replicated again among nodes of these supernodes.

Top and middle supernodes are for routing. Multipath routing is actively used when messages are forwarded. A lookup starts in parallel from all top supernodes that u knows and from the nodes associated with these supernodes. Then routing goes along multiple paths downwards in the butterfly. Each path is of logarithmic length. Some of them reach bottom supernodes that are responsible for the resource.

The CRN design requires $O(\log^3 N)$ state to keep routing tables. Routing takes $O(\log N)$ hops and $O(\log^3 N)$ messages in total (because of multipath routing). The resource replication factor is $O(\log N)$. The additional complexity aims at fault-tolerance, making a CRN overlay robust to churn and orchestrated attacks in dynamic environments. Even after an adversary deletes up to $N/2$ nodes, $(1 - \varepsilon)$ fraction of the remaining nodes have access to $(1 - \varepsilon)$ fraction of resources for a fixed parameter ε.

Although this supernode-based P2P architecture is hierarchical (and we consider such architectures in Chap. 7), the design solution concerns the flat node space. Regular nodes are distributed between supernodes independently on their capacity and other heterogeneous characteristics. Therefore, the routing role of a node is determined by the combination of supernodes the node belongs.

4.2.6 Mariposa

In contrast to butterfly-based designs with fixed discrete levels, the Mariposa design[1] applies the small-world principle for constructing a continuous variant of the Viceroy level-based topology. The circular $[0, 1)$ space (modulo 1) is used. Each node u estimates N and chooses randomly a range δ using the same harmonic pdf $h(x) = 1/(x \ln N)$ as in Symphony.

There are m' local links, 2 intermediate links, $2m$ long links and at most m global links per node $(m, m' \geq 1)$. Local links are to immediate clockwise successors. Intermediate links are to nodes responsible for keys $u + \log N$ and $u + \log N / \log m$. The intervals $[u - \delta, u]$ and $[u, u + \delta]$ are partitioned into m equal subintervals of length δ/m each, and u establishes a long link per subinterval. Global links are established if $\delta < 1/2$. The interval $[0, 1) \setminus [u - \delta, u + \delta]$ is partitioned into m equal subintervals of length $(1 - 2\delta)/m$ each, and u establishes a global link per subinterval.

Long links span nodes in two large ranges, clockwise and anticlockwise. Global links span nodes in u's opposite part of the ring. Continuous range δ plays the similar role as discrete level i in Viceroy; the smaller $1/\delta$, the larger hops u makes by long and global links. Greedy routing takes $O(\log N / \log m)$ hops with high probability.

4.3 Specialized Node ID Management

Principle 5 defines ID levels for node specialization in routing. DHTs designs from the previous section assume random assignment of IDs and their levels. As a result, a node cannot select its ID with desired specialization. Random assignment does not reflect the node heterogeneity and dynamics in real networks. It leads to load imbalance when some nodes have a higher load/capacity ratio than others. The routing performance degrades when a low-capacity node serves many lookups.

Similarly to the local adaptation schemes of Principle 3 in Chap. 3, nodes can dynamically adapt the overlay to their specialization. The basic approach is non-random selection of node IDs. It can be further extended with node migration when underloaded nodes move to overloaded areas of the overlay ID space. Specialized node ID management schemes were studied in [29, 30, 33, 39, 43] to keep routing performance and load balance within certain bounds. When a node joins it applies specific ID assignment to minimize the imbalance. When a node leaves or resource distribution changes dynamically participating nodes perform ID reassignment, changing their IDs to reduce observed imbalances.

---

[1]"Mariposa" means "butterfly" in Spanish.

4.3.1   ID Assignment

Consider a Chord-like DHT with a ring-based ID space S and the asymmetric space distance ρ. A node u is responsible for the keys between itself and its immediate successor $v = \operatorname*{argmin}_{w \in N, w \neq u} \rho(u, w)$,

$$S(u) = \{k \in S \mid 0 \leq \rho(u, k) < \rho(u, v)\}. \tag{4.1}$$

If resource keys are distributed uniformly in the ring S, then the fraction of resources that u is responsible for is proportional to $\rho(u, v)$.

Let node IDs be also taken uniformly in S. Although the expectation $E[\rho(u, v)] = 1/N$, with high probability the largest fraction [26] served by u is $\Theta((\log N)/N)$ and the smallest fraction [31] is $\Theta(1/N^2)$. Consequently, some nodes become higher or lower loaded than the average node since they are responsible $\Theta(\log N)$ times more or $\Theta(1/N)$ times less resources, respectively. Both estimates characterize the imbalance factor of basic consistent hashing.

If resource keys are distributed uniformly in the ID space then the size of zones $S(u)$ should be equalized for all nodes $u \in N$. If nodes differ in their capacity then the size of $S(u)$ should be proportional to the capacity of u. As was shown, the uniform selection of node IDs provides biased load per node for large N. A new node can apply a specific strategy to select its ID preserving a load-balanced distribution as much as possible. The basic scheme here employs the ability of multiple choice. A new node has multiple choices for its ID in the assignment, and a ranking model is applied to select the best ID.

Consider the ID assignment strategy from [27, 29] for a Chord-like DHT with the ring space $S = [0, 1)$. Nodes are assumed to have uniform capacity. The strategy designates the following points in S to be landmarks:

$$a_i = \frac{2i - 2^{\log_2 i} - 1}{2^{\log_2 i}} \quad \text{for } i = 1, 2, \ldots.$$

Intuitively, the first $N/2$ points are roughly equally spaced in $[0, 1)$.

A node u operates with several virtual IDs $\{u_j\}$, where $j = 1, 2, \ldots, c \log_2 N$ for a fixed constant $c > 2$. Participating in the system, u considers for each u_j the point $a_{i(j)} \in [u_j, v)$ with minimal index $i(j)$, where v is the active node preceding u_j in the space. Then u makes active u_{j^*} to be its current ID where $a_{i(j^*)} = \min_j a_{i(j)}$. In case of a tie, u makes active the ID that is closest to this $a_{i(j^*)}$. Although N can vary, each node u is responsible for interval in $[0, 1)$ of length at most $5/N$. That is, the imbalance factor is constant, compared with the logarithmic one when basic consistent hashing is used.

The strategy allows u to improve regularly its current position in S based on local knowledge of $\Theta(\log N)$ virtual nodes. Only one virtual node is activated at a given time, hence IDs are changed when load imbalance has been detected. In stable state, the global assignment is reached. When a node joins or leaves then $O(\log N)$ nodes have to change their IDs to return the system to the stable state. Consequently, $O(\log N / N)$ resource keys have to be reassigned between nodes.

4.3.2   Grouping Node IDs

The classifying model can be applied to form groups of several node IDs. In this Section, we briefly list strategies for load balancing based on simple groups of node IDs. Actually, the approach evolves to clustering; please refer to Chap. 5 for more details on this topic. Note that the ID assignment strategy [27, 29], which we considered above, is also based on specific groups of node IDs.

One of the first attempts to tackle the DHT load balancing problem was based on the consistent hashing approach [10, 26]. The hash table is divided into partitions. A partition is a contiguous subset of the resource key space, typically with specific left and right limits. For example, in the ring space of a Chord-like DHT, a partition is an arc, see (4.1). Each physical machine hosts a certain number of virtual DHT nodes. The hash table is distributed across the machines by assigning one partition per virtual node. Partitions have random size and they are assigned to virtual nodes randomly. For the balancing to be effective, each machine must host at least $\Theta(\log N)$ virtual nodes.

In bucket solution [36, 39] nodes are grouped into buckets, and each node can change its ID within the bucket to rearrange the load.

In load-balancing schemes of [43], each physical node uses several IDs (i.e., representing a group of virtual DHT nodes); virtual nodes migrate from heavily loaded physical nodes to lightly loaded ones.

4.3.3   ID Reassignment

When a node leaves, some existing nodes can change their IDs to rearrange the load or to preserve other useful characteristics like the routing performance. Such dynamic balancing the overlay topology becomes of high importance in adversarial scenarios when some node insertions and removals are done by an adaptive adversary [2]. When a legitimate node has detected imbalance, the node responses adjusting its ID for more appropriate specialization. This approach is called *node migration* [32].

In the multiple choice scheme above, the ID space is split evenly among nodes by selection of possible node IDs. The scheme ignores the actual resource distribution. In contrast, if a node changes its ID, moving to another position in the space, then, for given resource distribution, the system tries to distribute the resources among nodes adaptively.

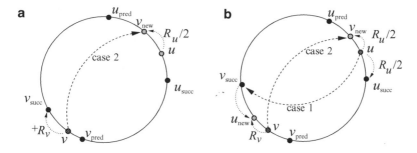

Fig. 4.3 Node migration when u and v discover load imbalance $R_v \leq \varepsilon R_u$. (**a**) Only case 2 happens when v_{succ} has reasonable load compared with u. (**b**) Case 1 happens in addition to case 2 when v_{succ} has high load

Consider the reassignment strategy for a Chord-like DHT from [27]. Let R_u be the number of resource keys u is responsible for. The basic requirement is that each node contacts random other nodes to check the load imbalance.

Let $0 \leq \varepsilon < 1/4$ be a constant. Every node u occasionally contacts a random node v. Let u_{pred}, u_{succ} and v_{pred}, v_{succ} be the immediate predecessors and successors of u and v, correspondingly. Note that $R_u \sim \rho(u_{\text{pred}}, u)$ and $R_v \sim \rho(v_{\text{pred}}, v)$. If $R_v \leq \varepsilon R_u$ then u and v have different load (the case $R_u < \varepsilon R_v$ is symmetrical), u is overloaded compared with v, and their IDs are updated as follows. This update reassigns resource keys, which u and v were responsible for.

1. If $R_{v_{\text{succ}}} > R_u$ then u selects a new position u_{new} in the middle of $(v_{\text{succ}}, v_{\text{pred}})$.
2. Select a new position $v_{\text{new}} \in (u_{\text{pred}}, u)$ such that the node v captures half of resources of node u.

Figure 4.3 illustrates the reassignment strategy. Case 2 is mandatory; it is depicted in Fig. 4.3a. The high load $R_u \sim \rho(u_{\text{pred}}, u)$ is halved. The first half $R_u/2 \sim \rho(u_{\text{pred}}, v_{\text{new}})$ is assigned to $v := v_{\text{new}}$, which moves to the new position in the arc $(u_{\text{pred}}, u_{\text{succ}})$. The second half $R_u/2$ remains at u. The condition $R_{v_{\text{succ}}} \leq R_u$ assumes that the current load of v_{succ} is acceptable; v_{succ} takes the additional load R_v, which is small compared with $R_{v_{\text{succ}}}$. As a result, the load of u and v become balanced.

The condition $R_{v_{\text{succ}}} > R_u$ of case 1 requires additional reassignment since the load of v_{succ} is high, see Fig. 4.3b. In this case, u moves from the arc $(u_{\text{pred}}, u_{\text{succ}})$, delegating v its role and assigning the second half $R_u/2 \sim \rho(v_{\text{new}}, u)$ to u_{succ} (in addition to the arc (u, u_{succ})). Now $u := u_{\text{new}}$ substitutes the migrated v in a middle point of the arc $(v_{\text{pred}}, v_{\text{succ}})$. The load of v_{succ} and u is equalized: u takes all R_v and a fraction of $R_{v_{\text{succ}}}$. Note that if case 1 happens often then it might be better to shift v to the middle in between its predecessor and successor without execution of case 2.

Dynamic manipulations with node IDs are expensive due to updating routing tables. A group of n nodes must maintain n routing tables, and migration incurs a significant cost. Active node migration leads to increasing churn in the overlay.

Additionally, security vulnerabilities appear since adversaries can affect the distribution of node IDs to destroy the network [9]. If node migration results in uneven distribution of node IDs in the space, then the basic assumption of the underlying DHT routing protocol is violated, and the routing performance degrades.

4.4 Distributing Resources Among Nodes

Node specialization arranges nodes into a global hierarchy. ID manipulation mechanisms adapt the overlay to node heterogeneity and dynamics. Nevertheless, another way exists for differentiating nodes for their specialization. Instead of changing IDs to appropriate specialization, resources are distributed according to the node specialization. Placing certain resources onto a specific set of nodes enables provisioning of those nodes to reflect demand, thus resulting in improved performance. For instance, computational-expensive resources are preferably placed at nodes with fast CPU; storage-expensive resources are at nodes with large disks.

Principle 6 (Resource distribution). *The system provides controlled resource distribution among nodes.*

Resource distribution can be controlled dynamically based on the *resource exchange* approach when the load is balanced through iterative resource exchange between nodes [15, 32].

Similarly to Principle 5, the classifying model can be applied for implementing Principle 6. Nodes with the same role form a group, and certain resource is assigned to a node of the appropriate group [28,50]. For example for load-balancing, nodes are grouped according to their available capacity [23, 44]. Administrative and geographical structures group nodes by organization membership or by location [40]. Different groups can be formed to reflect different levels of trust, locating important resources at high-trusty nodes. Another topical example is applications with considerable resource semantics, see Sect. 4.5.

Diminished Chord [28] is a version of Chord that allows any subset of nodes to jointly offer the same functionality. The design uses $O(\log n)$ additional neighbors per node in an n-nodes group and $O(n)$ additional data in the whole network (at nodes that are not necessarily in the group themselves). Routing a lookup for a node of a given group takes $O(\log N)$ hops.

Grouped Tapestry (GTap) [50] is a version of Tapestry that allows organizing nodes into many groups. It is similarly to Diminished Chord, but GTap has some advantages. At most one additional neighbor is required per node in an n-nodes group. Furthermore, GTap supports routing within a group, providing the path locality (see Principle 7 in Sect. 4.5 below).

SmartBoa [23] evolves the idea of large routing tables (see Sect. 3.6) such that the size of a routing table dynamically depends on the available bandwidth. Nodes are categorized into discrete levels according to their capacity. The lower the node's level is, the fewer neighbors it has to maintain. As a result, nodes are differentiated in routing; a node at a higher level can route a lookup more efficiently. In addition to node IDs, nodes have level labels and a node can change its level freely, enhancing the adaptation in heterogeneous environments.

HiGLOB [44] applies Principle 4 from Sect. 3.5 to reflect a global view of load distribution in the overlay. A node u constructs a histogram by partitioning the network into non-overlapping groups, each of which is connected to u by a neighbor. Each bucket in the histogram characterizes the average node load in the group. Based on this knowledge the node may redistribute the load when a new node joins the overlay and when an existing node becomes overloaded or underloaded.

One Hop Sites (1HS) [40] is a design for enterprise P2P overlays where each group (enterprise site) is a separate one-hop DHT. Resources are replicated between groups. The approach is opposite to Diminished Chord since the latter does not separate the overlay into independent DHTs. The design improves the resilience to network partitions.

Such a group hierarchy dynamically changes structurally (group construction/removal) and by content (node joins/leaves, resource allocation/exchange). It leads to overhead on both inter-group and intra-group levels.

4.5   Resource Semantics

Most of DHT designs we considered above resemble a balance tree in S [26]. Resources $r \in D$ are distributed near-uniformly among alive nodes $d \in S$ based on hashing $h : D \to S$. It destroys the application-aware semantics with own balance tree in the resource key space D, leading to inefficient resource distribution.

Resource semantics leads to evolution of Principle 6 to the principle of content and path locality in resource distribution among participating nodes.

Principle 7 (Content and path locality). *Resources are placed close to most of their users. Routing paths remain within a domain whenever possible, and a path between two overlay nodes of the same domain is also within this domain.*

Intuitively, if Alice works within an organization Foo, her resources are reasonable to place at nodes that belong to Foo. Each resource must be either explicitly placed on specific overlay nodes or distributed across nodes within a given resource domain. In some applications it is likely that documents produced within an organization are actively used by people at this organization. Another example is from [20]. A route from node explorer.ford.com to mustang.ford.com could pass through camaro.gm.com. This scenario people at ford.com might prefer to prevent.

Content locality allows application-specific data pre-fetching, enhanced browsing, and complex querying such as range queries, e.g., locating resources whose keys in D lie in a given range. Nodes store important resources within their organization, thus resulting in improved manageability and security.

An organization controls the administrative domain in which its resources reside; even when encrypted, data stored on arbitrary node outside the organization are susceptible to DoS and other types of attacks.

Path locality complements the content locality by improving manageability, security, performance, administrative control, and accountability of participating organizations. It increases availability and fault-tolerance. Even if an overlay domain becomes disconnected from the rest, nodes of this domain are able to reach important data (e.g., because of replication of popular data items in the domain). These principles are also helpful in the presence of NATs and firewalls that limit connectivity among hosts in different organizations.

Geometrical routing with zones (3.10) is an inspiration of *skip lists* for graphs, a known data structure to represent a balanced tree (see [5,20] and references therein). A skip list, however, can be constructed on top of D instead of S, hence preserving the semantic structure of D needed in Principle 7. Skip graphs [5,6] and SkipNet [20, 21] are DHT designs utilizing skip lists. They are similar, and we shall follow closer to the SkipNet design, which uses cyclic skip lists (rings).

There is no hashing of a resource name into numerical flat keys, and the closeness of related resources can be ensured in the topology. The original design assumes that each node represents only one resource item, hence node and resource IDs are the same. A node participates at multiple levels $i = 0, 1, \ldots, n = \lceil \log N \rceil$. Level $i = 0$ consists of all nodes ordered by their names, the root ring. There are at most 2^i rings at level i, and every node participates in one of them. A node in its ring of level i links to a neighbor (successor) that is about 2^i nodes away in the root ring. Consequently, the nodes in every ring are sorted according to their resource keys (e.g., lexicographically).

As shown in Fig. 4.4, this structure is similar to the Chord topology; an i-level ring consists of ith fingers. The fundamental difference is that Chord fingers skip over nodes in the numerical flat space. It immediately leads to greedy routing with $O(\log N)$ hops to find the node with closest name: searching first in higher levels (global routing) and gradually moving to lower levels (local routing). Note that a node also links to its predecessor in each ring.

The basic resource management scheme exploits ring IDs. Let u_{num} be a binary string generated at u. (It is a membership vector in Skip graphs and a numerical node ID in SkipNet.) Each ring is a double-liked list labeled by some binary string x. A node u belongs to ring x at level i if x is an i-bit prefix of u_{num}. If u_{num} is generated randomly for any u, then the number of nodes in each i-level ring is balanced.

Starting from the root ring, a new node u inserts itself in some ring at each level. Using greedy routing, u iteratively finds its immediate successor and predecessor at each its ring. Similarly, node departure involves iteratively removing a node from all of its rings. A background repair process maintains the correctness of ring memberships; the process is complicated due to the maintenance at several levels.

In practice, a node keeps several resources by incorporating node's name into its resource names. Resources with semantically close names are concentrated at the same node or at neighboring nodes. Hence each node holds a contiguous segment of resources. It supports Principle 7 of content and path

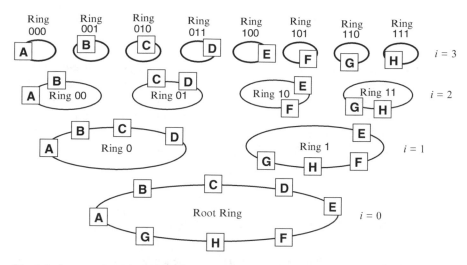

Fig. 4.4 An example of the ideal multi-level SkipNet with $N = 8$ nodes

locality For example, Alice's documents have names alice.foo.com/⟨docname⟩ for placing at node alice.foo.com. A path from node alice.foo.com to document mike.foo.com/song.mpg may go through node john.foo.com but not through mika.bar.com.

The basic skip lists scheme assumes the uniform resource distribution in D, otherwise nodes are subject to uneven load [6] (see also Sect. 4.6). To deal with this problem, various load balancing algorithms can be added, similarly to which were discussed in Sect. 4.3. However, the semantic ordering complicates load balancing since the order must be preserved in the system [32]. In the pairing algorithm [7], heavily-loaded nodes are placed next to lightly-loaded nodes, hence the former can migrate resources to their lightly-loaded neighbors. In the join algorithm [6], a new node performs a random walk, recording the load of each node it encounters; then it splits the load of the most heavily loaded of these nodes. In Skip Graphs++ [22], the probability that a node appears in higher levels of the skip lists structure is proportional to the capacity of the node, resulting in the load differentiation of heterogeneous nodes (cf. with SmartBoa [23] above).

Semantic ordering allows specific extensions to the basic design. Note that such extensions complicate even more the background repair processes [25]. Since nodes with close names are placed contiguously in the overlay, failures along domain boundaries do not completely fragment the overlay. Instead, they result in ring segment partitions. In the basic skip lists scheme, long-range neighbors at higher-level rings help to preserve the global connectivity. According to Principle 4 from Sect. 3.5 nodes can be augmented with additional links for better routing fault-tolerance and efficiency [14, 24].

Interesting mathematical schemes for constructing levels in skip list structure are proposed in [14]. They optimize the routing efficiency when the number of neighbors is differ than the original $n = \lceil \log N \rceil$ node state. Family trees [48] and rainbow skip graphs [16] are skip lists designs for $O(1)$ node state. Fault-tolerance properties were studied in [8, 16, 17]. NoN-routing [38] improves the expected query cost for searching skip graphs and SkipNet to $O(\log N / \log \log N)$, see also Sect. 3.5. An extension to multidimensional resource names can be found in Skip-webs [4] and pService [52]. SkipCluster [47] is a two-layer hierarchical overlay network derived from skip graphs and SkipNet to support both exact-match and range queries without consumption of extra memory space.

4.6 Non-uniform Resource Distribution

Resource semantics prohibit hashing $h : D \rightarrow S$, hence Principle 7 violates the assumption on uniform resource distribution. Load imbalances can happen when some nodes are responsible for more resources than other nodes. For example, if Alice from Foo stores essentially more documents than Bob from Bar, then the load of node foo.com can be higher than of node bar.com. Consider a DHT design that hierarchically applies Principle 6 for adapting the node workload to skewed resource distribution in S.

P-Grid [1, 11] inherits the idea of distributed trie [13] to combine advantages of structured and unstructured P2P systems into a DHT-like routing infrastructure. The trie is only virtual since it exists only via routing tables. Lookup paths are organized as a distributed trie to reflect the non-uniform resource distribution in S. Each node maintains a part of the trie, i.e., a tree for storing (binary) keys in which there is one neighbor for every common prefix. The key design choice is adaptive partition of the key space and assignment of the resultant zones to nodes such that to keep the load balance among nodes.

There is no maximal length for keys. Basic P-Grid assumes binary keys from $S = \{0, 1\}^+$. The space is recursively bisected such that the resulting zones carry approximately the same amount of currently available resources. Each zone is assigned to one or more nodes, so a certain replication level is possible. This P-Grid trie construction is implemented by pair-wise random interactions of nodes.

Bisecting S induces a canonical trie structure, a base for prefix-matching routing. A zone and its node u are identified by a bit sequence, a path from the root in the trie. For each position in the path, u maintains links to nodes that has a path with the inverse bit at this position. As a result, each leaf corresponds to a prefix; its routing table consists of randomly selected nodes responsible for keys with that prefix. This means that at each level of the trie the node has references to some other nodes that do not pertain to the node's subtrie at that level. Each node constructs its routing table such that it holds peers with exponentially increasing distance in the key space from its own position. In total, this enables efficient implementation of prefix-matching routing with $O(\log N)$ lookup performance.

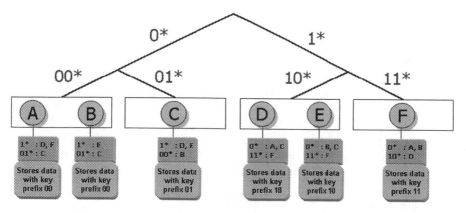

Fig. 4.5 The P-Grid bisection of the resource key space. The trie consists of six nodes responsible for four zones. Each zone is maintained by one or two nodes

A simple example of is shown in Fig. 4.5. The P-Grid trie consists of six nodes responsible for four zones, e.g., node A's path is "00" leading to two entries in its routing table: nodes D and F with paths "10" and "11" at the first level and node C with path "01" at the second level.

In routing a lookup for k, a node u first finds a leaf that shares a prefix with k. Then u forwards the lookup to an appropriate neighbor. The fundamental difference with prefix-matching routing of Pastry/Tapestry is that the distribution of prefixes in routing tables is not uniform; it adopts to the current resource distribution in S. In addition to the maintenance of routing table entries, a node can hash some resources for certain prefixes, hence also taking Principle 7 into account.

An alternative design that ensures balanced resources distribution among nodes can be found in [15]. They consider the concern of skew for resource distribution when only a few nodes are involved in the execution of most queries. Skew can be characterized by the imbalance ratio σ defined as the ratio of the loads of the largest and smallest resource partitions the nodes have in the system. In order to ensure that σ is small, resources may have to be moved from one node to another. On the other hand, an online load balancing algorithm should minimize the transfer cost (the number of items moved) subject to σ is bounded by a constant c, where the bound c is a tunable parameter.

The idea for load balancing is as follows. A node attempts to shed its load whenever its load increases by a factor $\delta > 1$, and attempts to gain load when it drops by the same factor. Formally, there is an infinite geometrically increasing sequence of thresholds $T_i = \lceil c\delta^i \rceil$ for all $i \geq 1$ and constant $c > 0$. When a node's load crosses T_i, the node initiates a load-balancing procedure—the threshold algorithm. Its basic operation is neighbor adjustment: a node moves a portion of its resources to its neighbor and thus attempt to balance out the load. The threshold algorithm guarantees $\sigma = \delta^3$ with a constant cost per resource item insert and delete. When the set of thresholds T_i is the Fibonacci numbers (with $T_1 = 1$ and $T_2 = 2$) then the algorithm guarantees $\sigma = \phi^3 \approx 4.24$, where $\phi = (\sqrt{5}+1)/2$ is the golden ratio.

4.7   Topology Evolution

One of the essential P2P properties is that the overlay graph can evolve as nodes
join and leave, the property is commonly referred as churn. Churn disrupts the
network topology, thus its maintenance becomes a central problem for structured
P2P networks. A good topology maintenance protocol should have the following
properties [35]. First, it should run efficiently, because joins and leaves happen
frequently in practice. Second, it should fully maintain the topology invariants,
because structured networks rely on the topology to provide efficient operations
such as lookups. Third, it should not restrict concurrency: trivial solutions such as
allowing only one join or leave at a time are not adequate because of the frequency
of joins and leaves.

In many DHTs, the ID length n has to be fixed before the installation. For instance
in greedy DHTs from Sect. 3.3, IDs are n-digit sequences in base b, yielding $N \le b^n$
and preventing easy dynamic enhancement for large sizes. The protocols become
suitable only for the sparse case when $N \ll |S|$.

Consider DHT designs that focus on adaptation of the topology to changing N.
Certain ideas are due to de Bruijn graphs, where the ID length and the number of
neighbors per node are independent (see Sect. 3.3). HiPeer [45] forms an overlay
with concentric multi-ring topology, see Fig. 4.6a for an example. The length of IDs
expands from the innermost ring to the outermost ring. The rings are evolutionally
constructed such that nodes on every inner ring from a de Bruijn network. When
nodes join or leave, the overlay expands or shrinks, and some nodes migrate from
one ring to another. As a result, inner ring nodes are more reliable than nodes on the
outer rings.

FISSIONE [34], Moore [18], BAKE [19] and SKY [51] emulate the Kautz
digraph structure. Similarly to the de Bruijn graphs, it has the logarithmic diameter
and constant degree properties while its congestion imbalance is lower. Node IDs
are finite sequences of digits in base $m + 1$. A sequence $u_{n-1} \cdots u_1 u_0$ must be a Kautz
string where any two consecutive digits are different, $u_{i+1} \ne u_i$. The ID length n may
vary, hence $S \subset \{0, \dots, m\}^+$. There are up to $N(n, m) = m^{n-1}(m + 1)$ IDs of a given
length n, since u_0 has $m + 1$ possibilities and all subsequent u_i have m possibilities.
Any resource gets its key $k \in S$ for large n. The node with the longest common prefix
is responsible for k.

A Kautz graph $\mathcal{K}(m, n)$ is similar to de Bruijn graph $\mathcal{B}(m, n)$ except that vertices
are labeled by Kautz strings, see Fig. 4.6b. A vertex $u_{n-1} \cdots u_0$ has m links to
$u_{n-2} \cdots u_0 x$ for $x \ne u_0$ and $0 \le x \le m$. In routing for k, the key prefixes incrementally
replace the suffixes in node IDs. A current node u finds its longest suffix that
appears as a key prefix and then forwards the lookup to resolve the next digit
in k. For example in Fig. 4.6b, lookup for 012021 from node 010 follows the path
$010 \to 101 \to 012$.

FISSIONE [34] emulates $\mathcal{K}(m, n)$ for $m = 2$ as its static topology. The design
uses key space partitioning into zones with complicated zone splitting and merging
algorithms.

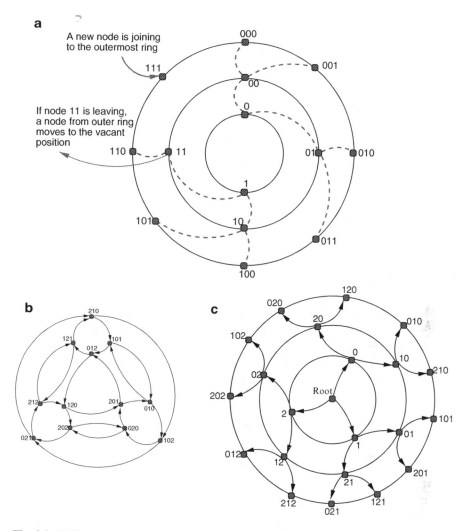

Fig. 4.6 (a) Node joins/leaves in the HiPeer concentric multi-ring topology. (b) Complete Kautz graph $\mathscr{K}(2,3)$. (c) Complete Kautz tree for $m = 2$ and $n = 3$

Moore [18] extends FISSIONE to a general Kautz graph $\mathscr{K}(m,n)$. A Moore overlay selects an initial n and starts with $N \in (N(n-1,m), N(n,m)]$ nodes forming the Kautz graph topology. Then the overlay either expands or shrinks to the range $(N(n-2,m), N(n-1,m)]$ or $(N(n,m), N(n+1,m)]$. Every node changes its ID appropriately by adding or removing one digit. The design requires global information for the topology maintenance.

BAKE [19] extends Moore with balanced Kautz tree structure. It is implemented on top of $\mathscr{K}(n,m)$, see Fig. 4.6c for an example. The idea is close to HiPeer [45] with its concentric multi-ring topology while nodes in BAKE are assigned only to vertices of the outermost ring. Each ring consists of nodes of the same level in the tree. The root has $m + 1$ children. Every other vertex has at most m children. Vertices of 0-ring are root's children $u = u_0$ for $0 \leq u_0 \leq m$. Vertices of i-ring ($i > 0$) are children $v = xu_{i-1} \cdots u_1 u_0$ of $u = u_{i-1} \cdots u_1 u_0$ for $x \neq u_{i-1}$ and $0 \leq x \leq m$. For a given n, all overlay nodes are assigned to vertices of $(n-1)$-ring. Each node keeps two additional ring links to its successor and predecessor. Topology expanding and shrinking leads to transition of overlay nodes from one ring to the next one.

Distributed Kautz graphs were proposed in [51] to support dynamic topology maintenance without "big" expanding and shrinking as in Moore and BAKE. The basic idea is to apply the "edge-vertex transition" of line graphs. Distributed Kautz graphs represent topology evolution as a sequence of graph transitions, starting from the initial topology and reflecting joins and leaves. SKY [51] is a DHT design based on those transitions. A general method (for arbitrary constant-degree graphs) of distributed line graphs was introduced in [49].

4.8 Summary

In this chapter we continued to build a set of design principles for applying hierarchical routing schemes in flat DHTs. The considered schemes appears in advanced DHT designs that operate with groups of nodes. Such grouping differentiates the role of each node in routing. It leads to global hierarchies in the system and becomes an unavoidable requirement in heterogeneous environments.

Principle 5 differentiates nodes by IDs, hence each node can manipulate with its ID in the selection of the role. Principle 6 differentiates nodes by placing resources to each node in accordance with the role. Principle 7 extends Principle 6 based on resource semantics, when application specifics define resource domains, and the role of each node is defined by the domain the node belongs to.

References

1. Aberer, K., Cudré-Mauroux, P., Datta, A., Despotovic, Z., Hauswirth, M., Punceva, M., Schmidt, R.: P-Grid: a self-organizing structured P2P system. SIGMOD Rec. **32**(3), 29–33 (2003). doi: http://doi.acm.org/10.1145/945721.945729
2. Abraham, I., Awerbuch, B., Azar, Y., Bartal, Y., Malkhi, D., Pavlov, E.: A generic scheme for building overlay networks in adversarial scenarios. In: IPDPS '03: Proceedings of 17th International Symposium on Parallel and Distributed Processing, p. 40.2. IEEE Computer Society (2003)
3. Abraham, I., Malkhi, D., Manku, G.S.: Papillon: greedy routing in rings. In: DISC '05: Proceedings of 19th International Conference on Distributed Computing. Lecture Notes in Computer Science, vol. 3724, pp. 514–515. Springer, Berlin (2005)
4. Arge, L., Eppstein, D., Goodrich, M.T.: Skip-webs: efficient distributed data structures for multi-dimensional data sets. In: PODC '05: Proceedings of 24th Annual ACM Symposium on Principles of Distributed Computing, pp. 69–76. ACM, New York (2005). doi: http://doi.acm.org/10.1145/1073814.1073827

5. Aspnes, J., Shah, G.: Skip graphs. In: SODA '03: Proceedings of 14th Annual ACM-SIAM Symposium on Discrete Algorithms, pp. 384–393. Society for Industrial and Applied Mathematics (2003)
6. Aspnes, J., Wieder, U.: The expansion and mixing time of skip graphs with applications. In: SPAA '05: Proceedings of 17th Annual ACM Symposium on Parallelism in Algorithms and Architectures, pp. 126–134. ACM, New York (2005). doi: http://doi.acm.org/10.1145/1073970.1073989
7. Aspnes, J., Kirsch, J., Krishnamurthy, A.: Load balancing and locality in range-queriable data structures. In: PODC '04: Proceedings of 23rd Annual ACM Symposium on Principles of Distributed Computing, pp. 115–124. ACM, New York (2004). doi: http://doi.acm.org/10.1145/1011767.1011785
8. Awerbuch, B., Scheideler, C.: Peer-to-peer systems for prefix search. In: PODC '03: Proceedings of 21nd Annual Symposium on Principles of Distributed Computing, pp. 123–132. ACM, New York (2003). doi: http://doi.acm.org/10.1145/872035.872053
9. Castro, M., Drushel, P., Ganesh, A., Rowstron, A., Wallach, D.S.: Secure routing for structured peer-to-peer overlay networks. In: Proceedings of 5th USENIX Symposium on Operating System Design and Implementation (OSDI 2002), pp. 299–314. ACM, Boston (2002)
10. Dabek, F., Kaashoek, M.F., Karger, D., Morris, R., Stoica, I.: Wide-area cooperative storage with CFS. In: Proceedings of 18th ACM Symposium Operating Systems Principles (SOSP '01), pp. 202–215. ACM, New York (2001). doi: http://doi.acm.org/10.1145/502034.502054
11. Datta, A., Girdzijauskas, S., Aberer, K.: On de bruijn routing in distributed hash tables: There and back again. In: IEEE P2P '04: Proceedings of 4th International Conference on Peer-to-Peer Computing, pp. 159–166. IEEE Computer Society (2004). doi: http://dx.doi.org/10.1109/P2P.2004.29
12. Fiat, A., Saia, J.: Censorship resistant peer-to-peer content addressable networks. In: SODA '02: Proceedings of 13th Annual ACM-SIAM Symposium on Discrete Algorithms, pp. 94–103. Society for Industrial and Applied Mathematics (2002)
13. Freedman, M.J., Vingralek, R.: Efficient peer-to-peer lookup based on a distributed trie. In: Revised Papers from 1st International Workshop on Peer-to-Peer Systems (IPTPS '01), pp. 66–75. Springer, New York (2002)
14. Fujita, S., Ohtsubo, A., Mito, M.: Extended skip graphs for efficient key search in P2P environment. In: ISPAN '05: Proceedings of 8th International Symposium on Parallel Architectures, Algorithms and Networks, pp. 256–261. IEEE Computer Society (2005). doi: http://dx.doi.org/10.1109/ISPAN.2005.45
15. Ganesan, P., Bawa, M., Garcia-Molina, H.: Online balancing of range-partitioned data with applications to peer-to-peer systems. In: VLDB '04: Proceedings of 30th International Conference Very Large Data Bases, pp. 444–455. VLDB Endowment (2004)
16. Goodrich, M.T., Nelson, M.J., Sun, J.Z.: The rainbow skip graph: a fault-tolerant constant-degree distributed data structure. In: SODA '06: Proceedings of 17th Annual ACM-SIAM Symposium on Discrete Algorithm, pp. 384–393. ACM, New York (2006). doi: http://doi.acm.org/10.1145/1109557.1109601
17. Guerraoui, R., Handurukande, S.B., Huguenin, K., Kermarrec, A.M., Le Fessant, F., Riviere, E.: Gosskip, an efficient, fault-tolerant and self organizing overlay using gossip-based construction and skip-lists principles. In: IEEE P2P '06: Proceedings of 6th International Conference on Peer-to-Peer Computing, pp. 12–22. IEEE Computer Society (2006). doi: http://dx.doi.org/10.1109/P2P.2006.19
18. Guo, D., Wu, J., Chen, H., Luo, X.: Moore: An extendable peer-to-peer network based on incomplete Kautz digraph with constant degree. In: Proceedings of IEEE INFOCOM'07, pp. 821–829. IEEE (2007)
19. Guo, D., Liu, Y., Li, X.Y.: BAKE: A balanced Kautz tree structure for peer-to-peer networks. In: Proceedings of IEEE INFOCOM'08, pp. 2450–2457. IEEE (2008)
20. Harvey, N.J.A., Jones, M.B., Saroiu, S., Theimer, M., Wolman, A.: SkipNet: a scalable overlay network with practical locality properties. In: USITS'03: Proceedings of 4th USENIX Symposium on Internet Technologies and Systems. USENIX Association (2003)

21. Harvey, N.J.A., Munro, J.I.: Deterministic SkipNet. Inf. Process. Lett. **90**(4), 205–208 (2004). doi: http://dx.doi.org/10.1016/j.ipl.2004.01.019
22. Hengkui, W., Fuhong, L., Hongke, Z.: Reducing maintenance overhead via heterogeneity in Skip Graphs. In: IC-BNMT '09: Proceedings of 2nd IEEE International Conference on Broadband Network & Multimedia Technology, pp. 638–642. IEEE (2009). doi:10.1109/ICBNMT.2009.5347832
23. Hu, J., Li, M., Zheng, W., Wang, D., Ning, N., Dong, H.: Smartboa: constructing P2P overlay network in the heterogeneous Internet using irregular routing tables. In: IPTPS '04: Proceedings of 3rd International Workshop on Peer-to-Peer Systems. Lecture Notes in Computer Science, vol. 3279, pp. 278–287. Springer, Berlin (2004)
24. Huang, X., Chen, L., Huang, L., Li, M.: Routing algorithm using SkipNet and Small-World for peer-to-peer system. In: GCC 2005: Proceedings of 4th International Conference on Grid and Cooperative Computing. Lecture Notes in Computer Science, vol. 3795, pp. 984–989. Springer, Berlin (2005)
25. Jacob, R., Richa, A., Scheideler, C., Schmid, S., Täubig, H.: A distributed polylogarithmic time algorithm for self-stabilizing skip graphs. In: PODC '09: Proceedings of 28th ACM Symposium on Principles of Distributed Computing, pp. 131–140. ACM, New York (2009). doi: http://doi.acm.org/10.1145/1582716.1582741
26. Karger, D., Lehman, E., Leighton, T., Panigrahy, R., Levine, M., Lewin, D.: Consistent hashing and random trees: distributed caching protocols for relieving hot spots on the world wide web. In: STOC '97: Proceedings of 29th Annual ACM Symposium on Theory of Computing, pp. 654–663. ACM, New York (1997). doi: http://doi.acm.org/10.1145/258533.258660
27. Karger, D.R., Ruhl, M.: New algorithms for load balancing in peer-to-peer systems. Technical Report LCS-TR-911, MIT (2003)
28. Karger, D.R., Ruhl, M.: Diminished Chord: a protocol for heterogeneous subgroup formation in peer-to-peer networks. In: IPTPS '04: Proceedings of 3rd International Workshop on Peer-to-Peer Systems. Lecture Notes in Computer Science, vol. 3279, pp. 288–297. Springer, Berlin (2004)
29. Karger, D.R., Ruhl, M.: Simple efficient load balancing algorithms for peer-to-peer systems. In: SPAA '04: Proceedings of 16th Annual ACM Symposium on Parallelism in Algorithms and Architectures, pp. 36–43. ACM, New York (2004). doi: http://doi.acm.org/10.1145/1007912.1007919
30. Kenthapadi, K., Manku, G.S.: Decentralized algorithms using both local and random probes for P2P load balancing. In: SPAA '05: Proceedings of 17th Annual ACM Symposium on Parallelism in Algorithms and Architectures, pp. 135–144. ACM, New York (2005). doi: http://doi.acm.org/10.1145/1073970.1073990
31. King, V., Saia, J.: Choosing a random peer. In: Proceedings of 23rd Annual ACM Symposium Principles of Distributed Computing (PODC '04), pp. 125–130. ACM, New York (2004). doi: http://doi.acm.org/10.1145/1011767.1011786
32. Konstantinou, I., Tsoumakos, D., Koziris, N.: Measuring the cost of online load-balancing in distributed range-queriable systems. In: IEEE P2P '09: Proceedings of 9th International Conference on Peer-to-Peer Computing, pp. 135–138. IEEE (2009)
33. Ledlie, J., Seltzer, M.I.: Distributed, secure load balancing with skew, heterogeneity and churn. In: Proceedings of IEEE INFOCOM'05, pp. 1419–1430. IEEE (2005)
34. Li, D., Lu, X., Wu, J.: FISSIONE: a scalable constant degree and low congestion DHT scheme based on Kautz graphs. In: Proceedings of IEEE INFOCOM'05, pp. 1677–1688. IEEE (2005)
35. Li, X., Misra, J., Greg Plaxton, C.: Maintaining the Ranch topology. J. Parallel Distrib. Comput. **70**(11), 1142–1158 (2010). doi: http://dx.doi.org/10.1016/j.jpdc.2010.06.004
36. Malkhi, D., Naor, M., Ratajczak, D.: Viceroy: a scalable and dynamic emulation of the butterfly. In: PODC '02: Proceedings of 21st Annual Symposium on Principles of Distributed Computing, pp. 183–192. ACM, New York (2002). doi: http://doi.acm.org/10.1145/571825.571857

37. Manku, G.S.: Routing networks for distributed hash tables. In: PODC '03: Proceedings of 22nd Annual Symposium on Principles of Distributed Computing, pp. 133–142. ACM, New York (2003). doi: http://doi.acm.org/10.1145/872035.872054

38. Manku, G.S., Naor, M., Wieder, U.: Know thy neighbor's neighbor: the power of lookahead in randomized P2P networks. In: STOC '04: Proceedings of 36th Annual ACM Symposium on Theory of Computing, pp. 54–63. ACM, New York (2004). doi: http://doi.acm.org/10.1145/1007352.1007368

39. Naor, M., Wieder, U.: Novel architectures for P2P applications: the continuous-discrete approach. ACM Trans. Algorithms **3**(3), 37 (2007). doi: http://doi.acm.org/10.1145/1273340.1273350

40. Risson, J., Harwood, A., Moors, T.: Stable high-capacity one-hop distributed hash tables. In: ISCC '06: Proceedings of 11th IEEE Symposium on Computers and Communications, pp. 687–694. IEEE Computer Society (2006). doi: http://dx.doi.org/10.1109/ISCC.2006.152

41. Saia, J., Fiat, A., Gribble, S.D., Karlin, A.R., Saroiu, S.: Dynamically fault-tolerant content addressable networks. In: IPTPS '01: Revised Papers from 1st International Workshop on Peer-to-Peer Systems, pp. 270–279. Springer, Berlin (2002)

42. Shen, H., Xu, C.Z., Chen, G.: Cycloid: a constant-degree and lookup-efficient p2p overlay network. Perform. Eval. **63**(3), 195–216 (2006). doi: http://dx.doi.org/10.1016/j.peva.2005.01.004

43. Surana, S., Godfrey, B., Lakshminarayanan, K., Karp, R., Stoica, I.: Load balancing in dynamic structured peer-to-peer systems. Perform. Eval. **63**(3), 217–240 (2006). doi: http://dx.doi.org/10.1016/j.peva.2005.01.003

44. Vu, Q.H., Ooi, B.C., Rinard, M., Tan, K.L.: Histogram-based global load balancing in structured peer-to-peer systems. IEEE Trans. Knowl. Data Eng. **21**(4), 595–608 (2009). doi: http://dx.doi.org/10.1109/TKDE.2008.182

45. Wepiwe, G., Simeonov, P.L.: A concentric multi-ring overlay for highly reliable P2P networks. In: NCA '05: Proceedings of 4th IEEE International Symposium on Network Computing and Applications, pp. 83–90. IEEE Computer Society (2005). doi: http://dx.doi.org/10.1109/NCA.2005.1

46. Xu, J., Kumar, A., Yu, X.: On the fundamental tradeoffs between routing table size and network diameter in peer-to-peer networks. IEEE J. Sel. Areas Commun. **22**(1), 151–163 (2004)

47. Xu, M., Zhou, S., Guan, J.: A new and effective hierarchical overlay structure for Peer-to-Peer networks. Comput. Commun. **34**(7), 862–874 (2011). doi: http://dx.doi.org/10.1016/j.comcom.2010.10.005

48. Zatloukal, K.C., Harvey, N.J.A.: Family trees: an ordered dictionary with optimal congestion, locality, degree, and search time. In: SODA '04: Proceedings of 15th Annual ACM-SIAM Symposium on Discrete Algorithms, pp. 308–317. Society for Industrial and Applied Mathematics (2004)

49. Zhang, Y., Liu, L., Li, D., Lu, X.: Distributed line graphs: A universal framework for building DHTs based on arbitrary constant-degree graphs. In: ICDCS 2008: Proceedings of 28th IEEE International Conference Distributed Computing Systems, pp. 152–159. IEEE Computer Society (2008)

50. Zhang, Y., Li, D., Chen, L., Lu, X.: Flexible routing in grouped DHTs. In: IEEE P2P '08: Proceedings of 8th International Conference Peer-to-Peer Computing, pp. 109–118. IEEE Computer Society (2008). doi: http://dx.doi.org/10.1109/P2P.2008.43

51. Zhang, Y., Lu, X., Li, D.: SKY: efficient peer-to-peer networks based on distributed Kautz graphs. Sci. China Ser. F Inf. Sci. **52**(4), 588–601 (2009)

52. Zhou, G., Yu, J.: pService: Towards similarity search on peer-to-peer web services discovery. In: Conference Advances in P2P Systems, pp. 111–115. IEEE Computer Society (2009). doi: http://doi.ieeecomputersociety.org/10.1109/AP2PS.2009.25

Chapter 5
Clustering

Abstract In the previous chapters we considered the dynamic differentiation of nodes and resources in flat P2P systems. This chapter shows that the design principles of the differentiation evolve to the clustering principle. It is the last conceptual step before the traditional hierarchy of HDHTs. We focus on reasons of clustering and consider local group formations possible on the top flat DHT overlays. Such formations eventually appear non-negligible from the point of view of the global overlay topology. They provide a mechanism to control the tradeoff between local and global routing.

5.1 Introduction

Many P2P designs deal with heterogeneity by organizing nodes or resources into explicit groups. Clustering is introduced in the system design to bring similar nodes together. It increases the scalability of the system and supports fault isolation. In the terms of performance, the routing distance depends on the similarity between nodes in the multi-hop path

$$u = w_0 \to w_1 \to w_2 \to \cdots \to w_{l-1} \to w_l = d.$$

If the nodes w_i are covered by a small set of clusters, then better performance can be achieved by improving intra-cluster routing (local routing). A node selects many neighbors from the same or similar groups. Global routing is required only for few hops in inter-cluster forwarding, and a node intentionally keeps less neighbors from far-away clusters.

Clustering aims at constructing generic system-level group structures, which are efficient for controlling load balance, proximity, administrative domains, data semantics and trust. In fact, "signs of clustering" appeared in Principles 1, 2, 3, and 4 of Chap. 3, where each node forms clusters of close nodes in its routing table. In graph theory, these characteristics are reflected in clustering and expansion

D. Korzun and A. Gurtov, *Structured Peer-to-Peer Systems: Fundamentals of Hierarchical*   111
Organization, Routing, Scaling, and Security, DOI 10.1007/978-1-4614-5483-0_5,
© Springer Science+Business Media New York 2013

coefficients [23,24]. For instance, Cayley graphs have good clustering properties [1] and can be applied in DHT designs [25, 32].

Further elements of grouping nodes appeared in Principles 5, 6, and 7 of Chap. 4. They adopt the heterogeneity by arranging the nodes appropriately to assign to them different responsibilities. These arrangements have higher impact on global overlay topology formation compared with clustering in local routing tables. The sum effect of local activity becomes higher, and the global topology embeds more complex hierarchical structures.

The extreme case is left for Chap. 7, where the role of groups in the topology becomes dominant. There, we shall see that each group can be assigned to one ore more nodes (supernodes), leading to multi-layer hierarchical architectures. In this chapter, we focus on explicit group formation that performs differentiation on local level (individually by a node), although the sum impact to the global network topology is non-negligible.

Clustering schemes appeared in early DHT designs where spontaneous groups of similar nodes are organized during the system lifetime. We considered preliminary techniques in Chap. 4. In virtual servers approach [9, 11, 14, 30, 38], several overlay node IDs are assigned to one physical machine, achieving better load balance. Proximity clustering [26, 27, 34, 43] aims at adaptation to the underlying network topology, leading to better routing performance.

Differentiation of resources applies resource clustering. Semantic clustering [39, 42] allows distributing resources among nodes based on the resource semantics. Similar resources concentrate in a given cluster, the distance between two nodes in the overlay is proportional to their resources dissimilarity in semantics. Topology formation with interest-based and utility-based clusters [3, 12, 20] is oriented to semantic similarity of participating nodes. The use of resource and node semantics allows advanced search queries. An initiator node uses a specification of requested resources instead of a flat resource key. A popular example is range queries, where the result is a contiguous set of similar resources.

Section 5.2 introduces the clustering principle and overviews possible application areas of clustering for structured P2P systems. We survey several popular clustering schemes and techniques they use in the definition of the node and resource similarity.

Section 5.3 considers cluster-oriented topology models. It begins with a conceptual model of virtual nodes that provides a two-level scheme of load balancing among active participants. The local knowledge problem prevents achieving a stable state with optimal clustering, different clustering algorithms perform approximated clustering, so we also consider the problem of false clustering, which can degrade the performance improvement of clustering.

Section 5.4 discusses a concrete case of clustering where the similarity between nodes is based on resources the nodes keep—the case of semantic clustering and close schemes. In these schemes, queries are routed to nodes who are interested in similar, in the semantic sense, information.

5.2 Clustering Principle

Principle 3 from Chap. 3 stated that each node can dynamically adapt locally to a global network hierarchy according to a metric τ. Principles 5, 6 and 7 from Chap. 4 supported topologies with specialized roles of nodes in routing. In general, node differentiation can be implemented with regard to a global network hierarchy.

Principle 8 (Clustering). *Let τ reflect a global network hierarchy. Nodes that are close with respect to $\tau(u,v)$ are also close with respect to $|u \rightarrow^+ v|$.*

In other words, if the global network hierarchy defines a set U of close nodes ($\tau(u,v)$ is small $\forall u,v \in U$) then they form an overlay cluster where each node can reach another in few hops.

An early example of applying clustering in P2P networks is CAP [16], a Cluster-based Architecture for P2P. Network-aware clustering technique was employed for content location and routing in unstructured P2P overlays, where the global hierarchy is determined by the underlying network topology, i.e., it reflects the domain-based Internet architecture. A cluster consists of nodes that are topologically close and under common administrative domain. The additional "cluster" level aims at scaling up query lookup and forwarding. The measurements based on real-life Gnutella traces showed performance improvement (in terms of successful lookup percentage and lookup path length) compared with the flat Gnutella system.

The CAP technique is an evolution of the local adaptation schemes, which we discussed in Sect. 3.4. In contrast to them, clusters form an additional level in the global overlay topology; a centralized server (called clustering server) performs clustering and cluster registration.

Let us survey several popular clustering schemes applicable in structured P2P networks. Models and concrete design solutions behind these schemes will be considered in subsequent sections.

The concept of "virtual nodes" for clustering $\Theta(\log N)$ virtual nodes (overlay node IDs) on a physical machine was introduced in [9,13]. It was developed initially for solving the DHT load balance problem, see also Sect. 4.3. With high probability a machine is responsible for a $O(1/N)$ portion of resource keys D, i.e., the load imbalance factor is reduced from $O(\log N)$ to $O(1)$. The concept is further refined in [11, 14, 38] by more careful virtual node selection and evolution in such ID-aware clusters. It reduces the maintenance cost at physical machines and adapts to heterogeneous capacities and high dynamics. A conceptual group-based model of virtual nodes is presented in [30] with load balance evaluation.

Virtual nodes are located at the same machine. Proximity clustering develops further this closeness idea, mainly for improving the routing latency [16, 26]. A node associates with a cluster based on a proximity metric τ and selects most neighbors from the cluster. Topologically-aware CAN construction [26] uses proximity clustering. The construction extends a discrete variant of PNS (see

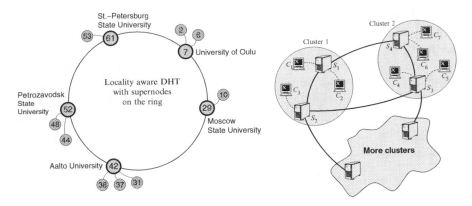

Fig. 5.1 (**a**) Close nodes are grouped into a cluster with a supernode and all supernodes form a Chord network. (**b**) In a cluster of nearby nodes, some supernodes (e.g., S_1 and S_2) act as local servers, to which other nodes (e.g., C_1 and C_2) are connected

Principle 3 in Chap. 3). Similarly, it allows selecting a next hop when forwarding a lookup, i.e., proximity clustering implements a discrete variant of PRS.

Proximity clustering often uses a global coordinate system. A node estimates physical distances to a few landmark nodes. An early example is topologically-aware CAN construction [26]. Such estimates form a local distance vector that characterize the position of the node. Similarity of distance vectors of two nodes provides a metric τ for clustering, e.g., the difference in landmark ordering that each of two vectors defines.

A distance vector can be hashed to a node ID space using a Hilbert curve or another space-filling curve that ensures preserving the closeness relationship [27]. Consequently, physically close nodes can construct close keys in the logical ID space. In [43], this mechanism complements PNS for selecting the long-range neighbors (expressways) that are physically close and situated near network access points. In [34], it allows locality-aware load balancing, see Fig. 5.1a for an illustration. Clustering schemes can follow Principle 4 (Chap. 3) using additional knowledge on the global network for more accurate cluster formation.

The idea of extending structured P2P designs with clustering semantically close nodes is introduced in several works [20, 39, 42]. A cluster is a group of nodes with similar content forming a semantic overlay [39]; it directly follows Principle 7 from Sect. 4.5. Another way of cluster formation is to place semantically related resources at nearby nodes, according to Principle 6 from Sect. 4.4. In particular, $\tau(u, v)$ can measure the dissimilarity in semantics between resources stored at u and v. As a result, many lookups can be resolved mostly with local routing. Classifications of the overlay content by categories [42] or communities [20] allow further improvement of the routing performance.

Space-filling curves are applicable in semantic clustering. Locality sensitive (or preserving) hashing constructs resource numerical keys. In the case of multiple attributes, several keys are constructed. A locating mechanism implements resource clustering to place the keys of semantically close resources to the same node or to a set of close nodes [18, 47].

According to Principle 5 from Sect. 4.2, Mercury [5] organizes nodes into groups to support multi-attribute range queries. One group is for each attribute. Similarly to the idea of grouped DHTs from Sect. 4.4, a node may belong to several groups. In addition to multi-attribute range queries, NR-trees [21] support k-nearest neighbor queries by clustering nearby nodes around the most powerful ones (Fig. 5.1b).

The existence of interest-based clusters is studied in [3]. Each node can estimate its preference of other nodes. The preference is a very general metric, e.g., for resource semantics (nodes keep similar content) or trust (nodes believe each other). Mutually preferred nodes form a cluster. In reciprocation strategies, nodes with low contribution to the P2P system become preferable for requesting more resources. The corresponding model was introduced in [15]. SkipStream [45] uses a similar scheme for on-demand streaming. Users are clustered in accordance with their playback offset and the resultant clusters build a clustered skip graph. Such a hierarchical overlay facilitates the streaming cooperation among heterogeneous nodes.

Principle 8 provides a natural base for hierarchical DHT designs. Nodes in each cluster assign a supernode (or a set of supernodes) according to their capacities. All supernodes form a higher-layer overlay. Therefore, clustering is the last step in the evolution of flat DHTs to HDHTs. As an example of the industry-level technology applicable in practical P2P network construction we refer to JXTA/JXSE project.[1] In particular, it defines the concept of PeerGroups to decompose the large number of nodes into manageable groups.

5.3 Cluster-Oriented Overlay Topologies

The previous section provided a retrospective of clustering schemes and their application areas for structured P2P systems. Let us consider several concrete designs with the focus on their reference models of clustering.

5.3.1 Virtual Nodes

Recall that the concept of virtual nodes for DHTs appeared initially [9, 13]. Each physical machine instantiates one or more DHT nodes—virtual nodes. Their IDs are taken randomly from the ID space S. A conceptual model of virtual nodes is introduced in [30]. It aims at dynamic balancing across a set of heterogeneous node.

---

[1] JXTA is a programming language and platform independent Open Source protocol started by Sun Microsystems for P2P networking in 2001. Official website is http://jxta.kenai.com/. In November 2010, Oracle officially announced its withdrawal from this project. It is currently migrating from Java.net to Project Kenaï.

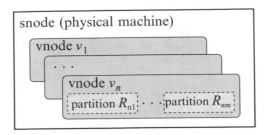

Fig. 5.2 The snode composition structure: the first level is for vnodes, the second level is for partitions a vnode is responsible in the resource key space. The sum size of partitions is approximately equal for each vnode

An overlay is organized as a set of physical machines: software nodes in [30] or simply snodes. They are active participants managing parts of the DHT. Each snode hosts several virtual nodes (or simply vnodes), see Fig. 5.2. In turn, each vnode is a set of partitions of the DHT. Therefore, each snode acts as a cluster of IDs and covers their sum responsibility area in the resource key space.

The model contemplates two levels of load balancing between participants: the first, supported by vnodes, provides coarse-grain balancing; the other one, based on the assignment of partitions to vnodes, provides fine-grain balancing.

In coarse-grain balancing, the number of vnodes per snode defines the snode enrollment level, which primarily depends on the amount of local computational resources an snode reserves for the DHT. Such amount is not necessarily static: for instance, a snode may start by reserving a certain amount of secondary storage for a DHT; later, that amount may change dynamically.

Fine-grain balancing is based on the number of partitions per vnode. This number is allowed to fluctuate between well defined bounds, during the creation or deletion of vnodes. Hence, all vnodes should be considered almost equal in terms of their load, while snodes may be heterogeneous in the capacity they provide to the system.

Let R_u be the fraction of D that a vnode u maintains. The model aims at minimizing a deviation of R_u from an ideal (average) share for all the vnodes in sum. As a result, every snode keeps the number of vnodes that its capacity supports. Every vnode is responsible for a similar share of the total load. The set of all vnodes is homogenous in terms of load balance; the set of all snodes is heterogeneous due to the tunable number of vnodes per snode.

If every node has global knowledge on the partition distribution over all the vnodes and their snodes, then join and leave operations can be performed efficiently to minimize the deviation. In addition, every snode is involved in the creation of every vnode. Consecutive creations of vnodes are executed serially, thus limiting the parallelism and reducing the scalability to a small number of snodes. For the design and evaluation details of the global approach please refer to [31].

In the local approach, all vnodes are partitioned into disjointed groups. Consequently, a group consists of many snodes and an snode participates in several groups. Within each group, balancing is based on the same algorithm used by the global approach, though restricted to the vnode set of the group. Local balancing actions may be performed concurrently at different groups. The number of vnodes in each group fluctuates between strict bounds, and the overall number of groups may change, as vnodes are created.

Each snode keeps knowledge on the local partition distribution over all snodes and vnodes for any group in which vnodes of the snode participate. The knowledge includes information about the number of partitions each vnode of the group maintains. Hence, the approach is a downsized version of the global one, an instance of the two-level decomposition. (We shall discuss it in more detail in Chap. 7 in the context of supernode model.) When the number of vnodes per group grows, the number of group decreases until the single global group, which contains all the nodes in the extreme case. Clearly, the quality of resultant balance for larger groups is better for the cost of maintenance performance (snodes joins and leaves, creation and removal of vnodes, creation, merge and split of groups).

5.3.2 *Proximity-Based Clusters*

In Brocade [46], a secondary overlay network of supernodes is used to improve the routing distance. These supernodes are the nodes that are situated near the network access points such as routers and gateways. Nodes in the primary overlay network establish direct connections with a supernode that is nearby, so forming a cluster. Supernodes advertise the nodes that are connected to them as objects served by the supernode in the secondary overlay. Routing from node u to d in the primary overlay involves three steps: (1) locating a supernode s_0 locally, (2) routing in the secondary overlay to the supernode s_1 that stores d, and (3) hopping from s_1 to d.

Although Brocade produces reasonable improvements from the flat overlay case, it pushes the problem to an auxiliary network of a smaller size. Brocade faces a dilemma in choosing the appropriate number of supernodes: if the number of supernodes is large, the logical overlay routing cost gets higher; if the number of nodes in the secondary network is small, a regular node needs to keep more state about the addresses of other nodes that are linked to the same supernode. This can be an issue in a dynamic environment. Some related models for this dilemma we shall consider in Chap. 7.

In eQuus [22], a Pastry-like DHT, nearby nodes form a clique in place of a single Pastry node. The goal is to make a DHT network locality-aware and highly resilient to churn. The eQuus network tolerates churn as long as there are surviving nodes in cliques. However, using cliques also entails the problem of maintaining the cliques, in addition to maintaining the overall topology. In particular, cliques have to be split (when cliques become too large due to joins) and merged (when cliques become too small due to leaves or crashes). In fact, eQuus applies two-layer hierarchical architecture, see more details in Chap. 7.

There are schemes that use landmarks to cluster nearby overlay nodes. Landmark clustering assumes that nodes close to each other are likely to have similar network

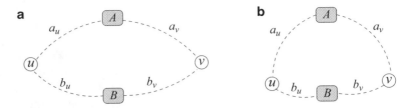

Fig. 5.3 False clustering with landmarks. Let $a_u = \tau(u,A)$, $a_v = \tau(v,A)$, $b_u = \tau(u,B)$, $b_v = \tau(v,B)$. Although u and v are distant, they have a similar ordering of landmarks A and B if $a_u \approx a_v$ and $b_u \approx b_v$. Cases (**a**) with $a_u \approx a_v > b_u \approx b_v$ and (**b**) with $a_u \approx a_v \approx b_u \approx b_v$ show that u and v have similar landmark orderings and similar landmark distance vectors. Note that the triangle inequality is not hold in the Internet, and the case $\tau(u,v) > a_u + a_v$ or $\tau(u,v) > b_u + b_v$ is also possible

latencies to a few landmarks spread evenly across the Internet. A pioneering work on landmark clustering is topologically-aware CAN construction [26]. Physically close nodes are clustered into logical vicinity during overlay formation. When a new node u joins the overlay, u finds such nodes that are close to u in the IP distance.

A distributed binning model supports this clustering scheme. Nodes partition themselves into bins such that nodes that fall within a given bin are relatively close to one another. The system provides a global set of well-known landmark machines. An overlay node measures its distance (round-trip time) to all landmarks, estimating the local distance vector. Arranging the distances, the node obtains its associated ordering of landmarks. Then the node independently selects the bin that corresponds to this ordering. Little effort is required for the landmarks to allow nodes to identify their bin by simple pinging.

Although this yields good performance improvements, it may cause significant imbalances in the distribution of the nodes in the CAN space that leads to hot-spots [7], e.g., there are bins with sparse and dense population. The clustering scheme is applied only when node join the system, so it does not handle properly changing network conditions. The distributed binning model is a coarse-grained approximation [33]. It may be inaccurate in differentiating close nodes—the false clustering problem, see illustrative examples in Fig. 5.3. Furthermore, an inadequate selection of landmarks can make landmark ordering completely useless.

To overcome the disadvantages of Brocade and topologically-aware CAN, expressway routing [43] constructs an auxiliary network (topology-aware expressway) for clustering nodes that observe the same latency behavior with a few chosen landmarks.[2] As in the previous designs, such an expressway overlay aims at approximation of the physical network topology, making performance independent of the network size N.

Consider the landmark clustering model used in the expressway design. Similarly to topologically-aware CAN construction, an expressway overlay uses a global

---

[2]Another variant, which [43] also considered, makes clusters from nodes belonging to the same autonomous system.

set of n landmark nodes. They can be part of the overlay itself or standalone. Each overlay node measures its network distance to the n landmarks, producing a landmark distance vector $(l_1, l_2, \ldots, l_n)$, a point in the Cartesian space (the landmark space). Intuitively, close nodes in terms of network latency distance τ have similar landmark vectors. To reduce false clustering, landmark vectors are used only in a pre-selection process to identify nodes that are potentially close to a new node. After that each node uses actual round-trip time measurements to other overlay nodes to identify the closest neighbors.

The expressway nodes independently determine their positions in the landmark space and publish their positions on the DHT. A node finds expressway nodes that are physically close to it by referring to this published information by using its own landmark vector as the key. The Hilbert curve (a particular instance of space-filling curves) is used to map landmark vectors to the DHT ID space. Importantly, this mapping preserves clustering properties, and close landmark vectors remain close in the ID space.

Expressway landmark clustering inherits shortcomings of landmark ordering [33]. Clustering with a common fixed set of landmarks is a coarse-grained approximation; it is not always effective in differentiating close nodes and is vulnerabile to landmark misplacements.

One of the key points of expressway construction is that network distance can be estimated without the need of direct measurements between overlay nodes. There exist network coordinate systems that embed nodes into some Cartesian space, similarly to the landmark space model above. An example is Vivaldi [10]. Latency between two nodes can be estimated using distance computation in that space. Consequently, a clustering scheme can exploit the coordinate model to clusters nodes, since nodes close to each other should have similar coordinates.

Although coordinate systems are useful in determining the latencies among nodes that have never communicated, there are also several important shortcomings [17, 35]. On the one hand, estimates are quite unpredictable. While many nodes obtain good estimates, a few ones can get extremely bad results. On the other hand, the network coordinate model assumes the triangle inequality, which is commonly violated due to Internet routing policies [10], see also the discussion for Fig. 5.3 above. Moreover, congestion or re-configuration of the underlying network can suddenly change the relative location of many nodes. Dynamics force nodes to recompute periodically their coordinates, leading to oscillations since a local decision making is hard to determine when it is convenient to update the coordinates.

TR-clustering [35] exploits traceroute paths instead of unstable measurement data such as round-trip-times. The traceroute tool makes clusters more stable, resilient to changing network conditions, though this supposes more effort to the nodes. The algorithm clusters together those nodes that share the same paths to a common set of landmarks. It is not based on Internet coordinate systems, attempting to form clusters with a low rate of falsely clustered nodes.

The key algorithm idea is exploitation of Internet routers with high vertex betweenness centrality, see also discussion in [24] on the role of this measure in models of complex networks. In the context of P2P overlays on top of the Internet, the betweenness centrality γ_A of a router A is the fraction of Internet paths between all pairs of nodes in N running through A (for simplicity we assume $A \notin N$),

$$\gamma_A = \frac{1}{N(N-1)} \sum_{u,v \in N, u \neq v} \frac{\sigma_{uv}(A)}{\sigma_{uv}},$$

where σ_{uv} is the number of shortest paths from u to v in the IP network and $\sigma_{uv}(A)$ is the number of them that run trough A.

This measure characterizes the influence of A over the flow of information between other nodes, assuming that information travels over the network by primarily following shortest paths. Traceroute probes are used to discover routers with high betweenness centrality. They are then selected to be landmarks for the overlay.

If a network contains clusters that are loosely connected to each other by a few intermediate routers, all the traffic is expected to go along these routers. As a result, they will report the highest vertex betweenness. By equipping the landmark set with these routers, the overlay is able to detect the clusters currently present in the network while partially revealing the structure of the Internet graph.

In contrast to other landmark-based algorithms, the TR-clustering algorithm allows its landmark set to be dynamic. Each cluster maintains an own set of landmarks and recomputes it based on traceroute probes of all nodes of the cluster. Selecting landmarks for the next round is followed by clustering, when each node individually selects its cluster based on landmark distances. The process is repeated until the either there are not more too dense clusters, or the desired clustering quality is achieved. The quality criterion is minimization of false clustering.

5.3.3 False Clustering

When distant nodes are clustered near each other the problem of false clustering appears, degrading the routing performance. The quality of existing clustering algorithms strongly depends on the quality of measurements of $\tau(u,v)$. In proximity clustering, clustering can be significantly inferior to the optimum when artificial data such as network coordinates are used. Moreover, the network coordinate model assumes that latency estimates satisfy the triangle inequality

$$\tau(u,v) \leq \tau(u,w) + \tau(w,v) \quad \forall u,v,w \in N,$$

which is not true in the Internet [10]. Violations of the triangle inequality lead to inaccuracies that rise the fraction of falsely clustered nodes.

Consider the clustering model from [35]. It represents an overlay network as a connected undirected graph $G = (N,E)$, where N represents the set of nodes and E the set of links. A clustering $C = \{C_1, C_2, \ldots, C_m\}$ of a network $G = (N,E)$ is a division of N into a finite number of disjointed sets such that $\bigcup_{i=1}^{m} C_i = N$ and $C_i \cap C_j = \varnothing$ $\forall i \neq j$. The sets C_i are called clusters. The total number of nodes in a cluster C_i is its size $|C_i|$. Denote C_u the cluster that contains the node u. If u and v are in the same cluster then $C_u = C_v$.

False clustering rate f is modeled as the fraction of falsely clustered nodes in a cluster. Intuitively, if a node contacts a randomly selected node from its cluster, then f is the probability of contacting a falsely clustered node. That is, false positives are taken into account: the quality of clustering degrades when a cluster contains a distant node.

Let $\hat{C}_u = C_u \setminus \{u\}$. Denote $\Gamma_v(l)$ the set of l closest nodes to u in terms of some metric τ, e.g., the round-trip time. For a node u, its individual false clustering rate is defined as

$$f_u = \frac{\hat{C}_u \cap \Gamma(l_u)}{|\hat{C}_u|} \quad \text{where } l_u = |\hat{C}_u|. \tag{5.1}$$

It measures the overlap between the $l_u = |\hat{C}_u|$ closest nodes to u and its cluster neighbors. The model assumes that the degenerated case $C_u = \{u\}$ is not possible.

Then the false clustering rate can be defined as the average over all nodes,

$$f = \frac{1}{N} \sum_{u=1}^{N} f_u. \tag{5.2}$$

A clear property is $0 \leq f \leq 1$, where values of 0 and 1 are assumed at best and at worst, respectively.

Model (5.1) does not account false negatives, i.e., the nodes in other clusters that should be included in C_u but not have been. Accounting false negatives requires that the optimal partitioning must be known in advance. It is a complicated problem even for static network and becomes more challenging in such a dynamic and large-scale environment as the Internet, where topology changes frequently due to detour paths and multi-homed nodes. That is, the false clustering rate does not depend on the benefit of any a priori structure and dynamics of the underlying network. The only knowledge is a distance metric τ, which allows defining the closest nodes.

Model (5.2) penalizes "dense" clusters than "sparse" clusters. It again relates the ration between false positives and false negatives. The sparser a cluster the more false negatives are, leading to the degenerated case $C_u = \{u\}$ $\forall u \in N$. If clusters become dense then false positives start playing a more important role in the quality of clustering. Therefore, (5.1) rewards clustering algorithms that attempt to create dense clusters.

Cluster validity is one of the most important issues of cluster analysis [4]. It aims at the evaluation of clustering results to give an indication of the partitioning that best fits a given data set. Although there are many traditional cluster validity indexes, they are not sufficient when one must deal with arbitrarily-shaped clusters. In contrast, (5.1) and (5.2) are based on no prior information about the geometry of the clusters. Consequently, the false clustering rate does not bias toward a particular underlying network shape.

Based on this model, [35] experimentally showed that false clustering is non-negligible for proximity-based clustering. It may impact negatively any proximity technique and have an important influence in solving the underlying network topology mismatching problem. Note that the model can also be applied for other clustering schemes where distance metrics τ are not related to proximity.

5.4   Cluster-Oriented Search Queries

5.4.1   Semantic Clustering

Content-based semantic-aware search is a challenging problem in P2P systems. For instance, conventional flat DHTs have no support for keyword searching. Although such DHTs can be extended, the search process may incur either huge traffic load for result intersection, or large overhead for multiple publication and update [8, 20, 47]. A possible solution is that resource keys are distributed over the nodes based on resource semantics, see also the discussion started in Chap. 4. The performance of locating content can be greatly improved by grouping peers interested in similar resources and routing their search requests within these groups. Clustering semantically close nodes together has been proposed in many studies, e.g., see [37, 39, 41].

5.4.1.1   Continuous Semantic Similarity Metric

The notion of *semantic overlay* was introduced in [39]. Such an overlay is a logical network where resources are published such that the routing distance between two nodes in the network is proportional to the dissimilarity in semantics of their resources. pSearch system [39] implements a semantic overlay using latent semantic indexing (LSI). It represents resources and queries as vectors in a semantic Cartesian space. Each element of a semantic vector corresponds to an abstract concept (e.g., a term in a common vocabulary) and the value shows the concept importance. The similarity between a search query x and a resource item y is measured as the cosine of the angle between their vector representations:

$$\text{Similarity} = \cos(x,y) = \frac{1}{|x| \cdot |y|} \sum_i x_i y_i.$$

Given a query, the problem of finding the most relevant resources is reduced to locating the resource vectors nearest to the query vector.

pSearch maps the semantic space to nodes in an overlay network and conduct efficient nearest-neighbor search in a decentralized manner. The CAN protocol is used where semantic vectors are keys in the CAN space. In the simplest case, the semantic space and CAN dimensions are equal. The lookup operation routes a query to the destination node, which is an instance of inter-cluster routing. Upon reaching the destination, the query is flooded to nodes within a radius r, determined by the similarity threshold or the number of wanted items specified by the requester. The latter step is an instance of intra-cluster routing for retrieval of a set of semantically close resources.

A close approach uses similarity indexes [2], based on the well-known technique of locality-sensitive hashing. Resources are hashed such that "similar" resources are much more likely to collide than dissimilar resources. A hash function is designed for a particular similarity metric of interest. At query time, the query is also hashed, and resources that are hashed to the same bucket are retrieved as answers. Some other cluster-based designs with the same idea of using a metric for estimating the distance between nodes sharing similar content can be found in [40, 47].

5.4.1.2 Discrete Semantic Classification

A different approach for semantic clustering was introduced in [20]. Clusters represent node communities, each shares similar resources. Resources are described with an ontology, which specifies formal classes of resources. The key assumption is that node's local resources fully represent interests of the node. That is, data ontology properties are used to classify nodes as well as search queries. Importantly that resource distribution is pure local: each node keeps own resources only, the storage is not delegated to other nodes.

When a new node joins the system, it first classifies its own resources, and then registers its major interests to the overlay. Specifically, an interest of the joining node is hashed into a key in the ID space and a responsible node is found by lookup operation. Then the new node would know other existing nodes sharing the same interest. The node contacts some of these existing nodes and joins their communities. In this way, nodes always connect to other nodes sharing common interests, and thus they can form a community overlay. In fact, this design is nearly a two-layer hierarchical DHT with all nodes connected into a flat DHT overlay on one layer and many community overlays on the other layer, see Chap. 7. Note that a node can participate in several community overlays at the same time.

An ontology-based DHT design was introduced in [28, 29]. It presents a structured P2P overlay network, which uses an upper ontology as a binary partition tree to organize nodes in a semantic space. The upper ontology is a static shared ontology with just "is-a" relationship between concepts (concept hierarchy) to annotate resources. The hash function is based on an upper ontology that partitions semantic space of the domain concepts. The partition tree organizes nodes in the semantic space as well as allows efficient routing of search queries. Each partition forms a community of nodes (cluster); the nodes are mirror of each other, maintaining the same resource information. The cost of a lookup query depends on depth of the shared ontology tree and is independent on the network size.

A similar approach is employed in [37] with the principle called interest-based locality. The principle posits that if a node has a particular piece of content that one is interested in, it is very likely that it will have other items that one is interested in as well. We also refer to [36] where the problem of ontology-based clustering in the context of P2P databases is discussed and experimentally evaluated. In particular, they contributed an architecture to group semantic similar peers within clusters and an incremental clustering process considering the dynamic aspect of P2P networks.

5.4.2 Interest-Based Clusters

Static semantic relationships between nodes and their resources can be discovered relatively easily. The biggest challenge is to build a network that maintains and exploits the discovered semantic structure. It is, however, is subject to dynamics, an evitable property of P2P systems. The behavioral patterns of participating nodes influence the semantic relations, thus affecting the cluster formation.

5.4.2.1 Resource Popularity

Non-uniform resource popularity leads to interest-based clusters when some nodes communicate between each other frequently, so forming a group. Queries are confined to members of groups (not necessarily disjoint) whose sizes are smaller than the network size. The non-uniform request patterns defines a global hierarchy. Based on it, the routing and node state complexity can be made a function of the cluster size, independent on the network size N.

Let nodes u and v frequently communicate with each other. Then making them neighbors ($u \in T_v$ and $v \in T_u$) improves the routing performance for the frequent lookups. The improvement could come possibly at the expense of an increase in the routing distance between some other pair of nodes that do not frequently communicate with one another. An effective P2P design should preserve that (1) the routing distance between nodes within a cluster scales with the cluster size and (2) the average routing distance between any pair of nodes is lower than that in the underlying flat DHT.

Consider the cluster-based design from [3], which implements this idea to optimize routing in a Chord DHT. Denote λ_{uv} the frequency of lookups from a node u for resources k located at v. Define a node v a preferred node for u, if $\lambda_{uv} \geq \lambda_0$, where λ_0 is a threshold constant. For each node u, its preferred nodes form a set P_u, which varies over time.

A request graph is $G_R = (N, E_R)$ where a direct edge $(u, v) \in E_R$ if an only if $v \in P_u$. Clearly, the connectivity of G_R depends on threshold λ_0; for larger λ_0 the connectivity becomes sparser.

A disjointed cluster is a subset $\mathscr{C} \subseteq N$ that satisfies the closure property $P_u \subseteq \mathscr{C} \ \forall u \in \mathscr{C}$. The notation $\mathscr{C}_u$ designates the cluster to which node u belongs. All nodes within a cluster share the common interest, distinct from other clusters. As a result, the set of nodes N is partitioned into several distinct clusters. Each cluster corresponds to a connected component in G_R.

Nodes with multiple interests could lead to the flat case when $\mathscr{C}_u = N \; \forall u \in N$. Identification of interest-based groups is still possible based on analysis of weakly linked clusters in G_R. Some of these clusters, however, are overlapped by nodes with multiple interests. The degree of overlap can be reduced by replication. Let $(u, v) \in E_R$ where u and v belongs to different clusters $\mathscr{C}_u$ and $\mathscr{C}_v$. (The number of such edges is small.) Then those resources of v that u is interested in are replicated on some nodes of $\mathscr{C}_u$. The edge (u, v) is removed shrinking P_u.

A cluster can optimize interest-aware communications by keeping the routing distance small among its members. The optimization of routing with respect to one interest group may conflict with that in the other interest group if there are nodes with multiple interests. Besides that, even if there is a tendency for nodes to cluster while serving their own interests, an outlier node in some cluster may frequently communicate with outliers in other clusters. When the improvement in routing for some interest group comes at the expense of a deterioration of routing distance for another interest group, the overall performance depends on the degree of overlap.

Consider the routing table of a node u. Let $m_c = \log_2 |\mathscr{C}_u|$ and $m = \log_2 N$. As in the basic Chord DHT, T_u includes the immediate successor on the ring. In contrast, other neighbors (fingers) are classified into m_c cluster neighbors and $m - m_c$ non-cluster neighbors. Cluster neighbors are selected from $\mathscr{C}_u$ with progressive distance steps 2^i for $i = 1, 2, \ldots, m_c$. Similarly, non-cluster neighbors are selected from $N \setminus \mathscr{C}_u$ with steps $2^{i - m_c}$ for $i = m_c + 1, m_c + 2, \ldots, m$. In fact, two sub-rings are formed, one within the cluster for local inter-cluster routing and the other for global routing.

In a lookup for k, if u knows that the destination is in $\mathscr{C}_u$ then u forwards the lookup to the best cluster neighbor. If u knows that the destination is outside $\mathscr{C}_u$ then u forwards the lookup using the best non-cluster neighbor. If u has no knowledge on the destination then u performs standard greedy routing, selecting the closest neighbor to the key.

In summary, each node u performs the following iterations.

1. Profile the requests to determine P_u.
2. Apply the replication protocol to prune P_u.
3. Apply the cluster identification protocol to determine $\mathscr{C}_u$.
4. Configure the routing table T_u.

In its pure form, the iterations should be synchronized with other nodes, since the construction involves the global request graph G_R. Nevertheless, u can apply only local knowledge to determine P_u and $\mathscr{C}_u$ approximately.

Theorem 5.1. *Consider a Chord DHT network extended with the above protocol. Let N be the number of nodes and $M \leq N$ be the mean cluster size. If ε is the fraction of times a lookup is forwarded to a non-cluster neighbor then the expected routing distance between random nodes u and d is*

$$\mathsf{E}[|u \to^+ d|] = O\left((1-\varepsilon)\log_2 M + \varepsilon M(1 - \frac{M}{N})\log_2 \frac{N}{M}\right) \qquad (5.3)$$

for large N and M.

Proof. Since each cluster forms own Chord sub-ring, the routing within the cluster takes $O(\log_2 M)$ hops. The expectation is $(1-\varepsilon)O(\log_2 M)$.

Consider a Chord ring of X nodes, each has $l < \log_2 X$ neighbors, i.e., the number of neighbors can be less than in the basic Chord. Let us prove that the routing distance is at most $Xl/2^l$ hops when the following two phases are used. At the first phase, the routing moves a lookup to the interval of size 2^l where the destination resides. It can be done by using the longest-range neighbors and requires at most $X/2^l$ hops. At the second phase, the standard Chord routing takes at most l hops to resolve the lookup within the interval.

For inter-cluster routing, $l = m - m_c$ neighbors is used for the sub-ring of $X = N - M$ nodes. Using the previous assertion, it takes at most

$$\frac{(N-M)(m-m_c)}{2^{m-m_c}} = \frac{(N-M)M(\log_2 N - \log_2 M)}{N} = (1 - M/N)M\log_2 \frac{N}{M}.$$

The expectation is $\varepsilon O\left(M(1 - M/N)\log_2(N/M)\right)$.   □

An interesting problem is finding $M^* = M(N,\varepsilon)$ that minimizes (5.3). Note that ε depends on the ratio M/N. In particular, $\varepsilon \approx 1$ if M is very small and $\varepsilon \approx 0$ if M is close to N.

For moderate cluster sizes $1 \ll M \ll N$, the improvement in routing performance is significant if ε is preserved small. That is $O(\log M)$ hops instead of $O(\log N)$. It means that u needs additional knowledge to determine for a given key there is $d \in \mathscr{C}_u$ responsible for k or there is no such $d \in \mathscr{C}_u$.

5.4.2.2 Social Rationality

Many P2P applications require that nodes behave altruistically in order to perform tasks collectively. This requirement is called *social rationality*: if a node has a choice of actions it should chose the action that maximizes the social utility (sum of all node utilities in the system). Individual nodes, however, even if they wish to follow the socially rationality, often will not have enough information to gauge the effects of their actions on others. Social rationality is also contrasted with individual rationality when nodes should select actions that maximize their individual utility. In addition, selfish or malicious nodes can get into the system, since P2P environments are typically designed for open.

The SLAC rule (Selfish Link-based Adaptation for Cooperation) provides a simple selfish re-wiring protocol that spontaneously self-organizes the network into internally specialized clusters [12]. Nodes within the cluster pool their interests, sharing tasks and working altruistically as a team although their individual behavior is selfish. In this design solution, the adaptation principles from Chap. 4 are evolved to offer dynamic specialization and respecialization if nodes recognize they could do better playing another role (and have the ability to do so).

Let u be engaged in some activity and B_u be some measure of its utility (or performance). The higher the value of B_u, the better the node performs in its target domain. The first assumption is that u tends to use its abilities to selfishly increase B_u in a greedy and adaptive way. The second assumption is that u can change its behavioral strategy (that is, change how it behaves at the application level) as well as drop and make links to nodes u knows. Third assumption is that u can discover other nodes randomly, compare its utility against those nodes, make links to them and their neighbors, and copy their strategies.

SLAC specifies how u should update its behavioral strategy and neighbors (routing table). Regularly, u compares its utility against a random node v. If $B_u < B_v$ then u copies behavioral strategy of v and "rewires" itself in the network by dropping all current neighbors from T_u and adding v and all v's neighbors to T_u.

Also, periodically and with low probability, u applies a kind of "mutation" operation. Mutation of the links involves removing all existing neighbors from T_u and replacing them with a single link to a node randomly drawn from the network. Mutation of the behavior (role) means changing the behavioral strategy randomly.

Note that the interactions between nodes are pair-wise between network neighbors. All the rewiring operations are symmetric: if node u makes a link $u \rightarrow v$ then v makes a link $v \rightarrow u$ and on the other hand if u drops a link $u \rightarrow v$ the link $v \rightarrow u$ has to be dropped as well.

If many nodes apply the SLAC rule then naturally some clusters are formed containing nodes of the similar utility. In fact, the rule uses the metric $\tau(u,v) = |B_u - B_v|$. The strategy is similar to BitTorrent, when nods with equivalent upload rates form a cluster in a file-sharing swarm. (Chapter 6 considers this topic in more detail.) Extensive experimental study [19] confirmed that the BitTorrent choking algorithm for neighbor selection facilitates the formation of clusters with nodes of similar upload bandwidth. Nodes of the same utility class prefer to upload to each other, with the occasional exception of random optimistic unchokes.

When applied in a suitably large population, over time, SLAC follows a kind of evolutionary process in which nodes with high utility tend to replace nodes with low utility with nodes periodically changing behavior and neighborhood. Experiments showed that there is no dominance of selfish behavior, as might be intuitively expected. A SLAC produced network is close to the social optimum although nodes behaved in a bounded rational and myopic selfish way.

Actually, a form of link-based incentive mechanism emerges. The focus is on the elimination of free-riding. Nodes make and drop links in the network to minimize the effects of others nodes' selfish behavior. Consequently, the topology itself reflects a network of cooperation.

Another social network-based design was introduced in [44]. Virtual social groups of nodes are formed using dynamic interrelationships (e.g., preferences, interaction history, and trust relationship). In fact, it builds a social network on top of the flat overlay to assist in exploring resources in the P2P overlay network.

The interactions between the nodes are used to incrementally build the social relationships between the nodes in the associated social groups. In such a P2P network, a search query is propagated along the social groups in the overlay social network.

PROSA [6] is a semantic P2P overlay network also inspired by social dynamics. In a PROSA overlay, nodes sharing similar resources are eventually connected with each other, naturally evolving to a small-world network. The PROSA topology dynamically changes, following changes in node preferences and attitudes: if a node gets involved in different topics, it will link to other nodes that provide resources in those topics, just making queries and waiting for responses.

5.5 Summary

A key challenge that is tackled with hierarchical routing schemes is the usability of a P2P system that consists of many heterogeneous nodes and keeps a huge amount of diverse resources. For scalability such a system must implement efficient and effective techniques for search and retrieval. Search techniques depend on overlay network topology and resource management, i.e., resource placement and resource indexes. As we showed the latter two problems can be solved based on mechanisms with grouping nodes and resources in the network.

This chapter finalized our exposition of the evolution of hierarchical routing schemes in flat DHT designs. Elements of grouping nodes, which appeared initially in Principles 5, 6 and 7, converge to system-level cluster structures. Eventually the evolution comes to Principle 8. The clustering mechanics lead to DHT designs which we called pre-hierarchical. It is the last step before pure HDHTs.

References

1. Akers, S.B., Krishnamurthy, B.: A group-theoretic model for symmetric interconnection networks. IEEE Trans. Comput. **38**(4), 555–566 (1989). doi: http://dx.doi.org/10.1109/12.21148
2. Bawa, M., Condie, T., Ganesan, P.: LSH forest: self-tuning indexes for similarity search. In: Proceedings of 14th International Conference World Wide Web (WWW '05), pp. 651–660. ACM, New York (2005). doi: http://doi.acm.org/10.1145/1060745.1060840
3. Bejan, A., Ghosh, S.: Self-optimizing DHTs using request profiling. In: Proceedings of 8th International Conference on Principles of Distributed Systems (OPODIS 2004). Revised Selected Papers. Lecture Notes in Computer Science, vol. 3544, pp. 140–153. Springer, Berlin (2005)
4. Bezdek, J.C., Pal, N.R.: Some new indexes of cluster validity. IEEE Trans. Syst. Man Cybern. B **28**(3), 301–315 (1998). doi: http://dx.doi.org/10.1109/3477.678624
5. Bharambe, A.R., Agrawal, M., Seshan, S.: Mercury: supporting scalable multi-attribute range queries. SIGCOMM Comput. Commun. Rev. **34**(4), 353–366 (2004). doi: http://doi.acm.org/10.1145/1030194.1015507

6. Carchiolo, V., Malgeri, M., Mangioni, G., Nicosia, V.: An adaptive overlay network inspired by social behaviour. J. Parallel Distrib. Comput. **70**(3), 282–295 (2010). doi: http://dx.doi.org/10.1016/j.jpdc.2009.05.004

7. Castro, M., Drushel, P., Hu, Y., Rowstron, A.: Exploiting network proximity in peer-to-peer networks. Technical Report MSR-TR-2002-82, Microsoft Research (2002)

8. Chazapis, A., Asiki, A., Tsoukalas, G., Tsoumakos, D., Koziris, N.: Replica-aware, multi-dimensional range queries in distributed hash tables. Comput. Commun. **33**(8), 984–996 (2010). doi: http://dx.doi.org/10.1016/j.comcom.2010.01.024

9. Dabek, F., Kaashoek, M.F., Karger, D., Morris, R., Stoica, I.: Wide-area cooperative storage with CFS. In: Proceedings of 18th ACM Symposium Operating Systems Principles (SOSP '01), pp. 202–215. ACM, New York (2001). doi: http://doi.acm.org/10.1145/502034.502054

10. Dabek, F., Cox, R., Kaashoek, F., Morris, R.: Vivaldi: a decentralized network coordinate system. In: Proceedings of ACM SIGCOMM'04, pp. 15–26. ACM, New York (2004). doi: http://doi.acm.org/10.1145/1015467.1015471

11. Godfrey, P.B., Stoica, I.: Heterogeneity and load balance in distributed hash tables. In: Proceedings of IEEE INFOCOM'05, pp. 596–606. IEEE (2005). doi:10.1109/INFCOM.2005.1497926

12. Hales, D., Edmonds, B.: Applying a socially inspired technique (tags) to improve cooperation in P2P networks. IEEE Trans. Syst. Man Cybern. A Syst. Hum. **35**, 385–395 (2005)

13. Karger, D., Lehman, E., Leighton, T., Panigrahy, R., Levine, M., Lewin, D.: Consistent hashing and random trees: distributed caching protocols for relieving hot spots on the world wide web. In: STOC '97: Proceedings of 29th Annual ACM Symposium on Theory of Computing, pp. 654–663. ACM, New York (1997). doi: http://doi.acm.org/10.1145/258533.258660

14. Karger, D.R., Ruhl, M.: Simple efficient load balancing algorithms for peer-to-peer systems. In: SPAA '04: Proceedings of 16th Annual ACM Symposium on Parallelism in Algorithms and Architectures, pp. 36–43. ACM, New York (2004). doi: http://doi.acm.org/10.1145/1007912.1007919

15. Korzun, D., Gurtov, A.: A local equilibrium model for P2P resource ranking. ACM SIGMETRICS Perform. Eval. Rev. **37**(2), 27–29 (2009). doi: http://doi.acm.org/10.1145/1639562.1639572

16. Krishnamurthy, B., Wang, J., Xie, Y.: Early measurements of a cluster-based architecture for P2P systems. In: IMW '01: Proceedings of 1st ACM SIGCOMM Workshop on Internet Measurement, pp. 105–109. ACM, New York (2001). doi: http://doi.acm.org/10.1145/505202.505216

17. Ledlie, J., Pietzuch, P., Seltzer, M.: Stable and accurate network coordinates. In: Proceedings of 26th IEEE International Conference Distributed Computing Systems (ICDCS '06). IEEE Computer Society (2006). doi: http://dx.doi.org/10.1109/ICDCS.2006.79

18. Lee, J., Lee, H., Kang, S., Kim, S.M., Song, J.: CISS: An efficient object clustering framework for DHT-based peer-to-peer applications. Comput. Netw. **51**(4), 1072–1094 (2007). doi: http://dx.doi.org/10.1016/j.comnet.2006.07.005

19. Legout, A., Liogkas, N., Kohler, E., Zhang, L.: Clustering and sharing incentives in BitTorrent systems. ACM SIGMETRICS Perform. Eval. Rev. **35**(1), 301–312 (2007). doi: http://doi.acm.org/10.1145/1269899.1254919

20. Li, J., Vuong, S.: Ontology-based clustering and routing in peer-to-peer networks. In: PDCAT '05: Proceedings of 6th International Conference Parallel and Distributed Computing Applications and Technologies, pp. 791–795. IEEE Computer Society (2005). doi: http://dx.doi.org/10.1109/PDCAT.2005.178

21. Liu, B., Lee, W.C., Lee, D.L.: Supporting complex multi-dimensional queries in P2P systems. In: ICDCS '05: Proceedings of 25th IEEE International Conference on Distributed Computing Systems, pp. 155–164. IEEE Computer Society (2005). doi: http://dx.doi.org/10.1109/ICDCS.2005.75

22. Locher, T., Schmid, S., Wattenhofer, R.: eQuus: a provably robust and locality-aware peer-to-peer system. In: Proceedings of 6th IEEE International Conference Peer-to-Peer Computing (P2P), pp. 3–11. IEEE Computer Society (2006). doi: http://dx.doi.org/10.1109/P2P.2006.17

23. Loguinov, D., Kumar, A., Rai, V., Ganesh, S.: Graph-theoretic analysis of structured peer-to-peer systems: routing distances and fault resilience. IEEE/ACM Trans. Netw. **13**(5), 1107–1120 (2005)
24. Newman, M.E.J.: The structure and function of complex networks. SIAM Rev. **45**, 167–256 (2003). doi: http://dx.doi.org/10.1137/S003614450342480
25. Qu, C., Nejdl, W., Kriesell, M.: Cayley DHTs — a group-theoretic framework for analyzing DHTs based on Cayley graphs. In: ISPA 2004: Proceedings of 2nd International Symposium on Parallel and Distributed Processing and Applications. Lecture Notes in Computer Science, vol. 3358, pp. 914–925. Springer, New York (2004)
26. Ratnasamy, S., Handley, M., Karp, R., Shenker, S.: Topologically-aware overlay construction and server selection. In: Proceedings of IEEE INFOCOM'02, pp. 1190–1199, vol. 3, IEEE (2002)
27. Ratti, S., Hariri, B., Shirmohammadi, S.: NL-DHT: A non-uniform locality sensitive DHT architecture for massively multi-user virtual environment applications. In: ICPADS '08: Proceedings of 14th IEEE International Conference on Parallel and Distributed Systems, pp. 793–798. IEEE Computer Society (2008). doi: http://dx.doi.org/10.1109/ICPADS.2008.32
28. Rostami, H., Habibi, J., Livani, E.: Semantic routing of search queries in P2P networks. J. Parallel Distrib. Comput. **68**(12), 1590–1602 (2008). doi: http://dx.doi.org/10.1016/j.jpdc.2008.06.005
29. Rostami, H., Habibi, J., Livani, E.: Semantic partitioning of peer-to-peer search space. Comput. Commun. **32**(4), 619–633 (2009). doi: http://dx.doi.org/10.1016/j.comcom.2008.11.020
30. Rufino, J., Alves, A., Exposto, J., Pina, A.: A cluster oriented model for dynamically balanced DHTs. In: IPDPS'04: Proceedings of 18th International Symposium on Parallel and Distributed Processing. IEEE Computer Society (2004)
31. Rufino, J., Pina, A., Alves, A., Exposto, J.: Toward a dynamically balanced cluster oriented DHT. In: Proceedings of Parallel and Distributed Computing and Networks (PDCN 2004), pp. 48–55. ACTA Press (2004)
32. Sánchez-Artigas, M., García López, P.: Echo: A peer-to-peer clustering framework for improving communication in DHTs. J. Parallel Distrib. Comput. **70**, 126–143 (2010). doi: http://dx.doi.org/10.1016/j.jpdc.2009.06.002
33. Sharma, P., Xu, Z., Banerjee, S., Lee, S.J.: Estimating network proximity and latency. SIGCOMM Comput. Commun. Rev. **36**, 39–50 (2006). doi: http://doi.acm.org/10.1145/1140086.1140092
34. Shen, H., Xu, C.Z.: Hash-based proximity clustering for efficient load balancing in heterogeneous DHT networks. J. Parallel Distrib. Comput. **68**(5), 686–702 (2008). doi: http://dx.doi.org/10.1016/j.jpdc.2007.10.005
35. Sínchez-Artigas, M., García-López, P., Gómez-Skarmeta, A.F., Santa, J.: TR-clustering: alleviating the impact of false clustering on P2P overlay networks. Comput. Netw. **52**, 3185–3204 (2008). doi:10.1016/j.comnet.2008.08.011
36. Souza, D., Pires, C.E., Kedad, Z., Tedesco, P., Salgado, A.C.: A semantic-based approach for data management in a P2P system. In: Transactions on Large-Scale Data- and Knowledge-Centered Systems III, pp. 56–86. Springer, Berlin (2011)
37. Sripanidkulchai, K., Maggs, B., Zhang, H.: Efficient content location using interest-based locality in peer-to-peer systems. In: Proceedings of IEEE INFOCOM'03, vol. 3, pp. 2166–2176 (2003). doi: http://dx.doi.org/10.1109/INFCOM.2003.1209237
38. Surana, S., Godfrey, B., Lakshminarayanan, K., Karp, R., Stoica, I.: Load balancing in dynamic structured peer-to-peer systems. Perform. Eval. **63**(3), 217–240 (2006). doi: http://dx.doi.org/10.1016/j.peva.2005.01.003
39. Tang, C., Xu, Z., Dwarkadas, S.: Peer-to-peer information retrieval using self-organizing semantic overlay networks. In: Proceedings of ACM SIGCOMM'03, pp. 175–186. ACM, New York (2003). doi: http://doi.acm.org/10.1145/863955.863976
40. Tirado, J.M., Higuero, D., Isaila, F., Carretero, J., Iamnitchi, A.: Affinity P2P: A self-organizing content-based locality-aware collaborative peer-to-peer network. Comput. Netw. **54**(12), 2056–2070 (2010). doi: http://dx.doi.org/10.1016/j.comnet.2010.04.016

41. Triantafillou, P., Xiruhaki, C., Koubarakis, M., Ntarmos, N.: Towards high performance peer-to-peer content and resource sharing systems. In: Proceedings of 1st Biennial Conference Innovative Data Systems Research (CIDR 2003) (2003)

42. Wan, Y., Asaka, T., Takahashi, T.: A hybrid P2P overlay network for non-strictly hierarchically categorized contents. In: CCGRID '08: Proceedings of 8th IEEE International Symposium on Cluster Computing and the Grid, pp. 41–48. IEEE Computer Society (2008). doi: http://dx.doi.org/10.1109/CCGRID.2008.10

43. Xu, Z., Mahalingam, M., Karlsson, M.: Turning heterogeneity into an advantage in overlay routing. In: Proceedings of IEEE INFOCOM'03, pp. 1499–1509, IEEE (2003)

44. Yang, S.J.H., Zhang, J., Lin, L., Tsai, J.J.P.: Improving peer-to-peer search performance through intelligent social search. Expert Syst. Appl. **36**(7), 10312–10324 (2009). doi: http://dx.doi.org/10.1016/j.eswa.2009.01.045

45. Yu, Q., Xu, T., Ye, B., Lu, S., Chen, D.: SkipStream: A clustered skip graph based on-demand streaming scheme over ubiquitous environments. In: ICPP '09: Proceedings of 2009 International Conference on Parallel Processing, pp. 269–276. IEEE Computer Society (2009). doi: http://dx.doi.org/10.1109/ICPP.2009.57

46. Zhao, B.Y., Duan, Y., Huang, L., Joseph, A.D., Kubiatowicz, J.D.: Brocade: landmark routing on overlay networks. In: IPTPS '02: Proceedings of 1st International Workshop on Peer-to-Peer Systems. Lecture Notes in Computer Science, vol. 2429, pp. 34–44. Springer, Berlin (2002)

47. Zhu, Y., Hu, Y.: Efficient semantic search on DHT overlays. J. Parallel Distrib. Comput. **67**(5), 604–616 (2007). doi: http://dx.doi.org/10.1016/j.jpdc.2007.01.005

Chapter 6
Local Ranking

Abstract Intrinsically, P2P networks are anonymous, dynamic, and autonomous. Participants have natural disincentives to cooperate because cooperation consumes their own resources and may degrade their own performance. Consequently, each node selfishly attempts to maximize own utility lowering the overall utility of the system. Avoiding this "tragedy of the commons" requires incentives for cooperation. This chapter considers design principles for providing incentives for cooperation in resource sharing among nodes. Fair resource sharing needs differentiation based on the contribution/consumption each node has in the system. We state the problem of local ranking in a P2P resource sharing system with multiple resources. It provides rank-based differentiation of competing nodes and their resources. Introducing ranks rewards cooperating nodes and punishes defect behavior. In this chapter, we focus on ranking models with local knowledge only; the class of models with global system knowledge is discussed later in Chap. 10.

6.1 Introduction

A P2P resource sharing system consists of nodes where each node is controlled by an independent entity and all nodes must cooperate as a self-organizing network with no central control. Cooperation is based on two operations: a node *provides* and *consumes* resources to and from other nodes, forming node's costs and gains. Concrete implementations include such popular systems as Amazon Simple Storage Service, P2PSIP, and BitTorrent (see Chap. 13).

By pooling together the resources of many participants, a P2P network provides a highly scalable platform for distributed resource sharing such as storage, content, bandwidth, or computing. The previous chapters primarily discussed the corresponding design principles of effective resource distribution. Being cooperative, participating nodes altogether achieve good system performance, scalability,

D. Korzun and A. Gurtov, *Structured Peer-to-Peer Systems: Fundamentals of Hierarchical Organization, Routing, Scaling, and Security*, DOI 10.1007/978-1-4614-5483-0_6,
© Springer Science+Business Media New York 2013

security, etc. This chapter focuses on design principles of cooperation in a P2P system, an essential issue when the environment is autonomous, anonymous, and dynamic.

Lack of cooperation can essentially degrade a P2P system [9, 20, 23, 35]. Due to autonomy, individual actions are voluntarily chosen and determined by independent and rational behavior. Nodes are intrinsically selfish; they attempt to maximize own utility lowering the overall system welfare. Individual consumption tends to come without proper contribution (freeriding). Anonymous nodes are less accountable for their activity and change their IDs with near zero cost (whitewashing). Due to dynamics, most interactions among nodes are one-time; nodes have no idea about other nodes' behavior history, except their current behavior, so operational decision-making is needed.

Cooperation among nodes can be supported with incentives [10]. A system incentive mechanism encourages every node to balance rationally when contributing its own resources and consuming external ones. In its simplest case, each node acquires and contributes equal amounts of resources, as used in BitTorrent [8] and its variants [5, 14, 16, 19, 29]. Even for this extremely simple case, stimulating participants to voluntarily provide resource is a very challenging task [26]. No exogenous organization can effectively enforce the punishment or reward. A rational node has to locally compute individual scores for its neighbors and resources. Based on this ranking the node operationally decides current provision and consumption strategies in its neighborhood.

The purest P2P assumption is that all communications are through neighbors. Any node $u \in N$ is assured only in a *local* view onto the network. In fact, u knows at a time instance t two metrics, $a_{vk} = a_{vk}(u,t)$ and $b_{vk} = b_{vk}(u,t)$, which are u's consumption and provision of resource k through neighbor v, respectively. The metrics aggregate the local history of u's direct observations. This assumption is known as bilateral exchange or direct reciprocity, see [4, 18, 20, 23]. In this chapter, we attempt to understand to which extent a node can rationally apply the scant local knowledge (a,b) for ranking in its neighborhood.

Section 6.2 considers the role of incentives and differentiation in P2P resource sharing. We study rank-based operational approach when a node computes separate local ranks for nodes and resources based on the current behavior of its neighbors. Then the node decides its provision and consumption in proportion to these ranks.

Section 6.3 discusses a generic linear model for P2P local ranking and its applicability for rank-based reciprocal incentives. We reduce the local rank computation to a LP problem and provide the condition of effective rank existence.

Section 6.4 relates other linear ranking model to prior work. We show that known deficit-based models are simply described as instances of this model. We also discuss techniques of such theories as economics, operations research, control theory, and game theory, which is widely applicable in modeling fair cooperation but have certain shortcomings when the local knowledge problem is essential.

6.2 Local Ranking Problem

Let each node observe activity of others directly and make own decisions based on this local knowledge. Decisions are operational since behavior of other participants is unpredictable, with no assumption on long-term invariants. In this section we consider operational decision-making in P2P resource sharing systems. We use BitTorrent-like (BT) systems as a reference case of single-resource exchange. It provides intuition and model assumptions for the general form of local ranking problem in multi-resource exchange.

6.2.1 Rank-Based Operational Decision-Making

Consider a P2P resource sharing system with a set N of nodes. Each node $u \in N$ has links to its neighbors $N_u \subset N$, see Fig. 6.1a. In turn, any u's neighbor has own set of neighbors, and the global network topology is formed. Assumption $|N_u| \ll |N|$ leads to the local knowledge restriction when u's directly observable knowledge about the entire system is partial [15, 24, 33]. Let R denote the index set of all system resources. For simplicity assume that each resource is of unlimited supply, e.g., video streaming. A node u consumes external resources $k \in R_u^+$ and provides local resources $k \in R_u^-$; let $R_u = R_u^+ \cup R_u^-$. Consumption and provision are through neighbors $v \in N_u$. Evolution is $N_u = N_u(t)$ and $R_u = R_u(t)$ in time t since u updates its neighbors and decides resources to provide/consume to/from the system.

Let $a_{vk}(t) \geq 0$ denote u's consumption of k through v and $b_{vk}(t) \geq 0$ be the provision of k to v. Both a_{vk} and b_{vk} are in the same units, e.g., consumption and provision rates. In the matrix form, $A(t)$ and $B(t)$ are positive ($|N_u(t)| \times |R_u(t)|$)-matrixes, where rows and columns change with t. They accumulate directly

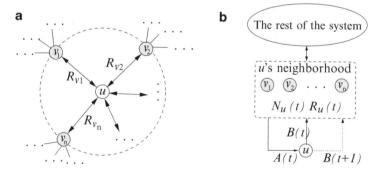

Fig. 6.1 Participation of a node u in the P2P resource sharing system: u's local perspective. (**a**) u makes exchanges (trades) with each neighbor $v \in N_u$ for resources $k \in R_v \subseteq R_u$. (**b**) u directly observes provision $A(t)$ from neighbors and knows their actual consumption $B(t)$, then u makes control decisions for expected consumption $B(t+1)$

Table 6.1 Decision-making at a node u in multi-resource consumption and provision. Ranks support the capacity allocation among resources and neighbors

Consumption: u is a client, $v \in N_u$ are servers	Provision: u is a server, $v \in N_u$ are clients
Selection of (k,v) for $k \in R_u^+$ and $v \in N_u$ is the consumption decision problem: which resource k and from which neighbor v to request.	Selection of (v,k) for $v \in N_u$ and $k \in R_u^-$ is the provision decision problem: to which neighbor v which resource k to provide.
Resource ranks $s = (s_k)_{k \in R_u}$	
Node ranks $r^+ = (r_v^+)_{v \in N_u}$	Node ranks $r^- = (r_v^-)_{v \in N_u}$
1. Allocate consumption capacity among resources $k \in R_u^+$ proportionally to s_k. 2. For given k, allocate its capacity share among neighbors $v \in N_u$ proportionally to r_v^+.	1. Allocate provision capacity among neighbors $v \in N_u$ proportionally to r_v^-. 2. For given v, allocate its capacity share among resources $k \in R_u^+$ proportionally to s_k.

observable participation history of the neighbors during $\tau \leq t$. For instance, the latest activity (short-term history) is used in BT-systems [5, 8, 14, 16, 19, 29, 32] and aggregation over a lengthy period (long-term history) is used in reputation mechanisms [2,7]. Each node applies a preferable accumulation method on its own.

Define u's local knowledge at time t the tuple $(N_u(t), R_u(t), A(t), B(t))$. It has strong practical reasons. Propagation of global knowledge is inefficient in large-scale and high-dynamic systems. Information from non-neighbor nodes is indirect, observed via intermediate nodes, thus becoming subject to trust constraints.

Consider discrete time periods $\{1, 2, \ldots\}$, indexed by t and possibly infinite. Let u decide its activity for period $t + 1$ based on the local knowledge collected in period t, see Fig. 6.1b. Decisions are on the operational level [13]: u performs actions in period t and then observes the feedback from neighbors in $t + 1$. Further we omit t in our notation when the context of repeated transitions $t \to t + 1$ is clear. The operational approach is reasonable when the system cannot be completely specified even as some types of stochastic processes. For instance, a neighbor may arbitrary refuse its rational or altruistic behavior, e.g., because of a bug or ill-will [17].

BitTorrent [8] is an example of a single-resource sharing system (file sharing), where the shared resource is bandwidth. Neighbors v consume from u in period $t + 1$ according with their provision to u in t. This is an instance of *bilateral exchange*; u operationally makes $b_v(t + 1)$ close to $a_v(t)$. Table 6.1 shows the general case when a node u performs two parallel iterative processes in $t + 1$. In provision, u iteratively selects (v,k) to provide v a portion of k, so having full control of $b_{vk}(t+1)$ for $v \in N_u$, $k \in R_u^-$. In consumption, u observes $a_{vk}(t + 1)$ for $v \in N_u$, $k \in R_u^+$ since it depends on v's provision strategy. Nevertheless, u can reduce a_{vk} by requesting less from v or u can stimulate increasing a_{vk} by more requests to v. Accordingly, u iteratively selects (k,v) to make a request for k to v.

These two processes result in sharing u's consumption and provision capacity between neighbors and resources. The bottom part of Table 6.1 shows two-phase rank-based sharing. Resource ranks s_k and node ranks r_v^+ and r_v^- are non-negative real numbers that quantify importance of $k \in R_u$ and $v \in N_u$ to u (local ranks or scores). The iterative rank-proportional allocation can be implemented in a round-robin fashion, typically with normalized ranks

$$\sum_{k \in R_u} s_k = 1, \quad \sum_{v \in N_u} r_v^+ = 1, \quad \sum_{v \in N_u} r_v^- = 1. \tag{6.1}$$

Further we assume normalization property (6.1) if the reverse is not stated explicitly.

Since u in consumption is primarily interested in resources, not from whom they consume, u decides which resource to request and only then from which neighbor. Similarly, in provision, u serves neighbors, allocating its capacity for them, and only then u decides which resources to provide to a selected neighbor. Now we can formulate the following principle for rank-based decision-making in multi-resource exchange.

Principle 9 (Rank-based multi-resource exchange). *A node u computes separate local ranks for nodes and resources. The ranks are used in two parallel iterative processes of provision and consumption according to Table 6.1. Provision of k from u to v is proportional to $r_v^- s_k$. Consumption of k from v to u is proportional to $s_k r_v^+$.*

6.2.2 Node Ranks in BT-Exchange

Consider BT systems [5, 8, 14, 16, 19, 29, 32] as a reference case for generic model assumptions on node ranks. A BT-system shares node bandwidth—bilateral single-resource exchange. In this exchange, resource rank s is needless ($s = 1$). The BT consumption rule is "consume as much as you can from any neighbor", thus node rank r^+ is also needless. The provision process in Table 6.1 becomes one-phase; u's upload bandwidth is allocated as $b_v = r_v^- b$ for $v \in N_u$.

A node u provides its upload bandwidth $b = \sum_{v \in N_u} b_v$ and consumes download bandwidth $a = \sum_{v \in N_u} a_v$. The incentives for u are straightforward: if u makes $b_v(t)$ low then v may reduce $a_v(t+1)$.

Definition 6.1 (BT incentives). There are incentives for u to cooperate with v if there exists $\delta_v \geq 0$ such that for any t

$$a_v(t) - b_v(t) \leq \delta_v, \tag{6.2}$$

where $d_v(t) = a_v(t) - b_v(t)$ is recent surplus of v.

That is, low provision b_v limits consumption a_v, up to $a_v \leq \delta_v$ for $b_v = 0$; high consumption a_v is possible only if u provides appropriately high b_v. Parameter δ_v is an upper imbalance bound that v tolerates. If $\delta_v = 0$ then u has no credit, and unit consumption requires provision in advance or immediate response; the case is difficult to realize in open network environment (exchange initialization, transfer latency, data retransmissions, etc.). If v is altruistic then $\delta_v = \infty$.

BT-exchange follows Principle 9 and can be easily described in terms of node ranks r_v^-. The provision process in period $t+1$ aims at making $b_v(t+1) = r_v^- b$. In original BitTorrent [8], u uploads to n top neighbors from N_u sorted by a_v, i.e., $r_v^- = 1/n$ for $v \in N_u^{best}$ and $r_v^- = 0$ otherwise. In proportional sharing [16,32], all neighbors are fed with $r_v^- = a_v/a$. Block-based BitTorrent [5, 14, 19] is more operational and immediately sets $r_v^- = 0$ if $a_v - b_v$ is lower a threshold.[1] FairTorrent [29] makes ultra operational decisions with one upload per period—to the neighbor with highest rate $a_v - b_v$, reducing the largest term in the sum surplus $\sum_{v \in N_u}(a_v - b_v)$.

The above decision-making aims at the rational use of (6.2) with uncertainty about δ_v. On one hand, $d_v < 0$ is unprofitable for u. On the other hand, u has to merely assume that δ_v is small, and $d_v > 0$ signals on possible violation in (6.2).

The traditional technique of operations research can be applied to this problem. Consider ranks r_v^- that minimize expected deviation from the balance $d_v(t+1) = 0$. Setting $b_v(t+1) = r_v^- b$ and approximating unknown in advance $a_v(t+1)$ with observed $a_v = a_v(t)$, we yield the following optimization problem (the mean-square deviation was selected).

$$
\begin{cases}
\sum_{v \in N_u} (a_v - r_v^- b)^2 \to \min \\
\sum_{v \in N_u} r_v^- = 1, \\
r^- \geq 0.
\end{cases}
\tag{6.3}
$$

The solution to (6.3) can be found by applying the Lagrange multipliers method and provides the following node ranks:

$$
r_v^- = \frac{a_v}{b} + \frac{b-a}{|N_u|b} \quad \text{if} \quad \min_{v \in N_u} a_v \geq \frac{a-b}{|N_u|},
\tag{6.4}
$$

where $\frac{b-a}{|N_u|b}$ is average relative skew of u's participation in the system. The condition in (6.4) is due to $r^- \geq 0$. If $a = b$ then $r_v^- = a_v/b$, coinciding with proportional sharing [16,32].

If u provides in sum less than it consumes ($a > b$) then there can be neighbors lower the average ($a_v < (a-b)/|N_u|$), and solving (6.3) becomes more complicated. Some BT-exchanges use the discrete heuristic—only highest d_v are important. If $a_v(t) \gg b_v(t)$ then u takes $r_v^- > 0$ to keep v's generosity and sets $r_v^- = 0$ otherwise. The case is similar to (6.3) but ranking is reduced to the best neighbors $N_u^{best} \subset N_u$.

---

[1]In general, u checks $\gamma a_v - b_v$ for exceeding the threshold, where $\gamma \geq 1$ is a design parameter; higher γ increases the overall system performance by sacrificing fairness towards high-capacity nodes [19].

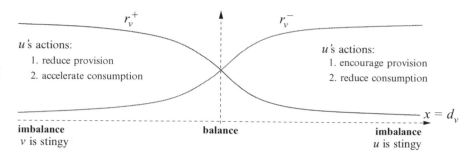

Fig. 6.2 Local single-resource exchange with $v \in N_u$. Let u have surplus $x = d_v$ for given v. Activity of other neighbors $w \neq v$ is fixed (in terms of d_w). Node ranks $r_v^-(x)$ and $r_v^+(x)$ behave monotonically

When u has limited consumption capacity then node consumption rank r^+ can be computed similarly to (6.3) and (6.4). The expected mean-square deviation $\sum_v (r_v^+ a - b_v)^2$ is minimized subject to $\sum_v r_v^+ = 1$ and $r^+ \geq 0$. The solution is symmetrical to (6.4):

$$r_v^+ = \frac{b_v}{a} + \frac{a-b}{|N_u|a} \quad \text{if} \quad \min_{v \in N_u} b_v \geq \frac{b-a}{|N_u|}, \tag{6.5}$$

and u consumes proportionally to its provision b_v.

Model (6.3) captures the property important for reciprocal incentives. If u minimizes deviation from the balance $d_v = a_v - b_v = 0$ then u supports preserving (6.2), even in unfavourable cases $\delta_v \approx 0$. Note that $r_v^- = r_v^-(a_v)$ in (6.4) and $r_v^+ = r_v^+(b_v)$ in (6.5) are non-decreasing functions on a_v and b_v, respectively.

The model has certain drawbacks. First, the conditions in (6.4) and (6.5) complicate the computation of optimal solution when $a \neq b$. In practice they are likely violated since a and b change unpredictably over the time: the same v can behave essentially differently, e.g., due to strategy mutation [12, 31]. Hence heuristics like "select some best neighbors to have non-zero rank" have to be employed. Second, dependences $r_v^- = r_v^-(a_v)$ and $r_v^+ = r_v^+(b_v)$ do not properly capture that high (low) values of a_v should be compensated with appropriate values of b_v. According to (6.4), u sets high r_v^- for high $a_v(t)$ regardless of $b_v(t)$. If u expects in (6.5) that $b_v = b_v(t+1) = r_v^- b$ then u sets $r_v^+ = a_v/a$, making r_v^- and r_v^+ independ on how generous or stingy u behaved on t. Simultaneous minimization of $\sum_{v \in N_u} (r_v^+ a - r_v^- b)^2$ is unhelpful since the sum consumption and provision "dissolve" individual activity.

Consider another approach for node rank modeling, which formalizes and generalizes the BT heuristics and incentives. Instead of fixing an optimization problem which ranks are solutions to, we assume that node ranks r_v^- and r_v^+ are monotone functions of $x = d_v = a_v - b_v$, reflecting imbalance between the actual provision and consumption as shown in Fig. 6.2.

Similarly to (6.3), node rank aims at reducing deviation from the balance: high $|d_v|$ leads to appropriate response. We do not require $r_v^+(x) = r_v^-(-x)$. For example, u may be more conservative in provision to stingy neighbors than in consumption, i.e., $r_v^-(x)$ decreases faster for $x \to -\infty$ than $r_v^+(x)$ for $x \to +\infty$.

6.2.3   Local Ranks: Nodes and Resources

In multi-resource exchange, surplus of v is a sum of its resource surpluses, $d_v = \sum_{k \in R_v} d_{vk}$. Resource compensation is possible since u operates with a set of resources. For example, u tolerates low provision a_{vk} from v because of high provision a_{vi} for $k, i \in R_u$ and $i \neq k$. Furthermore, cross-node resource compensation is possible when small a_{vk} are compensated with high a_{wi} for $w \neq v$.

In the multi-resource case, r^+ and r^- are functions of the surplus matrix $D = \{d_{vk}\}$ instead of the surplus vector $d = \{d_v\}$. Varying d_{vk} when other surpluses are fixed leads to appropriate changes in ranks to reflect the imbalance. The following definition models such behavior in terms of monotone functions, not necessarily contiguous (see also Fig. 6.2 for single-resource case).

Definition 6.2 (Local node rank). Let every d_{wi} be fixed abut $x = d_{vk}$. Then r^+ and r^- are *local node ranks* if and only if they satisfy the following properties for all $v \in N_u$ and $k \in R_v$.

Monotony: $r_v^-(x)$ is a non-increasing function and $r_v^+(x)$ is a non-decreasing function.

Marginality: $\lim_{x \to -\infty} r_v^-(x) = \lim_{x \to +\infty} r_v^+(x) = 0$ and $\lim_{x \to +\infty} r_v^-(x) = \lim_{x \to -\infty} r_v^+(x) = 1$.

The BT-heuristics for positive ranking the highest-providers agree with this definition (only r_v^+ is considered). For example, FairTorrent uses

$$r_v^+ = \begin{cases} 1, \text{ if } d_v = \max_{w \in N_u} d_w \text{ (in ambiguous case select one } v \text{ only)}, \\ 0, \text{ otherwise.} \end{cases}$$

If u operates as many independent BT-nodes, one for each resource, then there is no resource compensation. Each d_{vk} is balanced independently for each $k \in R_u$. In contrast, we aim at u that accounts all resources when cooperating with v.

Definition 6.3 (bilateral incentives for multi-resource exchange). There are incentives for u to cooperate with v if there is $\delta_v \geq 0$ such that for any t

$$\sum_{k \in R_u} (a_{vk}(t) - b_{vk}(t)) \leq \delta_v, \tag{6.6}$$

where $d_v(t) = \sum_{k \in R_u} (a_{vk}(t) - b_{vk}(t))$ is sum recent surplus of v.

Although the same resource can be used in exchanges with different nodes there is no neighbor to parasitize because of generosity of some other neighbors. The latter easily happens when u considers only the overall sum:

$$\sum_{v \in N_u} \sum_{k \in R_u} (a_{vk}(t) - b_{vk}(t)) \le \delta.$$

Similarly to BT-exchange, u realizes the incentives by balancing near $d_v = 0$ for all $v \in N_u$: high d_v likely violates (6.6) and low d_v is unprofitable for a rational node. Expected deviation is minimized by controlling $b_{vk}(t + 1)$ for $v \in N_u$ and $k \in R_u$. For each k reaction to observation $a_{vk}(t)$ may affect $b_{vk}(t + 1)$ as well as $b_{vi}(t + 1)$ for some $i \ne k$. Consequently, resource differentiation is needed to select appropriate resources for the reaction. Resource ranks $s_k \ge 0$ quantitatively implement the differentiation.

Resource rank needs different qualitative model assumptions than node rank. In the latter case, a neighbor v is both provider and consumer; each role receives its own rank. The simplest case for resource rank happens if any resource k has a single role being either *local* (u provides $k \in R_u^-$; $a_{vk} = 0$, $b_{vk} \ge 0 \; \forall v \in N_u$) or *external* ($u$ consumes $k \in R_u^+$; $a_{vk} \ge 0$, $b_{vk} = 0 \; \forall v \in N_u$).

A generalization is *transit resource*, which can be used for better trading opportunities in resource exchange. For example, assume u downloads file k from v ($a_{vk} > 0$); the file is not of interest of u itself; uploading k to w ($b_{wk} > 0$) allows u to download another file i from w ($a_{wi} > 0$). Hence k is transit resource that u uses to receive another resource i. This generalization reduces the gap between bilateral and multilateral exchanges, allowing multi-hop cyclic exchange paths, see more discussion in Chap. 10 and related work [18, 23]. A particular case of transit resource is bidirectional resource ($a_{vk}, b_{vk} > 0$), e.g., bandwidth in BT-systems.

For further analysis denote pure consumption and provision

$$\alpha_{vk} = \begin{cases} d_{vk} = a_{vk} - b_{vk}, & \text{if } d_{vk} > 0, \\ 0, & \text{otherwise.} \end{cases} \qquad \beta_{vk} = \begin{cases} -d_{vk} = b_{vk} - a_{vk}, & \text{if } d_{vk} < 0, \\ 0, & \text{otherwise.} \end{cases}$$

Clearly, $d_{vk} = \alpha_{vk} - \beta_{vk}$. Let $R_v^+ = \{k \mid \alpha_{vk} > 0\}$, $R_v^- = \{k \mid \beta_{vk} > 0\}$, and $R_v = R_v^+ \cup R_v^-$. The notation emphasizes the fact that R_v^+ consists of external-like resources (currently under u's consumption) and R_v^- consists of local-like resources (currently under u's provision) Then $R_u^+ = \bigcup_{v \in N_u} R_v^+$ and $R_u^- = \bigcup_{v \in N_u} R_v^-$. Although $R_v^+ \cap R_v^- = \varnothing \; \forall v \in N_u$ the set $R_u^+ \cap R_u^-$ can be nonempty.

Let $d_k = \sum_{v \in N_u} d_{vk}$ be surplus of k for u. Resource k is balanced when $d_k = 0$, i.e., neighbors provide in sum the same amount of k as they consume. The less $|d_k|$ the more preferable for u to consume or provide k. Hence s_k can be treated a balance closeness index and defined using the same technique of monotone functions as for local node ranks in Definition 6.2 above.

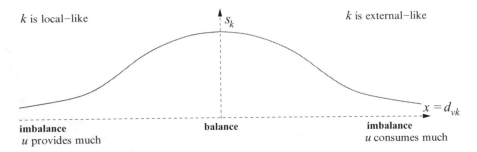

k is local–like s_k k is external–like

$x = d_{vk}$

imbalance   balance   imbalance
u provides much   u consumes much

Fig. 6.3 Qualitative behavior of resource rank s_k as a function of surplus $x = d_{vk}$ when other activity is fixed. If k is pure local or pure external then $s_k(x)$ is defined on $(-\infty,0]$ or $[-\infty,0)$, respectively. The function is neither required to be symmetrical $s_k(x) = s_k(-s)$ nor contiguous

Definition 6.4 (Local resource rank). Let every d_{wi} be fixed abut $x = d_{vk}$. Then s is *local resource rank* if and only if s satisfies the following properties for all $v \in N_u$ and $k \in R_v$.

Monotony $s_j(x)$ is a non-decreasing function on $(-\infty,0]$ for any $j \in R_v^-$ and $s_i(x)$
    is a non-increasing function on $[0,\infty)$ for any $i \in R_v^+$.
Marginality $\lim_{x \to -\infty} s_j(x) = \lim_{x \to +\infty} s_i(x) = 0$ for any $i \in R_v^+$ and $j \in R_v^-$.

The qualitative behavior is shown in Fig. 6.3. Intuitively, when d_{vk} becomes high then s prioritizes reduction in provision of any external-like resource $i \in R_v^+$, including k (since $k \in R_v^+$ if $d_{vk} > 0$). When d_{vk} becomes low then s prioritizes reduction in consumption of any local-like resource, including k (since $k \in R_v^-$ if $d_{vk} < 0$).

Normalized resource rank induces node ranks for every $v \in N_u$ as "sum rank" of resources available through v.

Let

$$r_v^+ = \sum_{i \in R_v^+} s_i, \quad r_v^- = \sum_{j \in R_v^-} s_j, \qquad (6.7)$$

Since s is normalized and $R_v^+ \cap R_v^- = \varnothing$ then $r_v^+ + r_v^- \leq 1$.

Theorem 6.1. *If* $R_v^+ \cup R_v^- = R_u$ *then* r_v^+ *and* r_v^- *in (6.7) are local node ranks (in terms of Definition 6.2).*

Proof. Consider a given local resource rank s as a function the surplus matrix $D = (d_{vk})$. Compute r^+ and r^- by (6.7). For $v \in N_u$ and $k \in R_u$ taken arbitrary consider variation $d_{vk} + \varepsilon$.

Case 1: $d_{vk} \geq 0$ and $\varepsilon > 0$. The variation of ε does not affect R_v^+. Summing $s_i(d_{vk}) \geq s_i(d_{vk} + \varepsilon)$ over $i \in R_v^+$ gives

$$r_v^+(d_{vk}) = \sum_{i \in R_v^+} s_i(d_{vk}) \geq \sum_{i \in R_v^+} s_i(d_{vk} + \varepsilon) = r_v^+(d_{vk} + \varepsilon).$$

Table 6.2 Symbol notation for the local ranking problem

Notation	Description
u, N_u	Node u locally observes the set N_u of all its neighbors.
R_u, R_u^+, R_u^-	Resources that u knows; $R_u = R_u^+ \cup R_u^-$, R_u, where $R_u^+ = \bigcup_{v \in N_u} R_v^+$ and $R_u^- = \bigcup_{v \in N_u} R_v^-$ are the sets of external-like and local-like resources, respectively. It is allowed $R_u^+ \cap R_u^- \neq \varnothing$.
R_v, R_v^+, R_v^-	Resources that u and its neighbor v exchange; $R_v = R_v^+ \cup R_v^-$, $R_v^+ = \{k \mid d_{vk}^+ > 0\}$, $R_v^- = \{k \mid d_{vk}^- > 0\}$, $R_v^+ \cap R_v^- = \varnothing$.
a_{vk}, b_{vk}	Metrics of u's consumption and provision of $k \in R_u$ through $v \in N_u$.
d_{vk}, d_v, d_k	Surplus counters: $d_{vk} = a_{vk} - b_{vk}$, $d_v = \sum_{k \in R_u} d_{vk}$, $d_k = \sum_{v \in N_u} d_{vk}$.
α_{vk}, β_{vk}	Clear u's consumption and provision of resource $k \in R_v$ through $v \in N_u$; $d_{vk} = \alpha_{vk} - \beta_{vk}$ such that $\min\{\alpha_{vk}, \beta_{vk}\} = 0$.
s_k	Local resource rank of $k \in R_u$, both for provision and consumption.
r_v^-, r_v^+	Local node ranks of neighbors $v \in N_u$, for provision and consumption, respectively.

Application of this inequality to the equations

$$r_v^+(d_{vk}) + r_v^-(d_{vk}) = 1 \quad \text{and} \quad r_v^+(d_{vk} + \varepsilon) + r_v^-(d_{vk} + \varepsilon) = 1$$

yields $r_v^-(d_{vk}) \leq r_v^-(d_{vk} + \varepsilon)$. Since $\lim_{x \to +\infty} s_i(x) = 0 \ \forall i \in R_v^+$ then

$$\lim_{x \to +\infty} r_v^+(x) = 0 \quad \text{and} \quad \lim_{x \to +\infty} r_v^-(x) = 1 - \lim_{x \to +\infty} r_v^+(x) = 1.$$

Case 2: $d_{vk} \leq 0$ and $\varepsilon > 0$. The proof that

$$r_v^-(d_{vk} - \varepsilon) \leq r_v^-(d_{vk}), \quad r_v^+(d_{vk} - \varepsilon) \geq r_v^+(d_{vk}), \quad \lim_{x \to -\infty} r_v^+(x) = 1, \quad \lim_{x \to -\infty} r_v^-(x) = 0$$

is symmetrical to the previous case.

Finally, the notation of local ranking problem is summarized in Table 6.2.

6.3 Linear Ranking Model

In single-resource exchange, such as BT-exchange, reciprocal incentives can be implemented with node ranks only. When nodes exchange multiple resources, resource ranks become essential for proper differentiation of nodes in accordance with their multi-resource activity. In this section we study a particular model of resource ranks. They are defined as solutions to a homogenous linear equation system. It states local exchange balance for all neighbors and resources. Computation of the ranks is reduced to an LP problem.

6.3.1 Linear Resource Ranks

Let a node u maintain surplus counters $d_{vk} = \alpha_{vk} - \beta_{vk}$ for all its neighbors v and resources k. Consider the following homogenous system of $n = |N_u|$ equations in $m = |R_u|$ non-negative unknowns.

$$\sum_{k \in R_u} d_{vk} s_k = 0, \quad v \in N_u. \tag{6.8}$$

Let $\|s\|_1 = \sum_{k \in R_u} |s_k| = \sum_{k \in R_u} s_k$ (i.e., the L_1 norm). If $s \neq \mathbf{0}$ (non-trivial) then $s/\|s\|_1$ is a normalized solution. We call *linear resource rank* any normalized solution to (6.8) or the trivial solution if (6.8) has no normalized solutions.

Given a linear resource rank s. Intuitively, if $s_k = 1/m$ then k is balanced. If $s_k < 1/m$ then the lower rank value is due to k excites the imbalance. If $s_k > 1/m$ then the higher rank value is due to other resources excite the imbalance.

Choose arbitrary $k \in R_u$ and then fix d_{vi} for all $i \neq k$. Coefficients $\{d_{vk}\}_{v \in N_u}$ are varied such that $x = d_k = \sum_v d_{vk}$ taking all real values. Consider a vector function $s(x)$ where $s(x)$ is a linear resource rank (selected arbitrary). The following theorem shows that linear resource rank approximates properties 1 and 2 of local resource ranks (see Definition 6.4).

Theorem 6.2. *Given $k \in R_u$ and $s(\cdot)$. For any $x = d_k \in (-\infty, \infty)$, if $x < 0$ then there is $y < x$ such that $s_k(y) \leq s_k(x)$; if $x > 0$ then there is $y > x$ such that $s_k(x) \geq s_k(y)$. Moreover,* $\lim\limits_{x \to -\infty} s_k(x) = \lim\limits_{x \to +\infty} s_k(x) = 0$.

Proof. It is sufficient to find $\varepsilon = y - x \neq 0$ such that $\Delta s_k = s_k(x + \varepsilon) - s_k(x) \leq 0$. Let $\varepsilon = \sum_{v \in N_u} \varepsilon_v$ where ε_v is the corresponding variation of d_{vk}. Since $s(x)$ and $s(x + \varepsilon)$ are solutions to (6.8) the following equalities hold for any $v \in N_u$:

$$\sum_{i \in R_u} d_{vi} s_i(x) = 0 \quad \text{and} \quad \sum_{i \in R_u} d_{vi} s_i(x + \varepsilon) + \varepsilon_v s_k(x + \varepsilon) = 0.$$

Summation over all v gives

$$\sum_{i \neq k} d_i s_i(x) + x s_k(x) = 0 \quad \text{and} \quad \sum_{i \neq k} d_i s_i(x + \varepsilon) + x s_k(x + \varepsilon) + \varepsilon s_k(x + \varepsilon) = 0.$$

Subtracting the first one from the other results in $\sum_{i \neq k} d_i \Delta s_i + x \Delta s_k + \varepsilon s_k(x + \varepsilon) = 0$,

$$\Delta s_k = -\frac{1}{x} \left(\varepsilon s_k(x + \varepsilon) + \sum_{i \neq k} d_i \Delta s_i \right), \tag{6.9}$$

$$s_k(x + \varepsilon) = -\frac{1}{\varepsilon} \left(x \Delta s_k + \sum_{i \neq k} d_i \Delta s_i \right). \tag{6.10}$$

Case 1. $s_k(x) = 0$. Then for any $\varepsilon \neq 0$

$$x\Delta s_k + \sum_{i\neq k} d_i \Delta s_i = \left(x s_k(x+\varepsilon) + \sum_{i\neq k} d_i s_i(x+\varepsilon) \right) - \left(\sum_{i\neq k} d_i s_i(x) - d_k s_k(x) \right)$$

$$= \sum_{i\in R_u} d_i s_i(x+\varepsilon) + \sum_{i\in R_u} d_i s_i(x) = 0.$$

Hence $s_k(x+\varepsilon) = 0$ by (6.10), and consequently $\Delta s_k = 0$.

If $s_k(x) > 0$ then take ε with $|\varepsilon| = \dfrac{1}{s_k(x)} \sum_{i\neq k} |d_i|$ and consider the further two cases.

Case 2. $s_k(x+\varepsilon) = 0$. Then $s_k(x+\varepsilon) \leq s_k(x)$ since $s(x) \geq \mathbf{0}$ by definition. Consequently $\Delta s_k \leq 0$.

Case 3. $s_k(x+\varepsilon) > 0$. If the expression in brackets in (6.9) is non-negative for $x, \varepsilon < 0$ and non-positive for $x, \varepsilon > 0$ then $\Delta s_k \leq 0$. Otherwise, let us prove by contradiction that none of the inequalities $\varepsilon s_k(x+\varepsilon) + \sum_{i\neq k} d_i \Delta s_i > 0$ for $x, \varepsilon < 0$ and $\varepsilon s_k(x+\varepsilon) + \sum_{i\neq k} d_i \Delta s_i < 0$ for $x, \varepsilon > 0$ is possible for the selected ε. First, the following upper bound is independent on ε:

$$\left| \sum_{i\neq k} d_i \Delta s_i \right| = \left| \sum_{i\neq k} d_i s_i(x+\varepsilon) - \sum_{i\neq k} d_i s_i(x) \right| \leq \sum_{i\neq k, d_i>0} d_i + \sum_{i\neq k, d_i<0} (-d_i) = \sum_{i\neq k} |d_i| = \text{const.}$$

Second, assume in (6.9) that $\Delta s_k > 0$ and $\varepsilon s_k(x+\varepsilon) < -\sum_{i\neq k} d_i \Delta s_i$ for $x, \varepsilon > 0$. Then

$$s_k(x+\varepsilon) < \frac{1}{|\varepsilon|} \left| \sum_{i\neq k} d_i \Delta s_i \right| \leq \frac{1}{|\varepsilon|} \sum_{i\neq k} |d_i|.$$

The same bound is true for the case with $x, \varepsilon < 0$. Substituting $|\varepsilon| = \frac{1}{s_k(x)} \sum_{i\neq k} |d_i|$ we yield the contradiction $s_k(x+\varepsilon) < s_k(x)$.

Finally, since $|\Delta s_i| \leq 1 \ \forall i \in R_u$ the absolute value of the expression in brackets in (6.10) is bounded by $\|d\|_1 = \sum_{i\in R_u} |d_i|$, which is independent on ε. Consequently, $s_k(x+\varepsilon) \to 0$ when $|\varepsilon| \to \infty$. $\square$

If $d_{vk} = 0 \ \forall v \in N_u$ then the resource k is inactive in the exchange, e.g., k is a new resource. System (6.8) provides no information for ranking k, and additional rule is needed. For instance, set $s_k = \varepsilon_0$, where $\varepsilon_0 > 0$ is a small default parameter. Inactive resources do not appear in (6.8), and further we assume that every $k \in R_u$ is active in (6.8).

Now consider a linear rank s with $s_k = 0$. Then the rank-proportional allocation ignores k. The knowledge $\{d_{vk}\}_{v\in N_u}$ is unused though $d_{vk} \neq 0$ for some v and k. Consequently, k does not influence to other ranks, as if k is not presented in (6.8). Hence, there is no differentiation compared with inactive resources. Furthermore, if there are several such k then there is no differentiation between them.

Let us call linear resource rank s *effective* if $s_k > 0 \; \forall k \in R_u$. Rejection of too imbalanced resources is possible when needed: if $0 < s_k < \varepsilon_0$ then k is worse than an inactive resource. Additionally, high s_i does not necessarily lead to high productivity with i, e.g., (a) u requests much i but the neighbors suddenly become stingy, (b) altruistic neighbors provide much i in spite of u's low response, or (c) u can provide much i but the neighbors request less. The capacity should be redistributed proportionally to ranks, including low-ranked resources since their sum share in R_u can be appreciable. In particular, it diminishes the problem of BT-exchange when many nodes are excluded from participation or underutilize capacity [25].

Given $v \in N_u$ we call *v-equation* the corresponding equation in (6.8). Rewrite each v-equation such that its unknowns appear with positive coefficients and consider the following linear system.

$$
\begin{cases}
\displaystyle\sum_{i \in R_v^+} \alpha_{vi} s_i = \sum_{j \in R_v^-} \beta_{vj} s_j, & v \in N_u, \\
s \geq 1.
\end{cases}
\tag{6.11}
$$

Given $i \in R_v^+$ and $j \in R_v^-$ define the ith and jth v-families of solutions:

$$
\begin{aligned}
\mathscr{H}_{vj} &= \left\{ h = \begin{pmatrix} \sigma \\ e_j \end{pmatrix} \; \Big| \; \sum_{k \in R_v^+} \alpha_{vk} \sigma_k = \beta_{vj}, \; \sigma_k \geq 0 \right\}, \\
\mathscr{H}_{vi} &= \left\{ h = \begin{pmatrix} e_i \\ \sigma \end{pmatrix} \; \Big| \; \alpha_{vi} = \sum_{k \in R_v^-} \beta_{vk} \sigma_k, \; \sigma_k \geq 0 \right\}.
\end{aligned}
\tag{6.12}
$$

Hence $\mathscr{H}_{vj}$ contains solutions to the v-equation such that $s_j = 1$ and $s_k = 0$ for $k \in R_v^- \setminus \{j\}$. Similarly, $s \in \mathscr{H}_{vi}$ is such that $s_i = 1$ and $s_k = 0$ for $k \in R_v^+ \setminus \{i\}$.

Further let vectors in $\mathscr{H}_{vj}$ and $\mathscr{H}_{vi}$ be of the same dimension $m = |R_u|$ as solutions to (6.8) or (6.11). Since the v-equation does not necessarily contain all unknowns of the whole system, s_k takes arbitrary non-negative values for $k \in R_u \setminus (R_v^+ \cup R_v^-)$.

Construct the following set of solutions to v-equation.

$$
\mathscr{H}_v = \left\{ s = \sum_{k \in R_v} \lambda_k h^{(k)} \; \Big| \; h^{(k)} \in \mathscr{H}_{vk}, \; \lambda_k \geq 1 \right\}.
\tag{6.13}
$$

We call a solution s to (6.8) a *v-effective* rank if and only if $s \in \mathscr{H}_v$. Any v-effective rank contains one $h^{(k)} \in \mathscr{H}_{vk}$, thus every resource k always "influences" to the rank with a level λ_k, where $\lambda_k \geq 1$ is a lower bound. The latter is uniform for all $k \in R_v$. Let us call $h^{(k)}$ *atomic contribution* of k.

Each v-family in (6.12) defines the set of all atomic contributions—a subset of the general solution to the v-equation.

Theorem 6.3. *The general solution to a v-equation in (6.11) is*

$$
\mathscr{H}_v^0 = \left\{ s = \sum_{k \in R_v^+ \cup R_v^-} \lambda_k h^{(k)} \; \Big| \; h^{(k)} \in \mathscr{H}_{vk}, \; \lambda_k \geq 0 \right\}.
$$

Proof. A linear combination $\sum_k \lambda_k h^{(k)}$ of solutions $h^{(k)}$ is also a solution to the v-equation. Let us prove now that any solution s belongs to $\mathscr{H}_v^0$.

Let $s = s^{(1)}$ be a non-zero solution with $s_{k_1}^{(1)} > 0$. Assume that $k_1 \in R_v^-$ since the case $k_1 \in R_v^+$ is symmetrical. Construct the solution $\tilde{s}^{(1)} = s^{(1)}/s_{k_1}^{(1)}$. Thus $\tilde{s}_{k_1}^{(1)} = 1$ and

$$\sum_{i \in R_v^+} a_{vi} \tilde{s}_i^{(1)} = b_{vk_1} + \sum_{j \in R_v^- \setminus \{k_1\}} b_{vj} \tilde{s}_j^{(1)}.$$

Since the right-hand side is non-negative and fixed, there exist non-negative $\sigma_i^{(1)}$ and $s_i^{(2)}$ such that $\tilde{s}_i^{(1)} = s_i^{(2)} + \sigma_i^{(1)}$ and

$$\sum_{i \in R_v^+} a_{vi} \sigma_i^{(1)} = b_{vk_1}, \quad \sum_{i \in R_v^+} a_{vi} s_i^{(2)} = \sum_{j \in R_v^- \setminus \{k_1\}} b_{vj} \tilde{s}_j^{(1)}.$$

Hence $\tilde{s}^{(1)} = s^{(2)} + h^{(k_1)}$ for some $h^{(k_1)} \in \mathscr{H}_{vk_1}$, and

$$s_k^{(2)} = \begin{cases} \tilde{s}_i^{(1)} - \sigma_i^{(1)}, & \text{if } k = i \in R_v^+ \\ 0, & \text{if } k = k_1 \\ \tilde{s}_j^{(1)}, & \text{if } k = j \in R_v^- \setminus \{k_1\} \end{cases}$$

is a solution to the v-equation.

If $s^{(2)}$ is non-zero then continue, find $s_{k_2}^{(2)} > 0$, let $\tilde{s}^{(2)} = s^{(2)}/s_{k_2}^{(2)}$, and construct $s^{(3)}$ such that $\tilde{s}^{(2)} = s^{(3)} + h^{(k_2)}$.

After $L \le |R_{uv} \cup P_{vu}|$ steps, we yield the required representation

$$\begin{aligned} s = s^{(1)} &= \lambda_1 \left(s^{(2)} + h^{(k_1)} \right) \\ &= \lambda_2 \left(s^{(3)} + h^{(k_2)} \right) + \lambda_1 h^{(k_1)} = \ldots = \sum_{l=1}^{L} \lambda_l h^{(k_l)}. \end{aligned}$$

where $\lambda_1 = s_{k_1}^{(1)}$, $\lambda_2 = s_{k_2}^{(2)} \lambda_1$, ..., $\lambda_l = s_{k_l}^{(l)} \lambda_{l-1}$. $\qquad\square$

In fact, the proof provides a stronger result that allows any of the following two decompositions

$$s = \sum_{j \in R_v^-} \lambda_j h^{(j)}, \quad h^{(j)} \in \mathscr{H}_{vj}, \quad \lambda_j \ge 0;$$

$$s = \sum_{i \in R_v^+} \lambda_i h^{(i)}, \quad h^{(i)} \in \mathscr{H}_{vi}, \quad \lambda_i \ge 0.$$

At each step of the iteration in the proof, select always $k_l \in R_v^-$ (resp. $k_l \in R_v^+$), since for a non-zero solution the right-hand (resp. left-hand) side of the v-equation is positive. Such decompositions of a v-equation solution in the basis $\{\mathscr{H}_{vk}\}$ are not unique and close to the notion of Hilbert basis for linear Diophantine equations [28].

To construct a v-effective rank we must select exactly one $h^{(k)}$ among many candidates in $\mathcal{H}_{vk}$ for each k. In Sect. 6.3.2 we shall show that this selection can be implicitly made using linear programming.

Each v-equation in (6.11) defines balance that supports our economic intuition. The sets R_v^+ and R_v^- are the opposite sides of "the market" between u and v. Variation in s_i for local-like resources $i \in R_v^+$ leads to proportional changes in s_j for external-like resources $j \in R_v^-$ and vice versa. For v-effective ranks, the following theorem provides quantitative estimates of this dependence.

Theorem 6.4 (Resource influence proportions). *Given* $s \in \mathcal{H}_v$, $i \in R_v^+$, $j \in R_v^-$.

1. *A contribution unit of i in s_i gives $h_k^{(i)}$ units to s_k to for all $k \in R_v^-$, where*

$$0 \le h_k^{(i)} \le \frac{\alpha_{vi}}{\beta_{vk}} = \frac{a_{vi} - b_{vi}}{b_{vk} - a_{vk}}.$$

2. *A contribution unit of j in s_j gives $h_k^{(j)}$ units to s_k to for all $k \in R_v^-$, where*

$$0 \le h_k^{(j)} \le \frac{\beta_{vj}}{\alpha_{vk}} = \frac{b_{vj} - a_{vj}}{a_{vk} - b_{vk}}.$$

Proof. Since these two case are symmetrical consider the first one only. From (6.13) we yield the representation

$$s = \sum_{l \in R_v \setminus \{i\}} \lambda_l h^{(l)} + \lambda_i h^{(i)} = s^{(i)} + \lambda_i h^{(i)},$$

where $s^{(i)}$ does not depend on the contribution of i. Application of (6.12) leads to the following component structure of s:

$$s_k = s_k^{(i)} + \lambda_i \begin{cases} 1, & \text{if } k = i, \\ 0, & \text{if } k \in R_v^+ \setminus \{i\}, \\ h_k^{(i)}, & \text{if } k \in R_v^-, \end{cases} \tag{6.14}$$

where $0 \le h_k^{(i)} \le \alpha_{vi}/\beta_{vk}$ for all $k \in R_v^-$ due to $\sum_{k \in R_v^-} \beta_{vk} h_k^{(i)} = \alpha_{vi}$. $\square$

Therefore, certain proportions

$$s_i : \left(s_k\right)_{k \in R_v^-} = 1 : \left(h_k^{(i)}\right)_{k \in R_v^-}, \quad \left(s_k\right)_{k \in R_v^+} : s_j = \left(h_k^{(j)}\right)_{k \in R_v^+} : 1$$

are in ranks for $i \in R_v^+$ and $j \in R_v^-$. Similarly to normalized ranks, the parameters λ_i and λ_j are relative. The mandatory influence of every k is a key point of our model. The rank normalization does not require every resource to influence the rank, hence there can be $s_k = 0$ for some k, which is impossible in our model.

The model differentiates the influence of resources in u's exchange with v. Every resource k has own set $\mathcal{H}_{vk}$ of atomic contributions. Each $h^{(k)}$ forms of an influence proportion of k to the opposite side of the resource market. Depending on the selection of $h^{(k)}$ the resource ranks of the opposite side grow differently. The sum growth, however, is preserved proportional either to α_{vk} or β_{vk}.

The term "effective" emphasizes that such ranks proportionally affect the decision-making processes for every resource, see Table 6.1 above. Qualitative characterization of incentives in the exchange between u and its neighbors is summarized in Table 6.3. If v changes its provision then u reacts in response and vice versa.

6.3.2 Reduction to Linear Programming

Let us analyze the role of the constraint $s_k \geq 1$ in (6.11). The following theorem shows that the constraint eliminates many ineffective ranks compared to a single v-equation.

Theorem 6.5. *Any v-effective rank $s \in \mathcal{H}_v$ is a solution to the linear system*

$$\begin{cases} \sum_{i \in R_v^+} \alpha_{vi} s_i = \sum_{j \in R_v^-} \beta_{vj} s_j, \\ s_k \geq 1, \quad k \in R_v^+ \cup R_v^-. \end{cases} \tag{6.15}$$

Proof. Let $s \in \mathcal{H}_v$, where $\mathcal{H}_v$ is defined in (6.13). Accordingly, s is a solution to the v-equation in (6.15), and

$$s = \sum_{i \in R_v^+} \lambda_i h^{(i)} + \sum_{j \in R_v^-} \lambda_j h^{(j)},$$

where $h^{(i)} \in \mathcal{H}_{vi}$ and $h^{(j)} \in \mathcal{H}_{vj}$. Take arbitrary $i \in R_v^+$. The definition of $\mathcal{H}_{vi}$ in (6.12) leads to $h_i^{(i)} = 1$, $h_k^{(i)} = 0$ for $k \in R_v^+ \setminus \{i\}$, and

$$s_i = \lambda_i + \sum_{j \in R_v^-} \lambda_j h_i^{(j)},$$

Since $\lambda \geq 1$ we conclude that $s_i \geq 1$. Symmetrically, taking arbitrary $j \in R_v^-$ we prove that $s_j \geq 1$. $\qquad \square$

The converse is not true. Although any solution to (6.15) can be presented as

$$s = \sum_{k \in R_v^+ \cup R_v^-} \lambda_k h^{(k)},$$

Table 6.3 Incentives by effective ranks: trends in terms of growth ($\nearrow$) and fall ($\searrow$)

Deviation in the exchange balance between u and its neighbor v			
u consumes less	u provides more	u provides less	u consumes more
$a_{vi} \searrow$ for a given $i \in R_v^+$	$b_{vj} \nearrow$ for a given $j \in R_v^-$	$b_{vj} \searrow$ for a given $j \in R_v^-$	$a_{vi} \nearrow$ for a given $i \in R_v^+$
Reaction of effective ranks			
$s_i \nearrow$, $r_v^+ \nearrow$ $r_v^- \searrow$, $s_j \searrow$ for some $j \in R_v^-$	$s_i \nearrow$ for some $i \in R_v^+$, $r_v^+ \nearrow$ $r_v^- \nearrow$, $s_j \nearrow$	$s_i \searrow$ for some $i \in R_v^+$, $r_v^+ \nearrow$ $s_j \nearrow$, $r_v^- \nearrow$	$s_i \searrow$, $r_v^+ \nearrow$ $r_v^- \searrow$, $s_j \nearrow$ for some $j \in R_v^-$
Result in the consumption process: the order in $\mathscr{C} = \{(k,w) \mid k \in R, w \in N_u\}$ is by (s_k, r_v^+)			
More requests for i to $w \in N_u$. More requests to v			Less requests for i to $w \in N_u$. Less requests to v
Compensating the i-loss of v among other neighbors.	Activating v and other neighbors for provision.	Deactivating v and its resources.	Compensating the i-growth of v among other neighbors.
Attempting $a_{wi} \nearrow$	Attempting $a_{vi} \nearrow$	Attempting $a_{vi} \searrow$	Attempting $a_{wi} \searrow$
Result in the provision process: the order in $\mathscr{P} = \{(w,k) \mid w \in N_u, k \in R\}$ is by (r_v^-, s_k)			
Decrease provision to v. Give lower priority to j	Increase provision to v. Give higher priority to j	Decrease provision to v. Give lower priority to j	Increase provision to v. Give higher priority to j
Penalizing v and its resources.	Stabilizing u's generosity.	Stabilizing u's selfishness.	Rewarding v and its resources.
Leading to $b_{vj} \searrow$	Leading to $b_{wj} \searrow$	Leading to $b_{vj} \nearrow$	Leading to $b_{vj} \nearrow$

there can be $0 < \lambda_k < 1$, see Theorem 6.3 above. Therefore, some solutions are ranks to which the resources influence less than the uniform bound $\lambda_k \geq 1$ dictates in (6.13).

Nevertheless, the accuracy is within multiplication to a constant. The following theorem states that system (6.11) is a proper model of the set of all effective linear resource ranks.

Theorem 6.6. *There exists a v-efficient rank $s = Cs'$ for any solution s' to (6.15) and a constant $C \geq 1$.*

Proof. It can be easily proved that if s' is a solution to (6.15), then $s' = \sum_k \lambda'_k h^{(k)}$ for some $\lambda'_k > 0$. If there is $\lambda'_k < 1$ then let $\lambda_* = \min\{\lambda'_k \mid \lambda'_k < 1\}$ and $C = 1/\lambda_*$. Otherwise, let $C = 1$. Hence

$$s = Cs' = \sum_{k \in R_v^+ \cup R_v^-} \frac{\lambda'_k}{\lambda_*} h^{(k)},$$

where $\lambda_k = \lambda'_k/\lambda_* \geq 1$.   □

As a result, there is no difference whether to use s' or s when computing the normalized rank. No further reduction of the search space is needed in (6.15).

Taking into account all neighbors of u, define a rank *effective* if it is *v*-effective for all $v \in N_u$. The set of all effective ranks is $\mathcal{H} \subseteq \bigcap_{v \in N_u} \mathcal{H}_v$. Note that for different v the same rank has different representations in (6.13). All effective ranks are among solutions to (6.11).

The definition of effective ranks is not oriented to computation. We now reduce it to a linear programming (LP) problem, where an optimization criterion (cost function) is used to find the best effective rank in the possibly infinite set of solutions to linear system (6.11).

Depending on a particular instance of the resource sharing problem some ranks account u's interests better. We can bound s_i since if (6.15) is solvable then there are solutions with arbitrary high s_k. Note that normalization preserves the proportions despite of how high the absolute values are. The following generic linear cost function provides the bound if $c > 0$:

$$\sum_{k \in R} c_k s_k \to \min. \tag{6.16}$$

The weights c_k allow additional problem-aware resource differentiation; the higher c_k the lower s_k.

Based on the point of view of market pricing, if many nodes provide resource $i \in R^+$ then its rank decreases. If many nodes consume resource $j \in R^-$ then its rank increases. In this case, c_i counts all nodes $v \in N_u$ such that $i \in R_v^+$ and c_j counts all v such that $j \in R_v^-$. These counters appear implicitly in node ranks:

$$\sum_{v \in N_u} r_v^+ = \sum_{v \in N_u} \sum_{i \in R_v^+} s_i = \sum_{i \in R} c_i s_i, \quad \sum_{v \in N_u} r_v^- = \sum_{v \in N_u} \sum_{j \in R_v^-} s_j = \sum_{j \in R} c_j s_j.$$

Table 6.4 Linear optimization criteria

Ref.	Cost	Interpretation
(C)	$\sum_{v \in N_u} r_v^+$	Higher priority in u's consumption to rare external resources since few nodes provides them.
(P)	$\sum_{v \in N_u} r_v^-$	Higher priority in u's provision to neighbors that consume resources of small variety.
(C+P)	$\sum_{v \in N_u} (r_v^+ + r_v^-)$	A trade-off in u's consumption/provision.
(C–P)	$\sum_{v \in N_u} (r_v^+ - r_v^-)$	A respectability threshold in u's consumption/provision: u minimizes its consumption and maximizes its provision.
(P–C)	$\sum_{v \in N_u} (r_v^- - r_v^+)$	A selfishness threshold in u's consumption/provision: u minimizes its provision and maximizes its consumption.

Table 6.4 shows possible linear cost minimization criteria based on these counters. Criterion (C+P) states a trade-off since neighbor consumption and provision ranks are bounded equally. Criteria (C–P) and (P–C) state thresholds of moving from this trade-off to the respectable and selfish direction, correspondingly. Obviously, they can lead to an unbounded optimization problem.

As in many optimization models for P2P resource sharing, this ranking model allows a generic function $f(r_v^+, r_v^-)$ that specifies the relation between r_v^+ and r_v^-. For instance, u can prioritize nodes $v \in N_u$ using mutual trust among them and use quadratic cost functions, similarly to [6]:

$$\sum_{v \in N_u} c_v (r_v^+ - r_v^-)^2 \to \min .$$

For every neighbor the consumption and provision ranks are equalized. The model, however, becomes nonlinear, hence increasing its computational complexity.

Instead of the use of homogenous linear constraints in (6.17) we can consider the equivalent linear programming (LP) problem:

$$\begin{cases} f = \sum_{k \in R_u} d_k \sigma_k \to \min \\ \sum_{k \in R_v} d_{vk} \sigma_k = -d_v, \quad v \in N_u, \\ \sigma_k \geq 0 \ \forall k \in R_u. \end{cases} \tag{6.17}$$

Any feasible solution σ to (6.17) yields a solution $s = \sigma + 1$ to (6.11) and vise versa.

The cost function f in (6.17) prioritizes resources by their surplus, and resources with high d_k will be of low rank. It agrees with Definition 6.4, and the optimization aims at better approximation of the rank monotony property than Theorem 6.2 for the general case of linear resource ranks.

One can construct the dual problem to (6.17), where the dual variables $-\rho_v$ take any real value.

$$\begin{cases} \phi = \sum_{v \in N_u} d_v \rho_v \to \min \\ \sum_{v \in N_u} d_{vk} \rho_v \le d_k, \quad k \in R_u. \end{cases} \quad (6.18)$$

According to the duality, $\rho_v = \frac{\partial f}{\partial d_v}$. It means that ρ_v arranges nodes by their influence to the resource-aware cost. Consequently, they can be used to compute node ranks, alternatively to (6.7) in Sect. 6.2.3.

6.3.3 Rank Existence

System (6.11) is not always solvable due to the constraint $s \ge 1$. Recall that it requires for every k to influence the rank. The balance equations are less restrictive since the number of nodes (the number of balance equations) is typically essentially less than the number of resources (the number of variables). LP methods require an initial solution for efficient computations. It concerns the more general problem than the solvability asking how we can construct a solution.

Consider the assumption (the left square bracket stands for logical OR)

$$\left[\begin{array}{l} |V| \le \bigcup_{v \in V} R_v^+ \quad \forall V \subseteq N_u, \text{ (X)} \\ |V| \le \bigcup_{v \in V} R_v^- \quad \forall V \subseteq N_u. \text{ (Y)} \end{array} \right. \quad (6.19)$$

Bound (6.19.X) states that the number of distinct resources u consumes is not less than the number of their providers. Bound (6.19.Y) is similar but for the number of distinct resources u provides to their consumers. In particular, if u provides (consumes) at least one distinct resource for any neighbor, then (6.19) holds. This case is typical in practice since nodes differentiate resources depending on neighbors. That is, (6.19) is true for a wide class of P2P resource sharing systems.

Let R_u^{ext}, R_u^{loc}, and R_u^{trn} be the distinct sets of external, local, and transit resources for node u, respectively. Then $R_u^+ \subseteq R_u^{\text{ext}} \cup R_u^{\text{trn}}$ and $R_u^- \subseteq R_u^{\text{loc}} \cup R_u^{\text{trn}}$. Rewrite balance equations (6.8) separating resource ranks $s = (x, y, z)$ to external $(x, i \in R_u^{\text{ext}})$, local $(y, j \in R_u^{\text{loc}})$, and transit $(z, k \in R_u^{\text{trn}})$ resources:

$$\begin{array}{l} \sum_{i \in R_u^{\text{ext}}} a_{vi} x_i + \sum_{k \in R_u^{\text{trn}}} a_{vk} z_k \\ = \sum_{j \in R_u^{\text{loc}}} b_{vj} y_j + \sum_{k \in R_u^{\text{trn}}} b_{vk} z_k, \quad v \in N_u. \end{array} \quad (6.20)$$

The following theorem shows that (6.19) is sufficient for existence of an efficient rank. A solution can be constructed explicitly, see the proof.

Theorem 6.7. *If (6.19) holds then (6.20) has a solution $s = (x, y, z)$ such that $x, y \ge 1$ and $z = 1$.*

Proof. Consider the case without transit resources. Then (6.20) is reduced to

$$\sum_{i\in R_u^{\text{ext}}} a_{vi}x_i = \sum_{j\in R_u^{\text{loc}}} b_{vj}y_j, \quad v\in N_u. \tag{6.21}$$

The unknowns x and y are disjunctive. We also assume that

$$\sum_{i\in R_u^{\text{ext}}} a_{vi}>0 \text{ and } \sum_{j\in R_u^{\text{loc}}} b_{vj}>0 \text{ for any } v\in N_u,$$

otherwise $y_j = 0$, $x_i = 0$, or the ranks are not constrained.

Let (6.19.X) be true. Apply Hall's theorem (the marriage theorem) to find a distinct representative $x_{i(v)}$ for each v-equation of (6.21). Set all other x_i to 1, splitting (6.21) to left-hand side independent v-equations

$$a_{vi(v)}x_{i(v)} = \sum_{j\in R_u^{\text{loc}}} b_{vj}y_j - \sum_{i\in R_u^{\text{ext}}} a_{vi},$$

which can be solved with $x\geq 1$ taking large enough $y\geq 1$. Any such y that makes the right-hand side greater or equal $a_{vi(v)}$ $\forall v\in N_u$ is appropriate.

Let (6.19.Y) be true. The case is symmetrical to the previous one; distinct representatives are selected among y_j and a solution can be found with $x,y\geq 1$.

Return to the general case of (6.20). Let (6.19.X) be true. (The case of (6.19.Y) is symmetrical.) Select distinct representatives $x_{i(v)}$ as above and set $z = 1$. We obtain the system

$$a_{vi(v)}x_{i_v} = \sum_{j\in R_u^{\text{loc}}} b_{vj}y_j + \sum_{k\in R_u^{\text{trn}}} b_{vk} - \sum_{k\in R_u^{\text{trn}}} a_{vk} - \sum_{i\in R_u^{\text{ext}}} a_{vi}, \quad v\in N_u.$$

Consider

$$C_v = \sum_{k\in R_u^{\text{trn}}} b_{vk} - \sum_{k\in R_u^{\text{trn}}} a_{vk} - \sum_{i\in R_u^{\text{ext}}} a_{vi} = \text{const.}$$

If $C_v < 0$ then taking large enough $y\geq 1$ results in a positive value in the right-hand side. The value can be made bigger than $a_{vi(v)}$, leading to $x_{vi_v}\geq 1$.

Summarizing the proof, a solution to (6.20) with $x,y\geq 1$ and $z = 1$ is constructed explicitly.   ☐

Consequently, (6.19) results in the solvability of (6.11). When (6.11) is unsolvable then there are too few distinct resources compared with the number of their providers or consumers. It means that some neighbors become indistinguishable from u's point of view. In this case, u can, for example, provide finer granularity in the resource distribution among neighbors, aggregate some neighbors into a bigger entity, or remove some neighbors from N_u.

6.4 Comparison with Other Resource Exchange Models

Design of incentives for P2P resource sharing has attracted much attention recently. In this section we overview known models for P2P exchange economies and compare the linear ranking model with them. We show that some approaches are reduced to the linear model. Also, there exist approaches for rank-based resource sharing that allows incorporation of the linear model.

6.4.1 Allocation of Common Resource

Each node u has common local resource to contribute to the system, e.g., storage space, CPU cycles, or bandwidth. As an example consider the bandwidth allocation problem; u distributes its upload bandwidth B among competing neighbors. Each neighbor v provides u with download bandwidth $a_v > 0$ and consumes upload bandwidth $b_v > 0$. There are the only local resource and the only external resource for any neighbor (one-to-one mappings $R_u^- \leftrightarrow N_u$ and $R_u^+ \leftrightarrow N_u$).

For upload bandwidth allocation u decides $r_v^- = f(a_v, b_v) > 0$ such that $\sum_{v \in N_u} r_v^- = 1$, where a_v and b_v are observed in the previous round. Then u proportionally assigns bandwidth $b_v' \sim r_v^-$ to v for the next round. The linear ranking model reduces the problem to the system of independent equations in unknowns $s = (x, y)$:

$$a_v x_v = b_v y_v, \quad x_v, y_v \geq 1, \quad v \in N_u,$$

where x_v is rank of v's provision and y_v is rank of v's consumption.

Let $a = \sum_{w \in N_w} a_w$ and $a = \sum_{w \in N_w} b_w$. A feasible solution is $x_v' = b_v$, $y_v' = a_v$, $v \in N_u$. Computing normalized ranks $s = (x, y)$ and using (6.7) yield

$$x_v = \frac{b_v}{a+b}, \quad y_v = \frac{a_v}{a+b}, \quad v \in N_u; \quad S^+ = \frac{b}{a+b}, \quad S^- = \frac{a}{a+b}.$$

Then the rank $r_v^- = a_v/a$ defines the upload bandwidth allocation proportional to v's contribution. The solution coincides with proportional sharing [16, 32], see also the discussion in Sect. 6.2.2.

Consider optimization criteria from Table 6.4 (on page 152). Criteria (C), (P), and (C+P) all lead to the following solution:

$$x_v' = \max\{\frac{b_v}{a_v}, 1\}, \quad y_v' = \max\{\frac{a_v}{b_v}, 1\}, \quad v \in N_u,$$

Criterion (C–P) defines a finite solution if and only if $b_v \geq a_v \ \forall v \in N_u$, and the best rank is

$$x_v' = \frac{b_v}{a_v}, \quad y_v' = 1.$$

Similarly, criterion (P–C) is valid if and only if $a_v \geq b_v \ \forall v \in N_u$, and

$$y'_v = \frac{a_v}{b_v}, \quad x'_v = 1.$$

Note that the parameters $\gamma_{uv} = a_v/b_v$ and $\gamma_{vu} = b_v/a_v$ are known as exchange ratios in bilateral models [4]. Let $\gamma_u^+ = \sum_v \gamma_{vu}$ and $\gamma_u^- = \sum_v \gamma_{uv}$.

If u is interested in local resource allocation only, then the model is simplified to $a_v x = b_v y_v$ for $v \in N_u$, where x is common for all neighbors and y_v differentiates only the provision of local resources. Similarly, if u is interested in external resource allocation only, then the model is $a_v x_v = b_v y$ for $v \in N_u$. The solution is

$$x' = 1, \quad y'_v = \frac{a_v}{b_v}, \quad r_v^- = \frac{a_v/b_v}{\sum\limits_{w \in N_u} a_w/b_w} = \gamma_{uv}/\gamma_u^-, \tag{6.22}$$

for local resource provision and, symmetrically,

$$x'_v = \frac{b_v}{a_v}, \quad y' = 1, \quad r_v^+ = \frac{b_v/a_v}{\sum\limits_{w \in N_u} b_w/a_w} = \gamma_{vu}/\gamma_u^+, \tag{6.23}$$

for external resource consumption.

Furthermore, if u is uninterested in differentiating b_v or a_v in the local or external allocation case, respectively. Then

$$r_v^- = \frac{a_v}{\sum\limits_{w \in N_u} a_w} = \frac{a_v}{a}, \quad r_v^+ = \frac{b_v}{\sum\limits_{w \in N_u} b_w} = \frac{b_v}{b}. \tag{6.24}$$

This solution coincides with (6.4) and (6.5) for $a = b$, see Sect. 6.2.2.

Yang and Veciana [35] studied symmetric peerwise proportional fairness using a global linear model. The system aims at resource allocation that follows the balance at time instance t,

$$b_v(t;u) = \frac{a_v(t;u)}{\sum\limits_{w \in N_u} a_w(t;u)} b(t;u) = \frac{a_v(t;u)}{a(t;u)} b(t;u), \quad v \in N_u,$$

where $a(t;u)$ and $b(t;u)$ is u's total download and upload bandwidth at t. The symmetric assumption is $a_v(t;u) = b_u(t;v)$ and $b_v(t;u) = a_u(t;v)$. Considering all nodes $u \in N$, the balance formally defines a global equation system in unknowns $x_{uv}(t) = b_v(t;u)$, and the fair resource allocation is its solution. In practice, exact solution of this system seems infeasible. Instead, u tries to reach the fair resource allocation by providing v with the upload bandwidth $b_v(u) = r_v^- b(u)$ for $r_v^- = a_v/a(u)$, hence directly following (6.24).

Note that [35] also formally found the importance of strictly positive allocations $x_{uv} > 0$. The linear ranking model explicitly requires strictly positive ranks, see the definition and interpretation of effective ranks in Sect. 6.3.

Ma et al. [21] analyzed bandwidth allocation with the game-theoretic approach. Neighbors bid for u's bandwidth. Each v suggests b_v to u, and u considers it the upper bound of the bandwidth assignment. The upload bandwidth allocation is proportional to a_v and b_v. The model suffers when neighbors misrepresent b_v strategically, see [34] and references therein.

Yan et al. [34] assumed that u computes ranks $r_v^- = f(a_v, b_v)$ knowing f in advance. The latter is increasing in a_v and decreasing in b_v. A node improves its ranking by contribution to the P2P system or deteriorates its ranking by provision. An utility function and a corresponding social welfare function are associated with each neighbor, where the ranks are parameters. Maximizing the minimum social welfare every v could obtain leads to the unique optimal allocation of the max-min fairness. The upload bandwidth that u allocates to v is a function of the ranks r_v^-. The linear ranking model can be incorporated into this model to provide the latter with ranks r_v^- in place of the function $f(a_v, b_v)$.

Many authors suggest a reputation system for differentiating nodes, see [22] and references therein. A node reputation statistically aggregates the past behavior of the node. Opinions from many other nodes are collected, and such indirectly observed knowledge leads to non-local models with corresponding scalability and trust issues. The linear ranking model can be applied with a reputation system, where the latter evaluates the input parameters a_v and b_v.

The neighbor selection problem is an instance of external resource allocation. Each node decides how actively to consume from each of its neighbors. Adler et al. [1] applied mathematical programming to study optimal rate selection for downloading from neighbors when the cost function is concave. The optimal rate is proportional to the maximum rate. In linear ranking model, it defines external resource allocation where b_v refers the available maximum rate. Consequently, u's download rates are proportional to r_v^+ for $v \in N_u$.

Habib and Chuang [11] considered the peer selection problem for P2P streaming. They considered a P2P media streaming system consisting of rational nodes who choose their contribution level in order to maximize their individual utility. The individual contribution level is converted into a score, which in turn is mapped into a percentile rank that determines the node rank (among other nodes in the system). Node selection depends on the rank ordering of the requesters and candidate suppliers. Contributable nodes are rewarded with high quality streaming sessions. The authors contributed a model of mapping node ranks to expected quality based on the rank-order tournament theory. Consequently, the linear ranks can be straightforwardly integrated into the streaming system in place of the score mechanism referred originally.

6.4.2   BT Systems

BitTorrent-like P2P file sharing systems implement bilateral tit-for-tat (TFT) strategies aiming in fair bandwidth allocation [5, 8, 14, 16, 19, 29, 32]. We show that they can be described with a particular instance of the linear ranking model for local resource provision.

BitTorrent TFT [8]

Each node u forms a set of the n_1 best neighbors v with the highest download rates a_v (v's provision history). Besides, u selects random n_2 "optimistic" neighbors (unchoking); the aim is in discovering new neighbors that would offer good downloads. During the recent time slot u uploads only to these $n_1 + n_2$ neighbors in a round-robin fashion. For downloading no specific selection strategy is used: u exploits all neighbors $v \in N_u$ consuming as much as possible.

It is an instance of model (6.24) for local resource provision regardless consumption history. Note that u needs the ranks for selecting the best neighbors to upload (ranking and then selecting nodes from the top). After the selection, u treats the selected neighbors equally (in a round-robin fashion), allocating the same upload bandwidth.

BitTorrent as Auction [16, 32]

A proportional share mechanism replaces the original BitTorrent unchoking algorithm. Each node u allocates its total upload bandwidth b to all its neighbors in proportion to the download rates from them,

$$b'_v = \frac{a_v}{\sum\limits_{w \in N_u} a_w} b = \frac{a_v}{a} b, \quad v \in N_u.$$

Wu and Zhang [32] showed that this mechanism allows converging quickly to a market equilibrium. Levin et al. [16] introduced the algorithms and implementation as well as showed with experiments that the mechanism improves system-wide performance. It is again an instance of (6.24) with $r_v^- = a_v/a$, which uses local resource provision regardless consumption history.

Block-Based TFT [5, 14, 19]

The strategy is similar to BitTorrent TFT except that u always checks the constraint

$$b_v - \gamma a_v \le \overline{d} \tag{6.25}$$

before uploading a data block to v, where γ and $\overline{d}$ are system parameters. If $b_v - \gamma a_v > \overline{d}$ then u postpones uploading to v. Bharambe et al. [5] proposed a scheme with $\gamma = 1$ in (6.25), i.e., $\overline{d}$ defines a threshold for the deficit $-d_v$. Liao et al. [19] suggested $\gamma \ge 1$ as a design parameter; when γ increases the overall system performance improves by sacrificing fairness towards high-capacity nodes.

It is an instance of model (6.22) for local resource allocation based on directly observable provision and consumption histories. Let u select n top neighbors with highest a_v at the beginning of each round, forming N_u^{best} as in the original BitTorrent TFT. During each round, a_v and b_v change due to activity of the neighbors. Ranks $r_v^- = \gamma_{uv}/\gamma_u^-$ in (6.22) differentiate the n top neighbors in accordance with (6.25), and u uploads a data block to $v \in N_u^{best}$ if

$$r_v^- \geq \left(1 - \frac{\overline{d}}{b_v}\right) / \left(\gamma \sum_{w \in N_u} a_w/b_w\right) = \frac{1 - \overline{d}/b_v}{\gamma \gamma_u^-}.$$

The top neighbors are still served uniformly: each $v \in N_u^{best}$ that satisfies (6.25) receives equal upload bandwidth from u. Ranks (6.22) allow a more intelligent strategy with proportional bandwidth allocation, operationally adapting to recent behavior of neighbors within each round.

FairTorrent [29]

Each node u uses a deficit counter $a_v - b_v$ for $v \in N_u$, a particular case of d_v in (6.8). The neighbors are sorted by their deficit counters. The top neighbor (highest deficit) is selected in every provision action. This sorting is similar to ranking by (6.22),

$$r_v^- = \gamma_{uv}/\gamma_u^- = (1 + d_v/b_v)/\gamma_u^-.$$

6.4.3 Plant-Like Systems

Let u consider the system as a plant from which u consumes $n = |R_u^+|$ resources $(R_u^+ = R_u^{ext})$ and to which u provides $m = |R_u^-|$ resources $(R_u^- = R_u^{loc})$. Direct differentiation is used for resources only, regardless through which nodes the access is. This kind of abstraction comes from control-theoretic approach [30].

In the linear ranking model the plant can be presented as the only neighbor (aggregate entity), and (6.8) is reduced to a single equation

$$\sum_{i \in R_u^+} a_i x_i = \sum_{j \in R_u^-} b_j y_j.$$

Setting all $\lambda_k = 1$ in (6.13) gives the following solution

$$s = \binom{x}{y} = \sum_{i \in R_u^+} h^{(i)} + \sum_{j \in R_u^-} h^{(j)} = 1 + \left(\frac{\sum_{j \in R_u^-} \sigma^{(j)}}{\sum_{i \in R_u^+} \sigma^{(i)}}\right).$$

Assume that $a_i = a$ for $i \in R_u^+$ and $b_j = b$ for $j \in R_u^-$. The symmetry yields

$$\sigma_i^{(j)} = \frac{b}{an}, \ \sigma_j^{(i)} = \frac{a}{bm}, \ \sum_{j \in R^-} \sigma_i^{(j)} = \frac{bm}{an}, \ \sum_{i \in R^+} \sigma_j^{(i)} = \frac{an}{bm},$$

and the ranks are

$$x_i = \frac{b}{(a+b)n}, \ y_j = \frac{a}{(a+b)m}, \ r^+ = \frac{b}{a+b}, \ r^- = \frac{a}{a+b}.$$

They differentiate between external ($i \in P_u$) and local ($j \in R_u$) resources. There is no differentiation of resources inside each group. This model allows u to control the sizes m and n of its two resource sets.

When a_i and b_j are non-uniform then the effective ranks provide further resource differentiation. In particular, the rank ($\sigma_i^{(j)} = \frac{b_j}{a_i n}, \ \sigma_j^{(i)} = \frac{a_i}{b_j m}$)

$$x_i \sim \frac{1}{a_i n} \sum_{j \in R_u^-} b_j = \frac{B}{a_i n}, \ y_j \sim \frac{1}{b_j m} \sum_{i \in R_u^+} a_i = \frac{A}{b_j m}$$

provides proportional allocation strategies, where $A = \sum_{i \in R_u^+} a_i$ and $B = \sum_{j \in R_u^-} b_j$.

Now consider the plant divided into several parts, each is accessed through a separate node. Introducing more neighbors constrains the rank stronger, and it results in more sophisticated differentiation. Instead of a single equation, we deal with the following system

$$\sum_{i \in R_v^+} a_{vi} x_i = \sum_{j \in R_v^-} b_{vj} y_j, \ v \in N_u.$$

Effective ranking follows two proportions: one is to the consumption and provision of different resources by a node v and another is to the total consumption and provision of different nodes. The v-equation defines the first proportion and the equation system as a whole captures the second one.

In the previous case (plant as a single aggregate neighbor), $x_i \sim B$ and $y_j \sim A$. Generalization is that several aggregate neighbors appear. Then

$$x_i \sim \sum_{v \in N_u} c_v^+ B_v, \ y_j \sim \sum_{v \in N_u} c_v^- A_v, \ A_v = \sum_{i \in R_v^+} a_{vi}, \ B_v = \sum_{j \in R_v^-} b_{vj},$$

i.e., proportionally to a linear combination of the overall interests A_v and B_v. The c_v^+ and c_v^- are v's shares in this proportion.

6.4.4 Network Games and Market Pricing

Considerable work has been appeared recently on applying the network exchange theory and graphical games for network resource sharing, see [4, 21, 34, 36] and references therein. The game-theoretic models are mostly focused on defining node utility functions $U(a(u),b(u))$, where consumption $a(u)$ and provision $b(u)$ define u's strategy. The sum of these utilities is the social utility to be maximized. Less attention is given to constraints that define feasible strategies. The constraints typically reflect node capacity bounds only. In contrast, constraints of the exchange balance is core of the linear ranking model, leading to essentially reduced search space for optimal strategies.

The work of Aperjis et al. [3, 4] is closest to the linear ranking approach. They considered the bilateral exchange balance where nodes charge each other for resources in a common monetary unit with settlement-free transactions. Node u charges node v a price $p_{uv} > 0$ per unit rate. Then each node u faces one exchange balance constraint for each neighbor v,

$$p_{vu} \sum_{k \in R_u^+} a_{vk}(u) = p_{uv} \sum_{k \in R_u^-} b_{vk}(u), \quad v \in N_u,$$

where $a_{vk}(u) = b_{uk}(v)$ is the rate at which u downloads resource k from v (or the rate at which v uploads resource k to u).

The model is close to linear ranking without transit ($R_u^+ = R_u^{\text{ext}}$, $R_u^- = R_u^{\text{loc}}$). In (6.8) and (6.11) the ranks $s_k = s_k(u)$ can be interpreted as the price of k at u. When $s_k = p_{vu} \ \forall k \in R_v^+$ and $s_k = p_{uv} \ \forall k \in R_v^-$ the models describe the same exchange balance. Note that the linear ranking model does not require globally synchronized prices p_{uv} and p_{vu}; different nodes u deal with own prices $s_k = s_k(u)$.

Aperjis et al. [4] also considered multilateral exchange balance. The system globally maintains one price per node. Let $p_v > 0$ be the price of node v. Then the balance is a single exchange constraint,

$$\sum_{v \in N_u} p_v \sum_{k \in R_u^+} a_{vk}(u) = p_u \sum_{v \in N_u} \sum_{k \in R_u^-} b_{vk}(u).$$

As in the previous model, if it is possible unambiguously set $s_k = p_v \ \forall k \in R_v^+$ and $s_k = p_u \ \forall k \in R_v^-$ then the linear ranking model states the same balance.

Aperjis et al. [3] considered the model with $p_k > 0$ (one price per resource) and the exchange constraint

$$\sum_{v \in N_u} \sum_{k \in R_u^+} a_{vk}(u)p_k \leq \sum_{v \in N_u} \sum_{k \in R_u^-} b_{vk}(u)p_k$$

It can be treated an aggregation model. Any global solution to the linear ranking model is a solution to the aggregation model.

The above game-theoretic models are focused on characterizing global equilibrium qualitatively, e.g., the existence of a point (a,b,p) such that maximizes the utility of each node u subject to its exchange balance constraints. In contrast, the linear ranking model aims at computing optimal ranks s for locally observable u's exchange (a,b).

6.5   Summary

We have discussed an approach for local P2P ranking and a corresponding analytic
model framework. In the approach, a node differentiates its neighbors in accordance
with their directly observed provision and consumption. Each node computes
ranks operationally, with no assumption on long-term invariants in strategies of
other nodes. Local rank computations are performed independently on the rank
computations at other nodes.

Multilateral exchange models take into account a system of complex reciprocal
relations between nodes and resources. If the balance for resource k in exchange
$u \leftrightarrow v$ has changed then it changes the strategies of other nodes for k and for some
other resources related to u and v. The presented linear ranking model describes
in (6.8) all exchange chains that pass through u. The model is a linear system of
balance equations. Its solutions are ranks that reflect imbalances in the resource
consumption and provision. The rank computation is reduced to an LP problem,
which allows efficient computation.

The ranks reflect shifts in load balance when nodes vary their interests in
resources. Balance shifts are mapped to effective ranks that provide incentives
for cooperation (see Table 6.3 on page 150). These incentives can be improved
depending on a given resource sharing problem. Various optimization criteria can
extend the model for selecting the best rank among effective ranks. The solvability
is preserved for typical P2P configurations.

This ranking model explicitly introduces transit resources into P2P exchange.
They are applicable if u has no local resources interested to its neighbors, diminishing the gap between bilateral and multilateral exchanges. In the model, the same
transit resource k simultaneously plays the role of local resource for some exchanges
$u \leftrightarrow v$ ($k \in R_v^-$) and the role of external resource for other exchanges $u \leftrightarrow w$ ($k \in R_w^+$).
The sum rank s_k aggregates this composite influence of k from all nodes with the
influence from other resources.

Many known exchange models treat consumption and provision symmetrically.
In our notation it means that $a_{vk}(t;u) = b_{uk}(t;v)$ and $b_{vk}(t;u) = a_{uk}(t;v)$. In
P2P networks the symmetry assumption is often violated due to network loss,
communication latency, and misrepresentation from other nodes. In local ranking,
each u measures $a_{vk}(t;u)$ and $b_{vk}(t;u)$ independently on other nodes.

As in many deficit-based models, we utilize both consumption a and provision b
as input parameters for ranking. Importantly that this model provides both consumption (r^+) and provision (r^-) ranks for neighbors. This explicit distinction of nodes
in the consumption and provision strategies leads to more flexible control when
variation of r_v^- does not immediately result in change of r_v^+ and vice versa.

References

1. Adler, M., Kumar, R., Ross, K.W., Rubenstein, D., Suel, T., Yao, D.D.: Optimal peer selection for P2P downloading and streaming. In: Proceedings of IEEE INFOCOM'05, pp. 1538–1549. IEEE (2005)
2. Aperjis, C., Johari, R.: Designing aggregation mechanisms for reputation systems in online marketplaces. SIGecom Exch. **9**, 3:1–3:4 (2010). doi: http://doi.acm.org/10.1145/1980534. 1980537
3. Aperjis, C., Freedman, M.J., Johari, R.: Peer-assisted content distribution with prices. In: Proceedings of ACM SIGCOMM Conference on emerging Networking Experiments and Technologies (CoNext '08). ACM, New York (2008). doi: http://doi.acm.org/10.1145/1544012. 1544029
4. Aperjis, C., Freedman, M.J., Johari, R.: Bilateral and multilateral exchanges for peer-assisted content distribution. IEEE/ACM Trans. Netw. **19**(5), pp. 1290–1303 (2011). doi: http://dx.doi. org/10.1109/TNET.2011.2114898
5. Bharambe, A.R., Herley, C., Padmanabhan, V.N.: Analyzing and improving a BitTorrent network's performance mechanisms. In: Proceedings of IEEE INFOCOM'06, pp. 2884–2895. IEEE (2006)
6. Bickson, D., Malkhi, D.: A unifying framework of rating users and data items in peer-to-peer and social networks. Peer-to-Peer Netw. Appl. **1**, 93–103 (2008)
7. Buchegger, S., Mundinger, J., Boudec, J.Y.L.: Reputation systems for self-organized networks. IEEE Tech. Soc. Mag. **27**, 41–47 (2008)
8. Cohen, B.: Incentives build robustness in BitTorrent. In: Proceedings of 1st Workshop on Economics of Peer-to-Peer Systems (2003)
9. DeFigueiredo, D., Venkatachalam, B., Wu, S.F.: Bounds on the performance of P2P networks using Tit-for-Tat strategies. In: P2P'07: Proceedings of 7th IEEE International Conference on Peer-to-Peer Computing, pp. 11–18. IEEE Computer Society (2007)
10. Feldman, M., Lai, K., Stoica, I., Chuang, J.: Robust incentive techniques for peer-to-peer networks. In: Proceedings of 5th ACM Conference Electronic Commerce (EC'04), pp. 102–111. ACM, New York (2004). doi: http://doi.acm.org/10.1145/988772.988788
11. Habib, A., Chuang, J.C.I.: Service differentiated peer selection: an incentive mechanism for peer-to-peer media streaming. IEEE Trans. Multimed. **8**(3), 610–621 (2006)
12. Hales, D., Edmonds, B.: Applying a socially inspired technique (tags) to improve cooperation in P2P networks. IEEE Trans. Syst. Man Cybern. A Syst. Hum. **35**, 385–395 (2005)
13. Harris, R.: Decision making techniques (1998). Accessed on 15 October 2012 http://www. virtualsalt.com/crebook5.htm
14. Jun, S., Ahamad, M.: Incentives in BitTorrent induce free riding. In: Proceedings of 2005 ACM SIGCOMM Workshop on Economics of Peer-to-Peer Systems, P2PECON '05, pp. 116–121. ACM, New York (2005). doi: http://doi.acm.org/10.1145/1080192.1080199
15. Korzun, D., Gurtov, A.: Survey on hierarchical routing schemes in "flat" distributed hash tables. Peer-to-Peer Netw. Appl. **4**, 346–375 (2011). doi: http://dx.doi.org/10.1007/s12083-010-0093-z
16. Levin, D., LaCurts, K., Spring, N., Bhattacharjee, B.: BitTorrent is an auction: analyzing and improving BitTorrent's incentives. SIGCOMM Comput. Commun. Rev. **38**, 243–254 (2008). doi: http://doi.acm.org/10.1145/1402946.1402987
17. Li, H.C., Clement, A., Marchetti, M., Kapritsos, M., Robison, L., Alvisi, L., Dahlin, M.: FlightPath: obedience vs. choice in cooperative services. In: Proceedings of 8th USENIX Conference Operating Systems Design and Implementation (OSDI'08), pp. 355–368. USENIX Association (2008)
18. Lian, Q., Peng, Y., Yang, M., Zhang, Z., Dai, Y., Li, X.: Robust incentives via multi-level Tit-for-Tat. Concurr. Comput. Pract. Exp. **20**, 167–178 (2008). doi:10.1002/cpe.v20:2
19. Liao, W.C., Papadopoulos, F., Psounis, K.: Performance analysis of BitTorrent-like systems with heterogeneous users. Perform. Eval. **64**, 876–891 (2007). doi:10.1016/j.peva.2007.06.008

20. Liu, Z., Hu, H., Liu, Y., Ross, K.W., Wang, Y., Mobius, M.: P2P trading in social networks: the value of staying connected. In: Proceedings of IEEE INFOCOM'10, pp. 2489–2497. IEEE (2010)

21. Ma, R.T.B., Lee, S.C.M., Lui, J.C.S., Yau, D.K.Y.: Incentive and service differentiation in P2P networks: a game theoretic approach. IEEE/ACM Trans. Netw. 14(5), 978–991 (2006). doi: http://dx.doi.org/10.1109/TNET.2006.882904

22. Mekouar, L., Iraqi, Y., Boutaba, R.: A contribution-based service differentiation scheme for peer-to-peer systems. Peer-to-Peer Netw. Appl. 2, 146–163 (2009). doi: http://dx.doi.org/10. 1007/s12083-008-0026-2

23. Menasché, D.S., Massoulié, L., Towsley, D.: Reciprocity and barter in peer-to-peer systems. In: Proceedings of IEEE INFOCOM'10, pp. 1505–1513. IEEE (2010)

24. Parkes, D.C., Cavallo, R., Constantin, F., Singh, S.: Dynamic Incentive Mechanisms. Artif. Intell. Mag. 31, 79–94 (2010)

25. Piatek, M., Krishnamurthy, A., Venkataramani, A., Yang, Y.R., Zhang, D., Jaffe, A.: Contracts: practical contribution incentives for P2P live streaming. In: Proceedings of 7th USENIX Symposium on Networked Systems Design and Implementation (NSDI 2010), pp. 81–94. USENIX Association (2010)

26. Rahman, R., Vinkó, T., Hales, D., Pouwelse, J., Sips, H.: Design space analysis for modeling incentives in distributed systems. SIGCOMM Comput. Commun. Rev. 41(4), 182–193 (2011). doi: http://doi.acm.org/10.1145/2043164.2018458

27. Ramachandran, K.K., Sikdar, B.: A queuing model for evaluating the transfer latency of peer-to-peer systems. IEEE Trans. Parallel Distrib. Syst. 21, 367–378 (2010). doi: http://dx. doi.org/10.1109/TPDS.2009.69

28. Schrijver, A.: Theory of Linear and Integer Programming. Wiley, New York (1986)

29. Sherman, A., Nieh, J., Stein, C.: FairTorrent: bringing fairness to peer-to-peer systems. In: CoNEXT '09: Proceedings of the 5th International Conference on Emerging Networking Experiments and Technologies, pp. 133–144. ACM, New York (2009). doi: http://doi.acm. org/10.1145/1658939.1658955

30. Wang, W., Li, B.: To play or to control: a game-based control-theoretic approach to peer-to-peer incentive engineering. In: Proceedings of 11th International Conference Quality of Service (IWQoS'03), pp. 174–192. Springer, Berlin (2003)

31. Wang, Y., Nakao, A., Vasilakos, A.V., Ma, J.: On the effectiveness of service differentiation based resource-provision incentive mechanisms in dynamic and autonomous P2P networks. Comput. Netw. 55(17), 3811–3831 (2011). doi: http://dx.doi.org/10.1016/j.comnet.2011.07. 011

32. Wu, F., Zhang, L.: Proportional response dynamics leads to market equilibrium. In: STOC '07: Proceedings of 29th Annual ACM Symposium on Theory of Computing, pp. 354–363. ACM, New York (2007). doi: http://doi.acm.org/10.1145/1250790.1250844

33. Xuan, P.: Techniques for robust planning in degradable multiagent systems. In: Scerri, P., Vincent, R., Mailler, R. (eds.) Coordination of Large-Scale Multiagent Systems, pp. 311–340. Springer, New York (2006). doi: http://dx.doi.org/10.1007/0-387-27972-5_15

34. Yan, Y., El-Atawy, A., Al-Shaer, E.: Ranking-based optimal resource allocation in peer-to-peer networks. In: Proceedings of IEEE INFOCOM'07, pp. 1100–1108. IEEE (2007)

35. Yang, X., de Veciana, G.: Performance of peer-to-peer networks: service capacity and role of resource sharing policies. Perform. Eval. 63, 175–194 (2006). doi:10.1016/j.peva.2005.01.005

36. Zhao, B.Q., Lui, J.C.S., Chiu, D.M.: Analysis of adaptive incentive protocols for P2P networks. In: Proceedings of IEEE INFOCOM'09, pp. 325–333. IEEE (2009)

Summary of Part II

This part has discussed hierarchical routing schemes existing in recent DHT designs where the assumption that nodes are homogeneous in their roles is fundamental. Such DHT designs are traditionally known as flat, in contrast to a hierarchical DHT. The discussion argues that hierarchies are implicitly inherited in flat DHT designs. The approach uses three basic arrangement models and their combinations, leading to a variety of hierarchies. Based on an arrangement the nodes can be structured onto groups. Recursive application allows further structuring within groups. It is a crucial step in constructing global hierarchy.

We sequentially built a set of design principles; each provides a base for applying hierarchical schemes. Chapters 3–5 focused on principles aimed at efficient routing. They assume that the participating nodes are cooperative. The logical structure of the principles is depicted in Fig. II.1, where the evolution levels from pure flat designs and to pre-hierarchical designs go from the top to the bottom. Principles at the same level are complementary. Each evolution level introduces more hierarchies in a design, achieving requisite performance and scalability properties.

Then in Chap. 6 we moved to the complementary problem—cooperation in P2P systems. It is a very broad topic covering issues that are beyond what traditional distributed systems have encountered. We paid attention to the incentive problem, which is undoubtedly required for fair cooperation in anonymous, dynamic, and autonomous P2P networks. The corresponding principles introduce local ranks to effectively reward cooperating nodes and to punish defect behavior. It brings more security and resistance to the whole system even if the local knowledge problem limits each node with direct observations in its neighborhood.

Hierarchical routing schemes and incentive mechanisms complicate DHT designs, especially the overlay maintenance. It leads to overhead, additional vulnerabilities and other shortcomings. Nevertheless, we argued that, given application demands, a flat design can be extended with locally constructed hierarchies. The extension aims at attaining a tradeoff between the protocol complexity and the required efficiency.

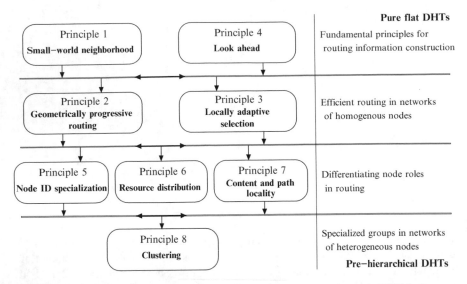

Fig. II.1 The logical structure of principles for hierarchical routing schemes in DHT designs

Part III
Beyond the Local Knowledge

Overview of Part III

The independence of dedicated infrastructure and centralized control are clear strength of P2P systems. On the other hand, these properties also become P2P weakness. To attain full efficiency, a network protocol must have complete information about all nodes who join the network. For instance, common assumptions of classical game theory are full-rationality of nodes, perfect knowledge of the game being played and all possible outcomes along with infinite computational time. Given these assumptions it is sometimes possible to analytically derive some equilibrium state for the game.

However, nodes of a P2P system can efficiently operate with local knowledge only. Achieving a globally optimal solution is too expensive or even irrelevant. For instance, the above classic game-theoretic assumptions do not hold in dynamic open P2P networks and the derivation of equilibria within dynamic topologies and changing populations is currently a hard problem for analytical techniques. Straightforward engineering solutions when a node collates accurate global statistics are non-scalable and impractical in any large and highly dynamic system.

Previously, Part II showed that careful structuring of local knowledge leads to reasonable approximations and a node can extrapolate its partial knowledge to the rest network. In this case it is not necessary to collect large amounts of information, or to undertake complicated and expensive distributed calculations, in order to implement the efficient protocol in a P2P network. Nevertheless, if node's capacity is high then acquisition of more knowledge about the global network can lead to further improvement. For instance, powerful nodes can keep bigger routing state or a node adjusts its routing table size in accordance with experienced churn and system population changes.

In this part, we consider methods for a node to collect additional knowledge about the global network. They provide tradeoffs between local and global knowledge. A base P2P protocol, which states default amount of local knowledge per node, can be appended such that appropriate nodes extend their local knowledge. This extension, as previously, follows some arrangement model to embed effective composition structure into the overall system of many nodes.

Chapter 7 describes further evolution of hierarchical routing schemes to hierarchical DHT (HDHT) architectures. Such architecture is a result of layering—nodes and their groups are arranged onto multiple layers. The stack of layers is decomposition of the network architecture along the vertical dimension. A layer consists of "close" groups that are structured along the horizontal dimension. Node role and responsibility depend on both group and layer which the node belongs to. The essence of HDHT design is that node's specialization is agreed with a global invariant—hierarchical architecture. This invariant extends node's local knowledge. We give taxonomy of HDHT architectures and make an overview of popular HDHT designs.

Chapter 8 introduces the cyclic routing (CR) method. It provides a systematic way for a node to collect stable and efficient overlay paths that run beyond neighbors. We consider cyclic paths; a cycle starts from the node to its neighbor, then runs through the overlay and returns to the source. On one hand, their construction is cheap since it is a result of performed lookups. On the other hand, cycles present global knowledge about available paths in the overlay. CR generalizes the idea of look-ahead approach to DHT routing, allowing a tradeoff between the look-ahead level and available node resources. It takes a place in between DHT routing using only neighbors (local) and optimal-path routing (global).

Chapter 9 focuses on the problem of describing P2P routes. It uses recent progress in the linear Diophantine analysis, which allows us to formulate the theoretical framework for P2P routing in general. We consider a Diophantine model, where a route aggregates several P2P paths that packets follow. Such aggregation is due to multipath routing as well as multiple sources and destinations. The model is based on abstract parallel process algebra, and we use a commutative context-free grammar to describe forwarding behavior of P2P nodes. A derivation in the grammar corresponds to a P2P route such that derivation initial and final strings define packet sources and destinations, respectively.

Chapter 10 targets the problem of fair cooperation in P2P systems. Known methods are based on trust and reputation models, which require accurate global statistics. We apply the same idea as in cyclic routing and propose a cyclic method for ranking P2P nodes based on available information of some cyclic paths. It can be thought of as a downsized version of such well-known global ranking algorithms as PageRank.

Chapter 7
Hierarchical DHT Architectures

Abstract Distributed Hash Tables (DHT) are presently used in several large-scale distributed systems in the Internet and envisaged as a key mechanism to provide identifier-locator separation for mobile hosts in Future Internet. Such decentralized structured data storage systems become increasingly complex serving popular social networking, P2P applications, and Internet-scale infrastructures. Hierarchy is a standard mechanism for coping with complexity, scalability, and heterogeneity in distributed systems. To address the shortcomings of flat DHT designs, many hierarchical P2P designs have been proposed over recent years. The last generation is hierarchical DHTs (HDHTs) where nodes are organized onto layers and groups. This chapter discusses concepts of hierarchical architectures in structured P2P overlay networks, focusing on HDHT designs. We introduce a framework consisting of conceptual models of network hierarchy, multi-layer hierarchical P2P architectures, and principles affecting the design choices. Based on the framework we provide taxonomy of existing P2P systems and thoroughly go over proposed hierarchical P2P alternatives.

7.1 Introduction

Designs of Internet-scale distributed systems evolve towards extremely complicated environments with a multitude of essentially heterogeneous and dynamic participants. It is a challenge that has been appearing in ubiquitous computing, Internet of Things, and other recent paradigms of computing. In this case, "flat" approaches lack in efficiency. System designs have to apply various differentiation and decomposition techniques. It results in hierarchical network architectures for organizing the participants, see, e.g. [3, 34, 35].

Structured P2P systems achieve their efficiency due to dynamic maintenance of rigorous network topology structure—its topology graph belongs to a class with well-defined invariant properties of connectivity [14, 21, 26, 36, 37, 43, 53]. Resources, functionality, and other types of responsibility are uniformly spread

D. Korzun and A. Gurtov, *Structured Peer-to-Peer Systems: Fundamentals of Hierarchical Organization, Routing, Scaling, and Security*, DOI 10.1007/978-1-4614-5483-0_7, © Springer Science+Business Media New York 2013

over all the nodes. It ensures low node degree and small network diameter, leading to modest node state and high routing performance. In this case, however, a P2P node knows only a part of the entire network—the local knowledge problem. To approximate the state of the rest network, the node extrapolates its local state based on invariant structural properties of the network topology. We studied these properties and corresponding design principles in Chap. 3.

The heterogeneity requires differentiation of nodes in a P2P system. Some of them are able to maintain bigger state becoming high degree nodes, which reduce the network diameter. On the other hand, there can be low-capacity nodes (e.g., running on mobile devices) that are not so active and resilient; their state and degree are minimal, which decreases the network connectivity efficiency. A system design adopts the heterogeneity by arranging the nodes such that they are subject to different responsibilities. Differentiation on local level (individually by a node) leads to pre-hierarchical P2P designs [26] that augment a flat P2P network with many locally-conducted hierarchies. We discussed them in Chap. 4.

Differentiation on the global level leads to hierarchical P2P networks. The network architecture considers an additional kind of entities—groups. Each group consists of nodes of similar responsibility. An example of group formation is clustering of proximity close nodes, the important design principle pointed in [27] and then further actively elaborated [26, 34, 51, 69]. Other examples are virtual servers [4, 12, 56, 59] when distinct P2P nodes reside at the same machine and semantic clustering [60, 64] when a group consists of nodes with semantically close content. We considered the clustering principle and its consequences in Chap. 5.

Layering is a further conceptualization step—nodes and their groups are arranged onto multiple layers [34, 35, 70]. The resultant stack of layers is decomposition of the network architecture along the vertical dimension. A layer consists of "close" groups, and the structure of groups on the same layer is along the horizontal dimension. Node role and responsibility depend on both group and layer which the node belongs to.

Layering supports traditional three-based hierarchies with the "divide-and-conquer" decomposition into disjointed parts. Leaf vertexes correspond to P2P nodes, non-leaf vertexes correspond to groups, and levels in the tree correspond to network layers. A typical example is administrative domain hierarchies, when a node belongs to an organization and the organization, in turn, belongs to a region. More general group-based hierarchies are also possible when relations between groups on different layers are not pure ancestor-descendant; some groups may be essentially overlapped and even nested. For example, the same node belongs to several groups, each group is responsible for a service, and the groups define nodes's service pool.

Section 7.2 introduces a framework of network hierarchy models, hierarchical P2P architectures, and principles affecting the design choices. We derive a set of conceptual models that describe global hierarchy in a structured P2P network, They cover possible hierarchies and support multi-layer hierarchical architectures using two design principles, which are dual. The disjoining principle provides node differentiation by forming low-overlapping groups. The nesting principle aims at functional differentiation when nodes with similar roles belongs high-overlapped groups.

Section 7.3 presents the taxonomy and essence of most known hierarchical P2P systems. A wide spectrum of existing hierarchical P2P designs is considered in a unified way, which our generic framework provides. In accordance with the above design principles we consider two classes of the hierarchical designs. The class with disjointed architecture uses the vertical dimension to differentiate the node responsibility by arranging heterogeneous nodes. The class with nested architecture uses the vertical dimension to arrange the overlay functionality: each layer performs the same function using different mechanisms.

Section 7.4 overviews basic performance models used for analysis of hierarchical P2P architectures. In contrast to flat designs, a hierarchical architecture has additional dimensions (groups, layers) that affect the fundamental lookup performance vs. node state tradeoff. Any design has to carefully set parameters along these dimensions to achieve reasonable performance bounds.

7.2 Conceptual Models

This section introduces conceptual models that cover possible hierarchies in decentralized networks. The models are further supported with a set of design principles for constructing P2P network hierarchy architectures. It provides a general framework that unifies the wide spectrum of existing HDHT designs with hierarchical multi-layer architectures.

7.2.1 Network Hierarchy Models

Kleinberg [24] considered three conceptual models for decentralized networks: grids, hierarchies, and set systems. The arrangement approach allows their modification to cluster-based hierarchies, tree-based hierarchies, and group-based hierarchies, respectively. Note that Kleinberg aimed at network models for analysis of decentralized search algorithms. We apply Kleinberg's results to describe system-level hierarchies appropriate for structured P2P networks.

7.2.1.1 Cluster-Based Model

The small-world phenomenon—connectivity by short chains of acquaintances—has become the subject of computer and social network analysis [5, 23], including P2P networks [37, 38]. Chapter 3 showed that many conventional DHT designs follow small-world models in constructing local routing tables (see also survey [26]). It results in geometrically progressive routing when the hop length $|u \rightarrow^+ v|$ is exponentially shorter than the space distance $\rho(u,v)$.

The basic scheme of neighbor selection is as follows. A node u divides its neighbors onto local and long-range ones (classifying model). Local neighbors of u are equal in terms of the space distance. Long-range neighbors are structured in u's vicinity (ordering model). The space distance metric provides ranks $\rho(u,v)$ for u to arrange other nodes v (ranking model).

Kleinberg's grid-based model is a simple example [23, 24]. The network is embedded in a two-dimensional $n \times n$ grid graph, defining all local neighbors for all nodes. Then a long-range link is established between nodes u and v with probability proportional to $[\rho(u,v)]^{-\alpha}$, where $\alpha \geq 0$ is a model parameter. Symphony [38] is the first DHT design that adapted this model for practical P2P settings.

The idea of distance has further evolved in DHT designs. Instead of the space metric ρ, composite distance metrics, such as routing distance τ, can be applied to reflect proximity of the underlying IP network, semantic closeness of resources kept at nodes, node service reputation, etc. The small-world phenomenon is then interpreted [24] as embedding a network into an underlying space with distance τ. Nodes tend to know their close neighbors in this space as well as to have contacts that span long distances. Establishing a link with a closer node is more likely.

This distance-aware neighbor selection leads to various hierarchical structures in the network as was shown in Chap. 4. The clustering principle evolves the distance idea further, see Chap. 5. The principle states that the routing distance τ must positively correlate with the space distance ρ. Consequently, given two groups of nodes

$$\mathscr{C}_{xr} = \{u \mid \rho(x,u) \leq r\}, \quad \mathscr{C}_{yR} = \{u \mid \rho(y,u) \leq R\}$$

for some landmarks $x, y \in S$ and positive scalars $r < R$. Then intra-group routing in $\mathscr{C}_{xr}$ is more efficient than in $\mathscr{C}_{yR}$. In other words, taking arbitrary $u, v \in \mathscr{C}_{xr}$ and $u', v' \in \mathscr{C}_{yR}$ we expect likely $\tau(u,v) < \tau(u',v')$. Note that this principle is continuous in nature.

P2P clusters are a result of the discretization with the classifying model. Close nodes become densely connected and form a discrete entity—the cluster with explicit space bounds. Figure 7.1 visually supports the intuition. Within a cluster each node can reach another in few hops. Therefore, a global structure of interconnected groups of close nodes appears, forming a two-layer hierarchy. In intra-cluster communication the routing distance τ is shorter than in inter-cluster communication.

P2P designs that follow the clustering principle were considered in Chap. 5 (see also surveys [26, 35, 43]). They are called "pre-hierarchical" in [26], "hybrid" in [43], or "planar group-based" in [35]. Initial designs have no mechanism for a cluster to be a decision-making entity, and nodes benefit only from the topological properties of cluster-based network structure. In this chapter we focus on the designs

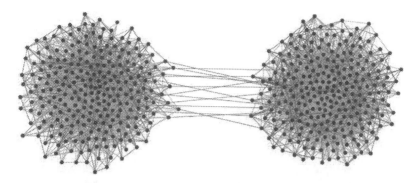

Fig. 7.1 The network with two clusters, where each cluster defines a community of 256 nodes. Nodes are densely connected within their community, in contrast to the poor inter-community connectivity. As a result, $\rho(u,v)$ is smaller within a cluster compared to the case when u and v belong to different clusters. The picture is from [63]

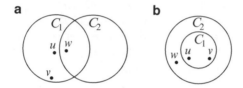

Fig. 7.2 When clusters C_1 and C_2 are not disjoint then there can be nodes u, v and w for which the distance relations contradict with (7.1): (**a**) partially overlapped clusters, (**b**) nested clusters

where each cluster (or a group of nodes in general) becomes a peer entity that can make own distinct decisions and actions in the network. We call such designs and their further generalization *hierarchical*; they are also known [35] as "partially centralized", "hybrid", "layered", or "multi-tier".

Principle 10 (Cluster-based model). *The hierarchy is two-layer with participant nodes on the bottom layer and clusters on the top layer. Nodes of the same cluster are densely connected; the inter-cluster connectivity is sparser.*

The model defines a hierarchy as a set of n interconnected clusters. It embeds the network such that its network structure ensures the following routing distance property

$$\left(\{\mathscr{C}_s\}_{s=1}^n, \tau\right),$$
$$\tau(u,v) \ll \tau(u,w) \text{ for } u,v \in \mathscr{C}, w \in \mathscr{C}', \mathscr{C} \neq \mathscr{C}'. \tag{7.1}$$

Each particular model defines its concretization of the relation "$\ll$". It must support the intuition that clusters are disjointed or almost disjointed, which is a consequence of the classification model. When clusters are essentially intersected then the intuition behind "$\ll$" suffers, see two examples in Fig. 7.2.

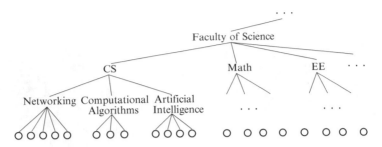

Fig. 7.3 A hierarchy of nodes at a university. *Circles* stand for P2P nodes. Each faculty consists of several departments. Each department has research groups with own nodes

Note that (7.1) captures the crucial role of hierarchy for the P2P routing dependability, e.g., see [54]. When $u, v \in \mathscr{C}$ (i.e., they are within one domain) then even if a node w of another domain $\mathscr{C}'$ is failed or if $\mathscr{C}$ is disconnected from the system, the nodes u and v are still able to communicate.

A simple formal model of cluster construction is a set of balls in the ID space S, e.g., in grid [24]. There is a set of landmarks $\{c_s\}_{s=1}^{n}$; they are points or dedicated nodes in S. Cluster $\mathscr{C}_s$ consists of all nodes u that satisfy $\rho(c_s, u) \leq R$ for a given radius R. The construction can be easily enhanced to construct disjointed clusters; if u belongs to several balls then a decision-making procedure assigns u in exactly to one cluster. In fact, the decision is about which c_s to associate each node with.

7.2.1.2 Tree-Based Model

The clustering principle allows constructing global two-layer hierarchies based on a distance metric. Another approach is to define a hierarchy in S more rigorously. In turn, the hierarchy induces space distance ρ in terms of the hierarchy tree. Then the hierarchy embeds the network, forming the hierarchy-aware connectivity structure with the resultant routing distance τ. This abstraction arises from classical hierarchical methods, known also as "divide and conquer".

An obvious example is location and administrative hierarchies. Nodes are categorized on lowest-level groups depending on which local area network they belong to. The upper level groups are defined in accordance with their scale, e.g., city, region, and state. Figure 7.3 depicts an example fragment of the university hierarchy where participant nodes belong to different administrative entities.

Kleinberg [24] described a formal network model based on a complete b-ary tree $\mathscr{T} = \mathscr{T}(b, N)$ with N leaves (hence $\mathscr{T}$ is of height $M = \log_b N$). Given leaves u and v, the hierarchy-induced distance $h(u, v)$ is the height of their lowest common ancestor in $\mathscr{T}$. Then a network of N nodes is constructed such that the probability of establishing a link $u \rightarrow v$ is proportional to $b^{-\alpha h(u,v)}$, where $\alpha \geq 0$ is a model parameter. As in the cluster-based model, shorter paths (routing distance τ) exist between nodes of the same group.

In general, an arbitrary tree $\mathscr{T}$ defines a hierarchy where N leaves correspond to nodes and other vertices are groups consisting of descendant groups and nodes. The distance metric is the tree distance as in the above model. In fact, it is a kind of introduction of more layers to two-layer cluster-based model (7.1); groups (non-leaf vertices) correspond to clusters, some of them are nested $\mathscr{C}_i \subset \mathscr{C}_j$ in accordance with the predefined hierarchy.

Principle 11 (Tree-based model). *The hierarchy is M-layer. The bottom layer $i = 1$ consists of N nodes. On the upper layer $i + 1$ each group consists of all nodes from its descendant groups on the layer i. Groups of the same layer are node disjointed. On each layer the inter-group connectivity is sparser than the intra-group connectivity.*

The tree-based model can be thought as iterative application of the cluster-based model. Each iteration results in the next layer, reflecting a higher scale level. Layer $i + 1$ consists of "clusters" for layer i. If $\{\mathscr{C}_{ij}\}_{j=1}^m$ are all groups of a fixed layer i then they satisfy (7.1). The model preserves the property of dense node connectivity within a group compared with the connectivity to nodes of other groups.

Further generalization is that there can be several distinct trees $\{\mathscr{T}_k\}_{k=1}^m$, reflecting the notion that nodes were simultaneously taking into account several "proximity" characteristics. For example, nodes in Fig. 7.3 can be distributed over many geographically different locations, and we should consider an additional area-location tree. Each $\mathscr{T}_k$ defines its own distance metric ρ_k. The resultant connectivity with routing distance τ must follow (7.1) on any layer of any $\mathscr{T}_k$. Note that in fact the set $\{\mathscr{T}_k\}_{k=1}^m$ defines additional m layers (superlayers) in the hierarchy.

7.2.1.3 Group-Based Model

Each of the above two models defines specific rules of group formation. Kleinberg et al. [22, 24] considered generalization where groups of the same network may be formed by means of different models, including at least cluster- and tree-based ones.

Assume that each node can belong to several groups. For example, a node u can belong to group C_1 (Helsinki Institute for Information Technology—administrative entity), group C_2 (powerful machines—performance level) and group C_3 (Europe—geographical location). As in the cluster- and tree-based models, nodes are more likely to be connected if they belong to the same group.

Kleinberg's group-based network model establishes the following properties using parameters $0 < \lambda < 1$ and $1 < \mu$.

(i) The full set N of all nodes is a group.
(ii) If $\mathscr{C}$ is a group of size $|\mathscr{C}| \geq 2$ and $u \in \mathscr{C}$, then there is a group $\mathscr{C}' \subset \mathscr{C}$ such that

$$\mathscr{C}' \neq \mathscr{C}, \quad u \in \mathscr{C}', \quad \lambda |\mathscr{C}| \leq |\mathscr{C}'| < |\mathscr{C}|.$$

(iii) For any set of groups $\{\mathscr{C}_i\}$ with a common node u,

$$\left|\bigcup_i \mathcal{C}_i\right| \leq \mu\sigma, \quad \text{where } \sigma = \max_i |\mathcal{C}_i|.$$

Property (i) ensures that for any subset of nodes there is a group such that it includes the whole subset. Property (ii) is a type of the "hierarchy balance" requirement when a group consists of subgroups of proportional size. For example, the tree-based model with a complete b-ary tree defines groups of b node-disjoint subgroups each, hence $\lambda \sim 1/b$. Property (iii) is a type of "bounded size growth" requirement; if groups has a common node then they are close in certain sense, so they cannot contain many distinct nodes. For example, the cluster-based model forms node-disjoint groups, hence a set in property (iii) always consists of one group.

For two nodes u and v, the induced space distance $\rho(u,v)$ is the minimum size of a group containing both u and v. Similarly to the previous models, it allows embedding the network into the hierarchy such that the connectivity structure positively correlates the routing distance τ with ρ. In Kleinberg's network model, the probability of establishing a link $u \to v$ is proportional to $[\rho(u,v)]^{-\alpha}$ for $\alpha \geq 0$.

Properties (ii) and (iii) motivate explicit separation of two dimensions in network hierarchy: vertical and horizontal. The vertical dimension defines layers and rules for nesting groups. Any hierarchy contains at least one chain of nested groups for any node u:

$$u \in \mathcal{C}_1 \subset \mathcal{C}_2 \subset \ldots \subset \mathcal{C}_m = N, \tag{7.2}$$

where each $\mathcal{C}_i$ belongs to a distinct layer and there is no $\mathcal{C}$ such that $\mathcal{C}_i \subset \mathcal{C} \subset \mathcal{C}_{i+1}$. In this sense, the tree-based model is pure vertical since the only chain of nested groups exists for a given node.

In the general case, there can be several chains for the same node, e.g., if the tree-based model uses several trees. The following requirement preserves the vertical structure "approximately nested". If $\{\mathcal{C}_i\}$ are arbitrary groups having a common node and any two of them do not belong to the same layer then

$$\left|\bigcup_i \mathcal{C}_i\right| \leq \mu(\max_i |\mathcal{C}_i|), \tag{7.3}$$

where $\mu(\cdot)$ is a monotone increasing function, defined by each particular model.

The horizontal dimension defines classification onto groups on the same layer i.

$$\bigcup_j \mathcal{C}_{ij} = N_i, \quad |\mathcal{C}_{ij} \cap \mathcal{C}_{ik}| \ll |\mathcal{C}_{ij} \triangle \mathcal{C}_{ik}| \ \forall j \neq k, \tag{7.4}$$

where $\triangle$ is the symmetric difference. In this sense, the cluster-based model is pure horizontal; its group distribution is always a partition of N, thus $\mathcal{C}_{ij} \cap \mathcal{C}_{ik} = \emptyset$.

In the general case, groups on the same layer may overlap. For example, when different research groups in Fig. 7.3 share some nodes. Requirement (7.4) preserves groups "low overlapped", where each particular model defines its concretization of the relation "$\ll$". Note that alternative chains in (7.2) can appear because of overlapping.

Node's neighborhoods on layer i are an example of overlapping groups $\mathscr{C}_{iu} = \{u\} \cup T_u^{(i)}$, where $T_u^{(i)}$ is u's routing table on layer i. If groups $\mathscr{C}_{iu}$ and $\mathscr{C}_{iv}$ are high overlapped then, from the interconnectivity point of view, they can be replaced with the cluster $\mathscr{C}_{i,uv} = \mathscr{C}_{iu} \cup \mathscr{C}_{iv}$.

The following principle summarizes the unified conceptual view on hierarchy construction.

Principle 12 (Group-based model). *A hierarchy is a set of interconnected groups (7.1). It spans both vertical and horizontal dimensions satisfying (7.2)–(7.4).*

Lloret et al. [35] introduced several types of group-based topologies appropriate for centralized and decentralized networks. That work focused on the topology graph and its role in the network efficiency improvement. They considered the case of disjointed groups, a particular instance of the general model of Principle 12. An essential finding they did is the importance of layering for scalable network structures. The optimization techniques for group formation algorithms and group distribution among network layers are considered by Lian et al. [34].

7.2.2 Layering Principle

Although pre-hierarchical P2P designs, see Part II and [26], utilize the clustering principle, an individual node does not necessarily identify itself a member of a certain global group as well as a group does not act as a decision-making entity in the network. The straightforward application of the clustering principle is continuous. There is no global group set fixed a priori. Each node u is able to compute the metric ρ, which numerically reflects the global network hierarchy. Values $\rho(u, \cdot)$ control u's connectivity to the network. Other nodes apply the same rules. As a result, clusters appear at the global network level, approximating the given hierarchy in actual overlay network topology.

Pre-hierarchical P2P designs follow the entire overlay agreement approach of flat P2P. All participants must globally agree on a set of protocols and parameter settings like the routing rules, size of routing tables, synchronization intervals, and replication strategy [45]. Optimal settings for these parameters depend on dynamic factors like churn rate, node failure probabilities, and fault correlation. They can be difficult to assess or estimate for a large overlay.

In the hierarchical approach, the overlay is a priori (by design) partitioned onto discrete layers, so forming the vertical dimension of network hierarchy. Many problems can be solved within a layer, hence the problem size and complexity are reduced compared with the entire overlay agreement approach.

In contrast to Kleinberg's hierarchy models, our conceptual models of network hierarchy emphasize the importance of vertical dimension. This requirement directly comes from practice; multi-layer architecture is a distinct characteristic property of any hierarchical P2P design, including HDHT designs [21, 48].

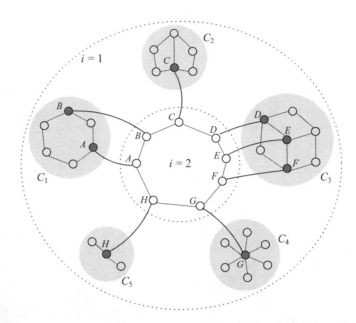

Fig. 7.4 Example of ordered two-layer architecture. Nodes are organized into clusters $C_1, \ldots, C_5$ on the *bottom layer*. They appoint supernodes $A, B, \ldots, H$ to form own overlay on the *top layer*. Each $A, B, \ldots, H$ is physically one node acting as two virtual nodes, one per layer

Principle 13 (layering). *There is a priori partitioning onto fixed discrete layers. Each node u identifies itself with a concrete layer when u performs its functions in the P2P system.*

The layering principle benefits from simplicity in the relation between the global network hierarchy and the actual overlay topology. A network layer can correspond to a domain in the global hierarchy. The maintenance cost is reduced by distributing the load among levels. Protocol and parameter settings are done within each domain. The principle first appeared in unstructured P2P systems [27, 70], then it was adopted in structured P2P designs.

A simple hierarchy is ordered two-layer (basic supernode model): the top layer of *supernodes* and the bottom layer of networks of *regular nodes*. This hierarchy is an instance of the cluster-based model, see Principle 10. Each supernode is assigned to a network of regular nodes and connects this network by proxying lookups on behalf of its regular nodes (a gateway to the top-level network). Supernodes are more available and powerful overlay nodes. For regular nodes this hierarchy reduces the impact of their short online times on the P2P system. Figure 7.4 shows an example network where nodes form clusters on the bottom layer; then each cluster maintains its own overlay and selects its supernodes to represent the cluster in the top layer overlay.

The layering principle is applicable for arbitrary finite number of layers to provide efficient scaling and organization. For instance, a design defines $M \geq 2$

layers to enable distinction between nodes with different capabilities, domains, and other global characteristics. The result is a hierarchy with multiple layers ordered among the vertical dimension.

Similarly to clustering in pre-hierarchical P2P designs, a multi-layer hierarchy influences the network topology, but the impact is higher. Nodes of the same domain form own network. Consequently, *hybrid P2P systems* that combine unstructured and structured topologies are possible [43, 57, 62]. A network on each layer runs own P2P protocol, either structured or unstructured. The anonymity is also improved (compared with flat designs); a network on a given layer is a black box for an external overlay node.

One disadvantage of layering is the assumption that a supernode can participate in any overlay. The assumption is not always true in IP networks since they have connectivity restrictions, e.g., due to NATs or firewalls.

7.2.3 Network Hierarchical Architectures

The layering principle introduces a stack of interconnected layers, structuring the network vertically. The horizontal dimension is for the intra-layer connection structure. A concrete design with composition of the vertical and horizontal dimensions defines *a hierarchical architecture*.

Artigas et al. [2] distinguished *vertical* and *horizontal* approaches for hierarchical architectures. The original definition focused on inter-overlay connection structure—a result of multiple routing tables that nodes maintain to participate in several overlays. This definition does not always provide a clear classification criterion, especially for the case of group-based hierarchy model. We modify the criterion accenting the role of layering and considering the relation between layers and their overlays.

Our further analysis brings two new design principles: *disjoining* and *nesting*. They focus on the relation between groups in the hierarchy. A group defines a role that group's nodes perform in the system. Layers and overlays are specific instances of the notion "group".

7.2.3.1 Vertical Approach

The characteristic property of this approach is that *every layer is a self-contained P2P overlay network*. Vertical architecture is a set (ordered or not) of M overlays. Each layer i overlay follows its own protocols. Layers may consist of different number of nodes, $N_i \leq N$. Any supernode u participates in several overlays, i.e., u is a gateway connecting them. Thus u has to maintain multiple routing tables,

$$u \in N_i \cap N_j, \quad T_u^{(i)} \cup T_u^{(j)} \subset T_u \quad \text{for some } i \neq j.$$

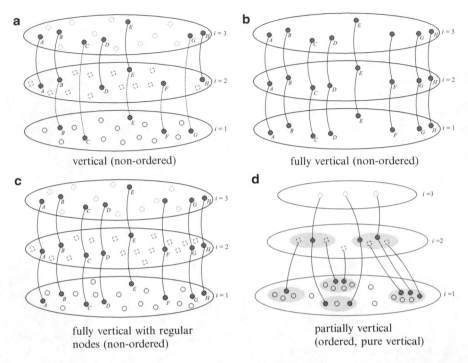

Fig. 7.5 Vertical architecture for $M = 3$ layers. A regular node (*unfilled circle*) belongs to exactly one layer. A supernode (*filled circle*) connects two or more layers. *Linked circles* represent one physical node acting as several virtual nodes due to layering. (**a**) Supernodes connect any combination of layers. (**b**) Any node is a supernode for every layer. (**c**) Regular node appears on one layer; supernode appears on all layers. (**d**) Tree- or group-based hierarchy; *grey ovals* are clusters of layer i overlay, which provides the P2P connectivity for all its nodes in spite of their clustering structure

In routing, u has to decide which of its overlays to use for a lookup. A regular node participates in exactly one overlay and forwards lookups to supernodes for inter-layer (global) routing.

Intuition behind vertical architecture is that the system partitioning is along the vertical dimension only. Within each layer one overlay network is presented, thus there is no specific horizontal partitioning. The division onto supernodes and regular nodes is for vertical glue between layers.

Vertical architecture benefits of the simplicity and straightforward implementation. For example, the ordered two-layer architecture is similar to shown in Fig. 7.4 earlier. All nodes participate in the bottom layer ($N_1 = N$), forming one big overlay network (e.g., flat DHT). The most responsible nodes become supernodes ($N_2 \leq N$), forming an additional overlay on the top layer (e.g., another flat DHT).

Increasing the number of layers provides more design opportunities, see Fig. 7.5 for illustration. Supernodes in Fig. 7.5a have two or three independent routing tables: F connects layers 1 and 2; C and G connect layers 1 and 3; A, D, H connect layers 2 and 3; B and E connect all three layers. When layers use essentially different P2P protocols (e.g., a mix of structured and unstructured protocols) the hierarchical architecture is hybrid.

In terms of the group-based model (Principle 12), each layer i defines own group N_i of nodes. The non-empty pair-wise intersection is due to supernodes,

$$N = \bigcup_{i=1}^{M} N_i, \quad N_i \cap N_j = S_{ij} \neq \varnothing \text{ for } i \neq j,$$

where S_{ij} consists of all supernodes participating in the overlays of layers i and j.

There are several alternatives for a supernode in (7.2). Requirement (7.3) can be satisfied, e.g., with the moderate number of layers $M \ll N$, then either approximately equal-sized layers or few ones that cover the majority of nodes.

Layers in vertical architecture can be *ordered* or *non-ordered*. The ordered vertical architecture applies layering to define different classes of the node responsibility or functional role in the global system. For instance, in the two-layer supernode model (see Fig. 7.4 on p. 180) the supernode overlay assures more efficient routing than the bottom overlay. A typical case is the decreasing number of nodes for the higher responsibility:

$$N_1 > N_2 > \cdots > N_M, \tag{7.5}$$

which is natural for nesting requirement (7.2). The pictorial idea is shown in Fig. 7.5d.

The non-ordered variant is suitable for *federated* architectures, when several domain-independent P2P overlays are combined. The case is illustrated in Fig. 7.5a, where a layer might represent a distinct administrative domain. A few supernodes belong to multiple domains. Layers are equal in their participation in the global system. The design must define rules for a supernode which layers it is associated with. Nesting requirement (7.2) becomes minor due to its reduction to $u \in N_i \subset N$ (for some i).

In certain sense, non-ordered vertical architecture is close to two-tier horizontal architecture (see Sect. 7.2.3.2): the bottom tier of regular nodes and the top tier of supernodes. Nevertheless, this architecture is vertical, and the distinctive property is that supernodes do not organize themselves into own overlay or a set of overlays.

The vertical approach allows *multi-hierarchical* architectures. Since every layer is a self-contained P2P overlay, the same procedure can be further applied to any layer to construct its own hierarchical architecture. For instance for multi-tree hierarchy (Sect. 7.2.1.2), each tree $\mathcal{T}_i$ corresponds to layer i, where the hierarchy is applied to all nodes $N = N_i$ and any node is a supernode for all M layers, see Fig. 7.5b. For federated architecture this recursive approach leaves much design freedom for the overlays that participate in the federation. Note that multi-hierarchy may combine layers with ordered and non-ordered architectures.

Ou et al. [48] further divided the vertical architecture class into *fully vertical* and *partially vertical*. In fully vertical architecture all N nodes are present on all M layers, as illustrated in Fig. 7.5b. Therefore, there is no need in dedicated gateway nodes. Since each layer is a full N-node overlay, lookup can be successively routed within any layer. Nevertheless, for the routing efficiency a node decides an appropriate layer i to bring into play for a given lookup.

Fully vertical architecture can be extended with regular nodes, as shown in Fig. 7.5c. Each supernode still must participate on all M layers, hence being a "universal gateway". Each regular node is presented within its domain (layer) only. In practice it can be because of the domain security policy does not allow such a node to be accessible directly from other domains, node capacity is low for maintaining multiple routing tables, or NATs and other IP-level restrictions prevent direct connectivity.

We use the term *pure vertical* for partially vertical architecture since it inherits the origin two-layer supernode model. The idea is depicted in Fig. 7.5d. The order of layers is essential. A node u participates in the overlays on layers $i = 1, 2, \ldots, m_u$, thus u does not appear on layers $i > m_u$. This solution is appropriate for highly heterogeneous environments. For example, a mobile device, which cannot maintain many routing tables, is presented only on the bottom layer.

The obvious merit of pure vertical architecture is that it naturally admits the cluster-based and tree-based hierarchy models with iterative layer construction. Nodes of the layer i overlay are organized into clusters $\mathcal{C}_{ik}$. A cluster assigns one or more its nodes to be supernodes for layer $i + 1$. If all supernodes of $\mathcal{C}_{ik}$ belong exactly to one cluster $\mathcal{C}_{i+1,j}$ on the next layer, then the hierarchy is tree-based. If supernodes of $\mathcal{C}_{ik}$ participate in different clusters on layer $i + 1$, then the hierarchy is group-based. Moreover, some nodes on layer i may ignore clustering and supernode assignment so far as the layer i overlay preserves its connectivity.

The known drawback of straightforward designs of vertical architecture is the overhead that supernodes afford in their routing table maintenance. The number of network connections can reach a big value for large N and M, worsening the scalability. For instance consider a federated network where its layer i overlay consists of $N_i = N/M$ nodes and uses a flat DHT like Chord [58]. A node has a routing table $T^{(i)}$ of $\log(N/M)$ entries. Since a supernode belongs to m layers it keeps $\sum_{i=1}^{m} |T^{(i)}| = m\log(N/M)$ entries. If the N-node network would be implemented with a single flat DHT then every node has routing table of size $\log N$. Let concretization of $M \ll N$ be $M^\alpha < N$ for $\alpha > 1$. The ratio

$$\frac{m}{\alpha} < \frac{m\log(N/M)}{\log N} < m$$

shows that the state overhead of a supernode is proportional to the number of layers the node participates in.

Note that this high overhead happens when the supernode maintains its routing tables independently. Contrary to the widespread opinion, vertical architecture does not always prevent lower overhead maintenance when routing entries are effectively reusable on different layers, as it first appeared in horizontal architecture designs (see the next section). In networks with low-overlapped layers, like federated P2P networks, the routing table maintenance, however, cannot efficiently benefit from the routing links reusability.

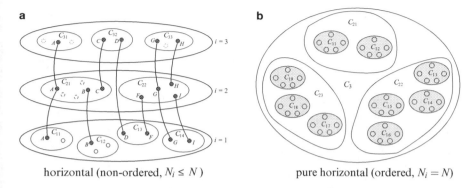

horizontal (non-ordered, $N_i \le N$)   pure horizontal (ordered, $N_i = N$)

Fig. 7.6 Horizontal architecture for $M = 3$ layers. (**a**) An inter-layer connected combination of overlays; there is no intra-layer connectivity between overlays. (**b**) Pure horizontal. *Bottom layer* $i = 1$ consists of the smallest overlays $C_{11}, C_{12}, \ldots, C_{19}$. Layer $i = 2$ merges them to the medium-sized overlays $C_{21} = C_{11} + C_{12}, C_{22} = C_{13} + \cdots + C_{16}, C_{23} = C_{17} + C_{18} + C_{19}$. Top-most overlay C_3 includes all N nodes

7.2.3.2   Horizontal Approach

The characteristic property of horizontal architecture is that, in addition to layering along the vertical dimension, *every layer is divided onto several disjointed overlays*, see Fig. 7.6. In this certain sense vertical architecture is a particular case of horizontal architecture with one overlay per layer. The term "horizontal" supports the intuition that the overlays of each layer spread out the horizontal dimension. In fact, partitioning a layer into overlays can be treated the application of the layering principle for the horizontal dimension.

As in vertical architecture, the horizontal approach allows non-ordered variants, see Fig. 7.6a. In contrast, they are not popular in P2P designs since the maintenance of the inter-layer connectivity between all overlays is complicated. Another property of the horizontal approach is that any layer i can cover all N nodes in sum ($N_i = N$), which is similar to fully vertical architecture.

Consider the pure variant of horizontal architecture: ordered stack of layers with $N_i = N$. Recall that in pure vertical architecture (Fig. 7.5d) every layer forms in exact one overlay, all N nodes participate in the bottom layer, the number of participating nodes is reduced in the overlay of the next layer, as shown in (7.5), and finally few supernodes participate in the top overlay. In contrast, pure horizontal architecture (Fig. 7.6b) partitions all N nodes into many small overlays on the bottom layer $i = 1$. The number of overlays is then reduced on every next layer $i + 1$ by merging layer i overlays, while keeping the same sum number of nodes on the layer. Finally, a single global overlay of N nodes appears on the top layer $i = M$.

This construction directly reflects the tree-based hierarchy (Principle 11), thus the pure horizontal architecture is appropriate for systems with explicit domain hierarchy, like shown in Fig. 7.3. Moreover, it allows the advantage that horizontal architecture typically provides: the sum routing table size is kept moderate. Instead of maintaining independent routing tables, a node u reuses its routing table entries in overlays on upper layers. It can be achieved with the telescoping scheme:

$$T_u^{(1)} \subset \ldots \subset T_u^{(i)} \subset T_u^{(i+1)} \subset \ldots \subset T^{(M)}. \tag{7.6}$$

The routing table size grows up with $i = 1, \ldots, M$. Any layer $i + 1$ overlay inherits and then extends the link structure of its layer i overlays (sibling networks). Additional links $\Delta_{ui} = T_u^{(i+1)} \setminus T_u^{(i)}$ are carefully chosen such that the size of every routing table remains small, comparable with a flat DHT, e.g., $|T_u^{(i)}| = O(\log N)$.

Further generalization allows partial inheritance, i.e., some entries of $T_u^{(i)}$ are not in $T_u^{(i+1)}$. For instance, if layers have unequal sets of nodes ($N_i \neq N_j$), then some neighbors from $T_u^{(i)}$ are absent on layer $i + 1$, thus they cannot be in $T_u^{(i+1)}$. The reuse requirement is reduced to keeping the symmetric difference $T_u^{(i)} \triangle T_u^{(i+1)}$ small. This way is also appropriate for non-ordered horizontal architectures architecture, where $T_u^{(i)} \triangle T_u^{(j)}$ should be kept small for $i \neq j$.

Contrary to the widely accepted opinion the routing entry reusability is not a distinct property of horizontal architecture. For instance, full vertical architecture designs can also benefit from supernodes that rationally maintain routing tables in the reusable manner. In fact, the reuse requirement is an instance of generic group-based principle (7.3) with groups $\mathscr{C}_{iu} = \{u\} \cup T_u^{(i)}$, leading high overlapped node's neighborhoods on different layers.

Since the requirement $N_i = N$ or at least $N_i \approx N$ is typical in horizontal architecture, the already mentioned disadvantage is that IP connectivity restrictions or high responsibility requirements can prevent participation of certain nodes in some layers. A more specific disadvantage is the unclear distinction among the load that different nodes take in the system. Node differentiation becomes difficult when almost every node participates in all layers.

One solution to mitigate the above disadvantages can be brought from pure vertical architecture. Each layer states the node responsibility level. A node u maintains its (nested) routing tables up to layer $m_u < M$, cf. (7.6). Communication with layers $m_u < i \leq M$ is through such nodes (supernodes for u) that maintain routing tables up to layer i at least. Hence, the node distribution among layers satisfies (7.5).

7.2.3.3 Disjoining and Nesting Principles

The group-based model can describe the core of any hierarchical architecture. A concrete architecture design further clarifies which relations exist among the groups, e.g., some groups are disjointed and some ones are nested. Some instances of such relations were considered above. Their analysis leads to two general design principles formulated in Table 7.1.

Table 7.1 Disjoining and nesting principles for hierarchical P2P architecture designs

Principle 14 (disjoining). *Nodes with different roles belong to disjointed or low-overlapped groups.*	**Principle 15 (nesting).** *Nodes with similar roles belong to nested or high-overlapped groups.*
Properties	
Classifying model: Groups are separated in accordance with their node population, responsibility, or functional role. *Low-overlapped groups:* If $u,v \in \mathcal{G}$ and $u \notin \mathcal{G}'$ then likely $v \notin \mathcal{G}'$. This tendency correlates with (7.4). *Coarse responsibility granularity:* The number of m-responsible nodes decreases rapidly with $m = 1,2,\ldots,M$. The number of layers M is typically small, hence providing few responsibility levels. Low-responsible nodes prevail, otherwise the hierarchy degenerates to a flat structure.	*Ordering model:* Higher layer groups inherit partially the node population and increment the responsibility and functional role. *High-overlapped groups:* If $u,v \in \mathcal{G}$ and $u \in \mathcal{G}'$ then likely $v \in \mathcal{G}'$. This tendency correlates with (7.3). *Fine responsibility granularity:* The number of m-responsible nodes decreases slowly with $m = 1,2,\ldots,M$. The number of layers M is can be big, hence allowing many responsibility levels and non-trivial nested structure (7.2). There are many high-responsible nodes, up to the totally supernode population.
Typical evidence in architectural design	
Layering	
Layers represent distinct domains; the overlap is low due to a small set of supernodes (compared with the major population). Popular in non-ordered architectures.	Every next layer increments the node responsibility. An m_u-responsible node u participates in layers $i = 1,2,\ldots,m_u \leq M$. Popular in ordered architectures.
Clustering	
Construction of the horizontal dimension. Each layer is composed of disjointed or low-overlapped groups by clustering similar nodes.	Construction of the vertical dimension. Some nodes of each cluster also appear on upper layers.
Multiple routing tables	
A node maintains layer-independent routing tables, leaving freedom for the intra-layer connectivity. Typical for vertical architecture.	A node maintains nested routing tables, allowing reuse of routing entries on different layers. Typical for horizontal architecture.
Node heterogeneity in load distribution	
A group acts as a collective entity. A high-responsible node affords essential capacity to represent the group on behalf of all its nodes.	A node increments its responsibility up to its own individually appropriate level.

Each principle captures an ideal case. Concrete hierarchical architecture designs are always subject to tradeoffs. The terms *low-* and *high-overlapped* groups show the possibility of deviation from the ideal cases. In fact, for a given architecture we can only say about the tendency: either it follows closer the disjoining principle or the nesting one.

Both principles state that a hierarchical architecture design reflects the node role differentiation and similarity by the construction of groups. The principles are opposite since they result in decreasing and increasing the group overlap, respectively. Nevertheless, they are applicable in a composition for both vertical and horizontal dimensions. They provide more understanding of the propositions we introduced in the group-based model and the layering principle.

The layering principle can be considered as an instance of the disjoining principle applied for constructing the vertical dimension of the hierarchy. Along the vertical dimension the groups can also follow the nesting principle and form a nested structure as it was formalized earlier in (7.2). Moreover, if groups from different layers are overlapped then their intersection is essential and the sum size is bounded with (7.3), an evidence of the nested principle. The horizontal architecture approach utilizes disjoining principle: all nodes of the same layer are partitioned into independent overlays, defining groups that follow (7.4).

The disjoining and nesting principles allow introducing another classification for hierarchical architectures onto *disjointed* and *nested*. This classification reflects the use of vertical stack of layers to capture the node responsibility level. The classification criterion is combinative; a given architecture cannot be pure disjointed or nested. Instead, it should be considered rather disjointed then nested and vice versa.

Disjointed architecture is based on conceptual separation of groups of different layers. Inter-layer overlapping is kept low. A node participating in several layers performs different roles on each its layer. The node responsibility is primarily measured with the number $1 \leq m \leq M$ of layers that the node connects (m-responsible node). Regular nodes are least responsible. The population size of m-responsible nodes decreases rapidly with growing m (e.g., exponentially).

The characteristic property is *the major population consists of low responsible nodes*. A common case is when many supernodes connect two layers only (adjacent layers in ordered architecture) even if $M > 2$. Since m is not large at least for majority of nodes, the responsibility scale has coarse granularity; the simplest case is binary classification: low-performance and powerful nodes. Note that if the major population becomes consisting of high-responsible supernodes then the node differentiation disappears, degenerating the hierarchy.

The architecture is suitable for non-ordered layers, where groups of different layers represent essentially diverse domains. Each layer consists of many regular nodes of its domain, making the separation of groups of different layers. Few supernodes are gateways, providing sparse inter-layer connectivity. Protocols on each layer are domain-aware, and this diversification separates roles the same supernode plays in overlay routing and maintenance on different layers.

In the ordered case the responsibility may correlate with the layer index $i = 1, 2, \ldots, M$ when nodes in N_{i+1} provide more capacity than nodes in N_i. Although it is similar to the nested case, disjointed architecture requires rapid layer size reduction in (7.5). Furthermore, some nodes from N_i may be not presented in N_{i-1}. Importantly that a higher layer provides more efficient routing mainly due to smaller network size and only partially due to higher node responsibility. The latter, in fact, is distributed among m layers, usually with no specific prioritization.

Even in the ordered case, groups of different layers are conceptually disjointed. A supernode on layer $i + 1$ acts on behalf of its group from layer i. In other words, a layer $i + 1$ supernode aims at group-aware decisions and actions; it can be replaced with another node of the group. Actually, it is evolution of the "virtual nodes" concept [4, 12, 59] when a physical node runs a group of virtual overlay nodes.

Layers are sparsely connected by a few supernodes; it makes disjointed architecture close to vertical one. However, disjointed architecture also allows the horizontal approach. See example schemes in Figs. 7.5a, c, d and 7.6a.

Nested architecture primarily arranges the overlay functionality among layers $i = 1, 2, \ldots, M$, where M can be large, allowing fine granularity of the node responsibility. For simplicity we assume the case of ordered hierarchical architecture. The bottom layer contains all N nodes and implements the basic function that every node must perform. Every next layer enhances the function with advanced mechanisms that more responsible nodes apply for better performance or for other improvements of the overlay operability.

Each node u appears on layers $i = 1, 2, \ldots, m_u \leq M$, participating in m_u networks in total:

$$u \in \bigcap_{i=1}^{m_u} N_i, \quad u \notin \bigcup_{i=m_u+1}^{M} N_i, \quad \forall u \in N,$$

where m_u depends on u's available capacity. Consequently, the following "nested" structure appears:

$$N_M \subset \ldots \subset N_{i+1} \subset N_i,$$

which generalizes (7.5). Each node can be intuitively thought "a column" in the pyramid of layers. In contrast to disjointed architecture, the size reduction between layers may be low.

The layer differentiation is due to routing criteria, not due to the population. A popular solution is that each layer aims at own routing scale, and higher layers require longer-range links for better global routing. They form expressways for lower layers. Actually, it is evolution of the "large routing table" concept when every node varies its routing table size depending on the capacity, see such flat DHT designs as SmartBoa [18], EpiChord [30], and Accordion [31].

The characteristic property is *the major population consists of high responsible nodes*. It is a counterpart of disjointed architecture. The number of high responsible nodes can be big and even comparable with the size of the whole population. This property does not degenerates the hierarchy since efficient routing applies an appropriate criterion within its layer depending on the current lookup state.

Nested architecture includes ordered variants of vertical and horizontal architectures. An obvious instance is fully vertical architecture (Fig. 7.5b) with diverse routing criteria on different layers. Also, pure vertical architecture (Fig. 7.5d) can be considered nested if each cluster may delegate many supernodes to the next layer.

Similarly, in horizontal architecture (Fig. 7.6b) the bottom layer consists of small-scale overlays. The node responsibility is restricted within a small overlay. Overlays on upper layers scale up in size, and the node responsibility level grows appropriately. This case is an example when any node participates in every layer.

7.3 Hierarchical DHT Taxonomy

This section overviews particular designs of HDHT architectures. Since a multitude of proposals has been appeared in the literature we cannot present here all of them. Nevertheless, we expect that our list is representative, and a proposal that is not in the list is close to one we described. The difference is non-principal in architectural terms, e.g., the ID space realization or the communication protocol details.

The disjoining and nesting principles usually appear in the same hierarchical architecture design in a composition. We first list the designs where the disjoining principle prevails, then we consider the designs with the prevalence of the nesting principle. Within each list the order is chronological, by the first publication of design (according to the best of our knowledge).

7.3.1 Disjointed Hierarchical Architectures

Disjointed architecture differentiates the node responsibility by arranging heterogeneous nodes along the vertical dimension. Typically, a coarse granularity scale is used, and the number of high responsible nodes is minor compared with the sum number of nodes from lower responsibility groups. Groups of different layers are separated—their node population, responsibility, or functional role are diverse. In particular, a two-layer design assigns a small fraction of nodes to the top layer where they act on behalf of their groups. The most of two-layer designs follow the basic ordered two-layer architecture shown in Fig. 7.4 from Sect. 7.2.2.

7.3.1.1 Hierarchical Systems by Garcés-Erice et al. [10]

The design adopts the two-layer supernode model, which was originally developed for unstructured P2P systems [27, 70]. Groups of the bottom layer consist of proximity close nodes. Each group forms an independent overlay, applying the horizontal architecture approach. Any overlay can use its own protocol (structured or unstructured) for intra-group routing, leading to a hybrid architecture design. On the top layer, all groups form a single overlay (e.g., Chord-based) for global routing.

Overlays on the bottom layer operate autonomously on each other. Within each group, one or more supernodes are selected to represent this group on the top layer. Supernodes are most powerful nodes, hence they can meet more responsibilities. In particular, a supernode maintains additional independent routing table for the top-layer overlay. The top overlay aims at efficient global routing due to the small overlay network size and good underlying network coverage.

When a node u joins the system, it must know ID of its group on the bottom layer. Hence u can contact any node existing in the system to locate a supernode of the group. If the group exists then u joins using the group overlay protocol. Otherwise, a new group is created with the only (super)node u.

The essential point is that an entire group acts as a virtual node in the top overlay; group's supernodes are only representatives and can be reassigned. Although the design allows most or even all nodes of a group to become supernodes, it leads to the degeneration when the bottom layer overlays disappear moving all routing functionality to the top layer.

This construction can be generalized to $M > 2$ layers. The higher layer the larger-scale routing its overlays provide. The top layer $i = M$ is a single global overlay. Each lower layer consists of many groups, and a group is a "node" in the overlay of the current layer. In a group of layer $1 \leq i < M$, nodes are classified into many regular nodes and a few supernodes. Supernodes represent the group on layer $i + 1$ maintaining the routing table for their group ID. Each regular node must know at least one supernode of its group. On layer $i + 1$, a supernode for a group of layer i either becomes a regular node or acts as a supernode again to represent its higher-level group on layer $i + 2$.

Starting at a regular node on layer $i = 1$, a lookup for key k sequentially goes to supernodes of groups on layer $i = 2, \ldots, M$ and reaches eventually the top layer. Then the lookup sequentially visits the groups responsible for k on layers $i = M, \ldots, 1$. Finally, on the bottom layer, the lookup is delivered to the destination node.

The hierarchy reduces the length of lookup paths compared with flat DHT of $O(\log N)$ routing complexity. For instance, in the $M = 2$ hierarchy with N_2 nodes in the top overlay, the reduction factor is $\log N / \log N_2$ if the top overlay and bottom-layer overlays provide $O(\log N_2)$ and $O(1)$ routing, respectively. It reasons the requirement $N_2 \ll N$.

The most routing and hierarchy maintenance load is pushed to supernodes. A certain mechanism is required for selecting and maintaining supernodes. Since several supernodes represent a single entity (group) in the higher-layer overlay, the latter needs to modify appropriately a conventional flat DHT protocol for the overlay.

The original $M = 2$ design was analyzed for the case of the Chord DHT on the top layer. Similar two-layer design was exploited in [33] for a hierarchical small-world system where the top layer is a single Symphony-based overlay and the bottom layer consists of group organized as independent Chord rings. Also this basic design was used in mDHT [29] that further emphasized one of the key postulates: an entire group forms a node in the top layer overlay, regardless which bottom-layer nodes of the group are recently its supernodes.

7.3.1.2 Kelips by Gupta et al. [16]

As in the hierarchical systems of Garcés-Erice et al. the design employs two layers: the bottom layer is for $N_1 = N$ nodes and the top layer is conceptual; it is for $N_2 \leq N$ node groups. Groups use own ID space $S_2 = \{0, 1, \ldots, N_2 - 1\}$, i.e., the total number of groups should be fixed a priori. A node associates itself with a group by hashing the node ID, yielding every group to be of size about N/N_2 nodes.

The connectivity structure of a group is closer to unstructured P2P topology: a node u knows most of nodes from its own group, leading $O(N/N_2)$ entries in u's routing table. For each other group, u stores additionally a constant-sized set of group nodes: $O(N_2)$ entries. Let R be the total number of resources in the system. Then u indexes R/N_2 resources, storing for resource r a pair (k_r, IP_r) with resource key and responsible node IP address.

On one hand, large values of N_2 reduce the memory for entries of own group nodes and the memory for resource indexes. On the other hand, the number of entries for other groups grows with N_2. The optimal tradeoff value is $N_2 = \Theta(\sqrt{N + R})$. Assuming $R = O(N)$, the optimal value leads to a routing table with $O(\sqrt{N})$ entries.

In a lookup, a querying node u hashes the resource name to the appropriate group ID and sends the lookup to the topologically closest node v that u knows for that group. Then v resolves the lookup by searching among its index and returns IP address of the responsible node. When v fails in resolving the lookup, then multi-hop (and multi-try) routing is enabled. Nevertheless, the probability of appearance of long lookup paths is very low since the routing state at nodes is highly redundant. As a result, the average number of hops is preserved within $O(1)$ hops.

Kelips is loosely structured and closer to unstructured P2P designs. The high routing efficiency requires the expensive maintenance (large routing tables and resource index) that uses gossip protocols for information dissemination, similarly to unstructured P2P overlays [22, 53]. It leads to $O(\sqrt{N} \cdot polylog N)$ cost, e.g., the expected convergence time for an event is $O(\sqrt{N} \cdot \log^3 N)$. Nevertheless, the $\sqrt{N}$-property of Kelips design can be used in more structured networks.

7.3.1.3 Structured Superpeers by Mizrak et al. [46]

Similarly to Kelips, each node maintains $O(\sqrt{N})$ local state to achieve $O(1)$ routing. The hierarchy is two-layer and follows the horizontal approach. Both layers use the same circular ID space. All N nodes are placed on the outer ring. Among them $N_2 = \Theta(\sqrt{N})$ high-capacity nodes are chosen to be supernodes; they create an additional ring—the inner ring—to provide fast global routing.

The outer ring is uniformly split into N_2 arcs such that the nodes of each arc form its own overlay on the bottom layer and assigns a supernode for the top layer. A bottom-layer overlay has the star topology: the supernode knows all nodes in its outer ring arc ($O(\sqrt{N})$ entries). In addition, any supernode maintains routing entries for all other supernodes ($\Theta(\sqrt{N})$ entries). As a result, the inner ring on the top layer is a single overlay with the fully-connected topology. This connectivity structure is close to unstructured P2P topology.

In a lookup for k, an initiator node sends the request directly to its supernode. If its arc includes k then the supernode locates the successor of k in the local routing table and returns the result (local routing). Otherwise, it forwards the lookup to the supernode who is responsible for the enclosing arc for k (global routing). That supernode locates the successor node in own routing table and returns the result. Even in the worst case, routing has constant cost $O(1)$.

In bootstrapping, first nodes of the system must provide an initial set of supernodes When the system evolves, an extra mechanism is needed for selecting new supernodes and keeping their amount equal $\Theta(\sqrt{N})$. A topology change of the inner ring is disseminated to all supernodes leading to $\Theta(\sqrt{N})$ maintenance traffic.

This two-layer design can be thought as a hybrid of hierarchical systems [10] and Kelips [16]. The architecture is disjointed since the difference between a regular node and a supernode is essential. In particular, a bottom layer overlay cannot function without its supernode.

7.3.1.4 OneHop by Gupta et al. [6, 15]

It is a three-layer hierarchy that supports one-hop routing with complete membership information at each node. Although memory cost allows a routing table containing millions entries per node, the network bandwidth is too expensive for straightforward dynamic membership maintenance algorithms. OneHop preserves efficient dissemination of the membership information by partitioning the update function among two additional layers.

OneHop ID space is circular, similarly to Chord. All N nodes form the fully-connected topology overlay on layer $i = 1$, leading to one-hop routing complexity by the cost of $\Theta(N)$ node state. Each node controls its immediate successor and predecessor using the Chord algorithm with periodic keep-alive messages. Additionally, a membership change can be detected with a lookup.

To maintain correct full routing tables, notifications of membership change events must reach every node within reasonable time. If a node sends an update notification to every other node, the bandwidth consumption becomes unacceptable. OneHop achieves efficient update dissemination by imposing additional layers.

Similarly to Mizrak's structured superpeers, the ID space is divided into $N_3 \ll N$ equal contiguous arcs (slices). A supernode (slice supernode) is assigned to each arc deterministically as the node that immediately succeeds the mid-point of the arc. Any regular node knows its slice supernode. When a new node has ID closer to the slice mid point then the node becomes the slice supernode. The slice supernode overlay on the top layer $i = 3$ has fully-connected topology: each supernode knows all other slice supernodes.

In turn, each slice is divided into $N_2 \ll N/N_3$ equal-sized units. For each unit its slice supernode assigns a unit supernode. The assignment is again deterministic: the successor of the unit mid point. Since a slice supernode knows all its unit supernodes, they form a star-topology cluster on layer $i = 2$. Any entry in a slice supernode routing table is marked either "regular node", "unit supernode", or "slice supernode".

When a node detects a membership change, it notifies its slice supernode. The latter collects all notification on a given time interval and then sends the message to every other slice supernode. Any slice supernode aggregates the information from other slice supernodes and periodically sends the aggregate message to its unit supernodes. A unit supernode piggybacks this information in keep-alive messages to its immediate successor and predecessor on the bottom ring.

A regular node u propagates the updates in one direction: if the update is from the predecessor, then u sends the update to the successor, and vice versa. This hierarchical approach imposes event dissemination trees with low redundancy; duplications of an update message may occur but they are rare.

OneHop is primarily attractive in small or low-churn systems. The authors recommended the OneHop design for systems with up to a few million nodes. OneHop does not differentiate node capacity: any node should be ready to afford additional responsibility if being occasionally assigned a supernode. Since the load imbalance between slice supernodes and regular nodes is high [32, 47], the design is inappropriate for a system with low-capacity nodes.

One-hop DHTs (in general, $O(1)$-hop DHT) are superior to multi-hop DHTs in stable and high-capacity network environments because of more efficient lookup bandwidth utilization. Risson et al. [55] proposed two hierarchical designs based on one-hop DHTs. The 1HS design (One Hop Sites) uses independent one-hop DHT overlays that can cooperate across protected sites and recover from network partitions between those sites. This hierarchical architecture is appropriate for a federation of large data centers. The 1HF design (One Hop Federation) arranges one-hop DHT overlays into a tree-based federation of regional hierarchies. Regional overlays have subtended organizational rings and organizational overlays have subtended site overlays. This hierarchical architecture is appropriate for global applications like name resolution or internet telephony.

7.3.1.5 Rings of Unstructured Clouds by Singh and Liu [57]

This two-layer hierarchy allows constructing hybrid P2P overlays that combine structured and unstructured topologies. The two layers provide the distinction of two desired properties: (1) structured P2P overlays (DHT rings) on the top layer are for efficient routing and (2) unstructured P2P overlays (clouds) on the bottom layer are for anonymity.

Flat DHT space is used for the global ring, e.g., the n-bit circular Chord space $S = [0, 2^n)$. All N nodes participate in the global ring forming a flat DHT overlay. A cloud is a small Gnutella-like network. To create a cloud, its name is hashed using multiple hash functions to find several nodes in the global ring. They connect to each other forming the cloud. Each becomes a rendezvous node, a kind of supernode acting as a cloud stakeholder. When a rendezvous node leaves the system, a new one that simply replaces the old one is found using the global ring DHT protocol.

Each node caches cloud names it sees in lookups. Hence a new node can always get a list of active clouds. A new node first joins the global ring and gets the list. Then the node hashes the cloud name to obtain a rendezvous node. The latter bootstraps the node into the cloud. Nodes controls the cloud size using a distant vector (distance from the rendezvous nodes); a cloud stops accepting new members if its distance limit is reached. The number of rendezvous nodes per cloud is small (system-wide parameter) compared with the typical cloud size, resulting in disjointed architecture.

The dynamic relation between a cloud and the services that its nodes provide uses rendezvous rings (R-Rings). An R-ring is an independent DHT overlay consisting of one rendezvous node from each cloud. A rendezvous node uses its cloud ID in the R-ring. Creating an R-ring follows the same DHT protocol as the global ring. There are multiple R-rings, one for each rendezvous node of every cloud. The R-ring structure introduces additional disjointed sub-hierarchy.

A lookup originates in a cloud and uses a random walk within the cloud. The originator sets random TTL and the lookup is forwarded to random neighbors until TTL $= 0$. The last node of the random walk becomes a crossover node. It preserves anonymity since a random node communicates on behalf of the originator.

If the lookup is tagged with a cloud name then the global DHT ring is used to locate a node of the target cloud. If the relation between clouds and services is dynamic the crossover node finds a rendezvous node of its cloud. Then the lookup runs over the R-ring to locate a rendezvous node of the target cloud. The rendezvous node broadcasts the lookup in its cloud. Then a responsible node replies to the crossover node, which forwards the reply back to the originator. Note that a random walk can be applied also on the target cloud side for service provider anonymity.

Random walks decrease the routing performance. Each rendezvous node has to maintain an additional routing table for the R-Ring. The load due to node rendezvous and crossover responsibility is uniformly distributed among N nodes, regardless of their heterogeneity.

The rendezvous mechanism is extremely important for hierarchical P2P architectures. It protects overlays in the hierarchy from object mobility when a node or resource moves to another overlay. Risson et al. [54] proposed a rendezvous abstraction for a wide class of hierarchical P2P architectures. Overlays of the hierarchy stores additional records implementing a location information plane that manages globally unique, persistent, semantic-free identifiers.

7.3.1.6 Chordella by Zoels et al. [77, 79]

They analyzed the generic two-layer horizontal hierarchical architecture when every bottom-layer overlay delegates exactly one supernode to the top layer. The top level uses a flat DHT to form the global ring of supernodes; the analysis applies the Chord DHT for the reference case. On the bottom layer each supernode manages its own overlay of regular nodes, which exploit the supernode as a proxy.

A unique l-bit ID is associated with each node, either supernode or regular one. The resource ID space reflects the two-layer hierarchy: a resource item has a $2l$-bit key with first l bits being independent on the second l bits. The first part of a key identifies the responsible group and the second part determines the responsible node within that group.

Routing is two-phase: global and local. First, a lookup for k resolves the responsible group finding in the top-layer overlay the supernode by a DHT lookup for the first l bits of k. Second, the lookup determines the responsible node within that group using the second l bits of k.

There are three alternative connectivity structures for overlays of regular nodes: fully-meshed (FuMe) when every node is connected to all other nodes of its overlay, single-connection (SiCo) when every regular node is connected to its supernode only (i.e., star topology), and DHT-based when a flat DHT forms each overlay. Note that the analysis was limited with the pure case when all bottom-layer overlays follows the same connectivity structure.

In the FuMe and SiCo structures, any group's supernode knows all members of its group, and local routing is one-hop. In the DHT-based structure, local routing is $O(\log N/N_2)$ if the Chord DHT is used. Note that a node always contacts its supernode, even if the responsible node is within the same group.

The analysis considered the load efficiency that such hierarchical architectures provide compared with flat DHTs. It is especially important for heterogeneous environments when many nodes cannot provide much capacity to the system. They focused on the optimal ratio between supernodes and the total number of nodes in the system.

- The optimal supernode fraction $0 < \alpha < 1$ (i.e., $N_2 = \alpha N$) that minimizes the total operation costs subject to bearable load on supernodes.
- Bottom-layer overlay connectivity structures that allow small α, reducing the required number of supernodes, which have to take the most overhead of the hierarchy.

The analysis concentrated on $\alpha \leq 25\%$ since for higher values the hierarchy degenerates. The numerical evaluation showed that $\alpha \geq 10\%$ for typical network settings. Note that small values of α reflect the characteristic tendency of disjointed architecture.

The design is supported with a cost model that shows the average number of sent and received messages for regular nodes and supernodes. The model leads to a distributed algorithm that dynamically adjusts α to a cost-optimal value. If the network load is low, then a supernode can demote itself to a regular node. If the experienced load is high, then a supernode promotes a regular node to a supernode.

The evaluation showed that the SiCo structure is superior to the FuMe and DHT ones, in terms of the tradeoff between minimizing the total network traffic and avoiding to overload the highest loaded supernodes.

The design was extended in [78] with a load balancing algorithm for supernodes. It aims at assigning an appropriate number of regular nodes (or other types of the load) to every supernode. Every supernode keeps information on the load level of some $O(\log N)$ other supernodes. A new node is assigned to such a supernode that currently has a low load level. When a supernode leaves, its regular nodes have to rejoin the system. In summary, it is a step towards self-adoptable hierarchy with the fixed number of layers.

SA-Chord [52] is an instantiation of the Cordella design. The goal is to move the routing load entirely to a small set of the most capable nodes. These N_2 supernodes are "routing nodes"; only they may forward lookups. The $N_1 = N - N_2$ regular nodes are "non-routing nodes"; they are lookup sources and destinations only. The top layer (routing ring) is a modified Chord DHT and the bottom layer follows the SiCo connectivity structure. A regular node always requests its supernode for a lookup. Then the routing ring resolves the lookup to the responsible supernode. The latter finally forwards the lookup to the destination regular node.

SA-Chord uses the flat circular Chord ID space. There is no need in hierarchical IDs since regular nodes does not participate in routing. A supernode knows all regular nodes on the arc clockwise from itself to the closest supernode. The routing ring is a Chord extension. Instead of one finger per power-two interval, SA-Chord allows k fingers, where $k \geq 1$ is a system-level parameter. This routing table redundancy improves the routing ring performance to $O(\log_{2k} N_2)$ hops per lookup compared with $O(\log_2 N_2)$ of the basic Chord. Additionally, proximity neighbor selection is applied to reduce routing latency.

7.3.1.7 HONet by Tian et al. [62]

The design uses the basic two-layer hierarchy and constructs on top of it a hybrid P2P system (see another hybrid two-layer design in Sect. 7.3.1.5). The bottom layer consists of many topologically-based clusters of small size, each maintains own overlay with its supernode (cluster root node). All cluster roots form a single global overlay (the core network) on the top layer.

Any overlay is DHT-based and has own independent ID space. De Bruijn network is used for the reference case. A node (resp. resource) is identified by its cluster root ID in the core network and its node ID (resp. resource key) in the local overlay. Hence a full ID is a pair (c, k).

A cluster root is chosen as the most stable member of the cluster. HONet uses a coordinate system with a group of well-known landmark nodes. Cluster coordinates are coordinates of its root; they are stored in the core network DHT. Space-filling curves, which preserve locality, are used to map multidimensional coordinates to the one-dimensional core network ID space (inspired by work [69]).

When a node joins, it searches in the core network a nearby cluster using its own coordinates mapped to the core network ID space. If no cluster root is found within a predefined cluster radius, then the node becomes a root of a new cluster. Otherwise, the node joins the existing cluster.

Hierarchical routing is two-phase, similar to the previous designs. If a lookup for (c,k) is started in a cluster $c' \neq c$ then the global routing is performed in the core network, and the lookup moves from one cluster to another. The lookup reaches a root node of the target cluster using c. Then the lookup continues on the bottom layer, i.e., in the local DHT overlay using the key k.

For better routing efficiency the design employs elements of unstructured P2P topology. Each node, in addition to its DHT-based routing table, creates random links to nodes of other clusters if the node capacity allows. The link construction uses a random walk algorithm when a node initiates a message with a finite TTL. A node u forwards random walk messages to a neighbor v with probability proportional to f_v, where f_v is a generic fitness metric that characterizes v's available capacity and the performance of link $u \rightarrow v$. When a node receives a random walk message with TTL $= 0$, it can provide the inter-cluster link.

A node publishes information about its inter-cluster links in its local cluster DHT using cluster IDs as the keys. That is, if a node u is responsible for a key c in the cluster $c' \neq c$ then u knows all nodes in its cluster c' that keep random links to members of the cluster c. Hence u can serve as a local reflector to c to accelerate global routing of lookups for (c,k). If a node v from c' does not know a random link to a node from c then v forwards the lookup to the local reflector u based on local DHT routing. The reflector forwards the lookup either to a local node that knows a random link to the cluster c or to the cluster root to perform hierarchical routing. When the lookup enters another cluster using hierarchical routing, it can try fast inter-cluster routing again.

Note that this solution applies the small-world principle when a routing table consists of short-range and long range neighbors [23, 26]. The same global routing acceleration method was proposed in Cyclone [1, 2] (see Sect. 7.3.2.10). Although the design applies the same cluster-based model as the hierarchical systems of Garcés-Erice et al., the architecture is closer to vertical one; all clusters tend to form a global (unstructured) overlay on the bottom layer where clusters are connected using random links.

7.3.1.8   Content-Based Hierarchy by Zoels et al. [76]

The design applies content-based hierarchy to support efficient topic-based search. The hierarchy reflects a content category tree, thus following the tree-based hierarchy model. The architecture is multi-layer ordered horizontal. It is set up by an arbitrary number of independent Chord rings as topic-spaces(it can be generalized

Fig. 7.7 Example of content-based multi-ring hierarchy [76]. *Filled circles* correspond to supernodes; each maintains its own ring for a subtopic and acts as a regular node (*unfilled circle*) in the parent ring. Topic content is distributed over all ring nodes. Path $A \rightarrow B \rightarrow C$ corresponds to qualified name "/root/germany/munich/airport"

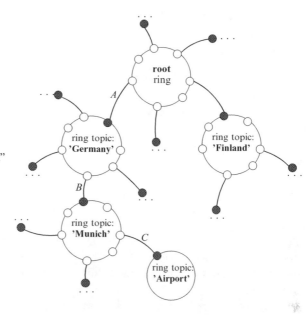

to any flat DHT). Each ring corresponds to a vertex in the tree and stands for an individual topic, divided, if needed, onto subtopics with own rings on the lower layer. The root ring is the only overlay on the top layer.

A ring is initiated and managed by its "ringmaster" (supernode). Initiating a new ring, its supernode informs the parent ring supernode and remains a regular node in the parent ring. A supernode is always aware of its parent ring and all own child rings. Most of ring nodes are regular; they belong to that ring only. Recall that it is characteristic tendency of disjointed architecture. Figure 7.7 depicts an example.

There are two types of resource names: qualified and unqualified. A qualified name refers to the exact path in the content category tree (e.g., "/root/germany/munich/airport") and identifies the corresponding rings. An unqualified name is a keyword to lookup the resource in a Chord ring (keywords are hashed to the ring ID space).

The design allows a dynamic structure adaptation when users can create and remove their own topic spaces. It requires, however, some discipline among users; they should work with the ring respective to a given topic. Alternatively, an extra mechanism can be employed, e.g., a bootstrap server that is aware of the global system structure.

The design is much inspired by multi-ring hierarchy [45] (see Sect. 7.3.2.8). The latter, however, is closer to nested architecture; many nodes can participate in multiple rings. Also [45] is oriented to administrative domain hierarchies.

7.3.1.9   Hierarchy-Adaptive Topology by Zhang et al. [71]

The design is dynamically adaptable implementation of pure vertical architecture. The number of layers M is not fixed and is changed self-adaptively according to current value of N. The tree-based hierarchy model is applied for bottom-up construction of the layers.

The ID space is d-dimensional torus with Cartesian coordinates, the same as in CAN [50, 51]. Space operations are inherited from CAN. Node ID determines a zone (hyper-rectangle) in the space, and the node is responsible for all keys that are points in its zone. Neighbors are nodes whose zones overlap along $d - 1$ dimensions and abut along one dimension. Zones can be split and merged, which are the core operations in this hierarchical design.

The layer construction uses bottom-up and top-down directions. The former is cluster-based and takes into account the underlying network proximity. The latter introduces tree-based partition of the ID space. The construction is dynamic and depends on the current value of N. There are certain parameters and rules that control the size of clusters, and hence the number of nodes on each layer and the number of layers.

In the bottom-up direction, layer i nodes are grouped into clusters according to proximity in the underlying IP network. Each cluster selects its supernode among the most powerful nodes; it will represent the cluster on the layer $i + 1$. Any node (child node) has direct link to its supernode (parent node). Furthermore, the design requires a node to keep links to all its ancestor nodes on upper layers (ancestor routing table). Each supernode also maintains links to all its immediate children (child routing table).

In the top-down direction, the entire ID space is partitioned among all the layer $i + 1$ nodes according to the CAN space partition operation. The common layer $i + 1$ overlay is formed using neighbor connections (neighbor routing table). The zone that a node u maintains on layer $i + 1$ is further divided between u's children, which form a proximity cluster on layer i.

To join the system a new node u finds the proximity closest node and becomes a member of the corresponding cluster on the bottom layer. If all other nodes are far from u then u forms own cluster. Being a member of a cluster, u may become a supernode by some election procedure. After that u joins clusters on upper layers. When u leaves the system its role is delegated to another node.

If a lookup key k is within the current zone of a node u, then the lookup already reaches the destination. Otherwise, u forwards the lookup to the ancestor node whose zone is the smallest one that covers k (bottom-up routing first). Then the lookup iterates through the child nodes whose zones cover k until the destination node. Routing to neighbors on the same layer is used when ancestor nodes are failed or busy.

Nodes on higher layers have more responsibilities since they handle more lookup requests. A node can maintain fast inter-layer links (shortcuts) that reduce the load of higher layer nodes and accelerate global routing. The idea is similar to HONet but instead of random walks the design applies the expressway method of eCAN [66,73] (Sect. 7.3.2.2), which is based on the nesting principle. A node u creates a additional shortcut to a grandchild of a neighbor of u's ancestor on some layer.

The key points of this hierarchical design are (1) proximity neighbor selection and (2) longer hops in the ID space on higher layers. The first one is due to clustering when nearby nodes in the underlying network become neighbors in the overlay. The second point is due to geometrical reduction of the number of nodes with $i = 1, 2, \ldots, M$, and hops between neighbors become larger distance leaps.

7.3.1.10   Chord^2 by Joung and Wang [19]

The design is a variation of the basic ordered two-layer architecture. The aim is at reducing overlay maintenance overhead for regular nodes, which compose the major population. A Chord ring of all N nodes is on the bottom layer (the regular ring) and a Chord ring of $N_2 \ll N$ supernodes is on the top layer (the conduct ring). The costly part of the maintenance related to Chord fingers is pushed to the conduct ring, since supernodes are assumed to be the most powerful and stable nodes.

A regular node knows at least one supernode. Regular nodes proactively (periodical checks) maintain their successor and predecessor links. When a node u detects node joining or leaving, u reports to a supernode, and the conduct ring becomes responsible for the notification of all affected nodes. Supernodes receive join/leave information and reactively[1] send the notifications. Receiving an update notification, a regular node entirely updates its routing table (all finger links). Therefore, regular nodes do not perform the costly proactive finger maintenance of the basic Chord.

The system starts from its regular ring. Nodes collect the uptime information (or other metric for supernodes). Each node estimates uptimes of other node and recommends the most stable nodes to be supernodes. (Various reputation-based election mechanisms can be also introduced.) Once a node is promoted to be a supernode, it constructs an additional ID for the conduct ring and joins this ring following the basic Chord protocol. A variant when IDs in both rings concise is possible.

The conduct ring DHT stores information about regular ring topology. Each regular node u inserts to the conduct ring a link object (u, u_{succ}) where u is used as a key and u_{succ} is u's successor. Similarly, u inserts a finger object (v, u) for every neighbor $v \in T_u$. Now, whenever a node w leaves the regular ring, a supernode knows which nodes have neighbor links to w and notifies these nodes to replace w with w's successor in their routing tables.

When a node joins the regular ring, its routing table of size $O(\log N)$ can be constructed based data from the conduct ring, the total cost is $O(\log N_2 \cdot \log N)$ compared with $O(\log^2 N)$ in the basic Chord. When a node leaves the regular ring,

---

[1] It is in contrast to Chord with its proactive routing table maintenance when every node periodically calls the stabilization and fixfingers procedures.

either it informs a supernode or its predecessor detects the departure. The cost takes $O(\log N)$ messages to update routing tables in affected regular nodes. In the basic Chord, the cost of node leaving is included to the maintenance cost.

A regular node periodically check only their successors and predecessors; each round takes $O(1)$ messages compared with $O(\log N)$ in the basic Chord, where a round affects all neighbors. The conduct ring maintenance is the same as in the basic Chord: the costs of node joining and maintenance are equal, each takes $O(\log^2 N_2)$ messages per operation. Since supernodes are stable, the maintenance cost can be reduced by increasing the interval of the periodical checks. Additionally, each supernode stores $O(\frac{N}{N_2})$ link objects and $O(\frac{N \log N}{N_2})$ finger objects.

Note that the Chordella design also allows Chord rings on both layers. It, however, follows the horizontal architecture approach with many Chord rings on the bottom layer. In contrast, Chord^2 is a vertical architecture design with a single Chord ring per layer. From this point of view, Chord^2 is closer to HONet, but the latter has less structured topology on the bottom layer.

Tanta-ngai and McAllister [61] applied the similar idea for separating the local and global routing onto own layers (cf. Brocade and eCAN in Sect. 7.3.2). The bottom layer is a flat Chord ring in $S = [0, 2^n)$. On the top layer, an auxiliary supernode ring (the expressway) uses finer arc-partition granularity than two-power intervals of Chord. A supernode u has a link to the closest supernode in every arc

$$[u + ap^i, u + (a+1)p^i), \quad a = 1, 2, \ldots, p-1,$$
$$i = 0, 1, \ldots, \lceil \log_p 2^n \rceil - 1,$$

where $p \geq 2$ is an integer parameter (expressway forwarding power). More capable nodes become stronger connected via the expressway ring, aiming at better global routing. Instead of periodic updates of the Chord protocol, a reactive event-based notification mechanism is used for maintaining the expressway ring. The maintenance is cheaper assuming that supernodes are relatively stable.

7.3.1.11   Hierarchical DHT-Based Overlay Networks by Martinez-Yelmo et al. [39–41]

It is a variant of the basic ordered two-layer hierarchical architecture with the focus on global multimedia applications where many inter-operating domains exist for users or content. The particular reference case is P2PSIP, where SIP service is based on P2P manner and P2P signaling traffic is carried by SIP messages.

The two-layer hierarchy directly reflects the application domain division, following horizontal architecture. A domain corresponds to an exclusive group of nodes (multimedia participants); they form an independent overlay network (domain overlay) using any flat DHT. Each group is represented by at least one supernode on the top layer. The basic design uses one supernode per domain. Supernodes are dedicated entities allowing regular nodes (e.g., devices like a mobile phone) to be of lower load. These $N_2 \ll N$ supernodes form the global interconnection overlay using a flat DHT.

The interconnection overlay acts like a directory service among the different domains, and each supernode publishes there its location information and information about its domain overlay. There is a common packet format to assure interoperability between different domains. The supernode selection and update mechanisms must be integrated in the maintenance protocol of each DHT; they can be based on the strategy from [44].

Similarly to many hierarchical designs, ID prefixes and suffixes are used for identifying objects on different layers. A hierarchical node ID is composed by a suffix ID (l_s bits hashed from its domain name, e.g., "example.com") for intra-domain communication and a prefix ID (l_p bits hashed from its full name, e.g., "user@example.com") for identifying the domain. For each resource, a node stores a resource tuple containing the resource hierarchical ID, the resource full name, and the resource itself.

In a lookup for a hierarchical resource key k, a node u compares its own prefix with k's prefix. If they are different then u forwards the lookup to its supernode. Otherwise, intra-domain routing happens. As in Chordella and the hierarchical systems of Garcés-Erice et al., the reduction factor of a regular node routing state is $\log N / \log N_2$ if DHTs with $O(\log N)$ state are used in domain overlays. Supernodes have to support N/N_2 times more bandwidth and queries. In contrast the mentioned designs, where all the work is delegated to supernodes, a regular node forwards a lookup to its supernode only when inter-domain communication is needed.

Another hierarchical design for P2PSIP was introduced by Le and Kuo [28]; it does not use domain-based classification and is primarily oriented to node physical capacity, which varies essentially in heterogeneous environments. There are $M \geq 2$ layers, where M is fixed in advance assuming that N does not vary significantly (it does not prevent some nodes to be temporarily inactive). Each layer forms own Chord-based overlay, following vertical architecture. The design clusters N heterogeneous nodes according with their capacity and forms the population of each layer such that the number of nodes is geometrically reduced with $i = 1, 2, \ldots, M$.

An analytical model of M-layer hierarchical P2P architecture for Internet telephony systems was introduced in [17]. The horizontal approach with disjointed architecture is applied with the system-level parameters $C > 1$ and $K > 1$. On layer i its nodes organize into non-overlapped groups of size C using an arbitrary clustering mechanism. A group is an overlay network based on any P2P protocol. Each group delegates one supernode for the upper layer $i + 1$ and controls at most K groups from the lower layer $i - 1$. Notably that such a group can include nodes that do not belong to any lower layer—an additional sign of disjointed architecture. Although each group can be a DHT-based overlay with $O(\log C)$ intra-group routing, the inter-group routing algorithm is closer to unstructured P2P networks when a lookup is duplicated up to $K + 1$ directions: K or less groups on the lower layer and supernode's group on the upper layer.

The model hierarchy varies with the number of active nodes N. Nodes join the system according to a non-stationary Poisson process with arrival rate $\lambda(t)$ and stay active for i.i.d. random time, hence the problem is reduced an $M_t/G/\infty$ queue. Then the average population size is estimated for sinusoidal arrival rate and exponential sojourn time (both assumptions are typical for telephony systems). In the steady state it yields periodic behavior of $N = N(t)$ as well as of the lookup performance. Although the model can be applied for the two-layer design of Martinez-Yelmo et al. and the M-layer design of Le and Kuo, the assumption on immediate hierarchy reorganization with varying N and the flooding-like routing algorithm make the model less realistic.

7.3.1.12 HSN by Guisheng et al. [13]

It is a two-layer hierarchy with the small-world ring of supernodes on the top and many clusters on the bottom, hence following the horizontal architecture approach. The basic idea is close to structured superpeers of Mizrak et al. All N nodes participate in the global ring on the bottom layer. It is partitioned into several clusters, each one is an independent overlay. A cluster assigns a supernode (cluster head), and the N_2 supernodes form own overlay on the top layer.

In the global circular ID space, a cluster covers an arc in between two consecutive supernode IDs. The start (clockwise) node u is the cluster supernode; other consequent nodes on the arc are regular, forming the set $\mathscr{C}_u$. Note that the supernode is a predecessor for all its cluster nodes. The final node does not belong to this cluster; it is a supernode of the next one. Every node in a cluster knows its cluster arc. Consequently, a node always can determine whether inter- or intra-cluster routing is needed.

Short links connect regular nodes within a cluster. Each regular node keeps short bidirectional links to its n_{short} consecutive successors. That is, any regular node also knows n_{short} its consecutive predecessors. In addition, a short link connects a regular node with its supernode. Intra-cluster communication uses greedy routing, which is similar to the basic Chord algorithm. Since the arc size is kept small a mechanism like Chord fingers is not very essential in the local routing and the simple use of consecutive successors is adequate. Moreover, the bidirectional property allows anti-clockwise forwarding, thus improving the greedy routing performance.

When a regular node determines that inter-cluster routing is needed, the node forwards the lookup to the supernode. The latter decides and fixes the routing direction (clockwise or anticlockwise) for subsequent intra-cluster routing. After that the intermediate supernodes on the path do not change the routing direction.

Long links connect supernodes in a ring for inter-cluster routing. Each supernode u keeps n_{long} long bidirectional links (in addition to its $|\mathscr{C}_u|$ short links). The neighbor selection algorithm follows the small-world principle: neighbors should appear on all distance scales; the probability of closer neighbors is higher. Actually, the design uses the algorithm similar to Chord fingers: the ring is partitioned onto power-two intervals and one long link is selected from each interval. For $n_{\text{long}} = \Theta(\log N_2)$ or higher it leads to $O(\log N_2)$ routing in the supernode ring. For lower n_{long} the estimate of $O(\text{polylog} N_2)$ hops is valid [5, 24].

A mechanism for cluster split and merge is needed to keep every cluster within the given bounds to the number of nodes. The protocol preserves the inequality

$G_{min} \leq |\mathscr{C}_u| \leq G_{max}$ for system-level constants G_{min} and G_{max}. The reactive routing maintenance is on the top layer; when a message passes through a supernode, it piggybacks its node ID and IP address to the message. The proactive routing maintenance is on the bottom layer; every regular node periodically sends heartbeat messages to its neighbors.

The connectivity structure of the supernode ring is close to the Chord protocol: a discrete implementation of the small-world principle that leads to logarithmic routing [26]. From this point of view, the use of any DHT with logarithmic routing leads to comparable performance values. The shown experiment routing outperformance over the flat Chord DHT is achieved partially because of the bidirectional routing. Note that there exist several bidirectional modifications of Chord with the similar outperformance over the basic Chord, e.g., see [8, 42].

7.3.1.13 GTPP by Ou et al. [48]

GTPP (general truncated pyramid P2P) architecture is a generic design scheme for tree-based P2P hierarchies. On layers $i = 1, 2, \ldots, M - 1$, nodes are grouped into several disjointed overlays with own P2P protocols.[2] Each overlay assigns one supernode to be a gateway to the next layer. Finally, a single overlay appears on the top layer $i = M$.

From the vertical perspective, the overlay structure consists of multiple trees rooted by the nodes of the top overlay, which make the hierarchy looking like a truncated pyramid (see also Fig. 7.5d in Sect. 7.2.3). The number of nodes on each layer decreases exponentially with $i = 1, 2, \ldots, M$, assuming that the majority of nodes have relatively weak capability and the much powerful nodes are in the minority.

A lookup is sequentially forwarded bottom-up until it reaches the top overlay. The latter delivers the lookup to the closest supernode. Then the lookup sequentially goes top-down until the responsible node is found.

The analysis showed that GTTP decreases the expected routing latency though the path hop length can be higher than in a flat DHT. The GTPP layering provides certain rules for load allocation to heterogeneous nodes: the higher the layer the more load its nodes have. GTPP architectures with $M = 2, 3$ layers are most reasonable in practical settings.

7.3.2 Nested Hierarchical Architectures

Nested architecture arranges the overlay functionality along the vertical dimension. Each layer performs the same function using different mechanisms. Typically, the

---

[2]For the reference case, GTPP utilizes Chord for all overlays.

arrangement is incremental, and a node selects its "function stack" with the basic function on the bottom and subsequent enhancements on upper layers. Processing an operation a node decides which layer suits best and applies that layer mechanisms for the operation. In particular, a two-layer design often employs the top and bottom layers for global and local routing, respectively.

7.3.2.1   Brocade by Zhao et al. [74]

To the best of our knowledge, Brocade is the earliest HDHT design. The hierarchy is two-layer, constructed according with vertical architecture. The primary overlay of all $N = N_1$ nodes is on the bottom layer and the secondary overlay of $N_2 \leq N$ supernodes is on the top layer. Tapestry is a flat DHT to implement both overlays. The design, however, does not prevent the use of other flat DHTs. Although the overlays use the same ID space, they are constructed independently, and a supernode has two independent IDs.

On the bottom layer, proximity close nodes are grouped together and get assigned a supernode for the secondary overlay. A supernode provides shortcuts across distant domains in the underlying IP network. The key evidence of nested architecture is that shortcuts form an additional mechanism for better global routing. The secondary overlay is optional, and global routing, although less efficient, is possible my means of the primary overlay only.

If $N_2 \approx 0$ then the overlay degenerates to a flat DHT network. If $N_2 \approx N$ then almost all nodes diversify routing: local links in the primary overlay and long-range links in the secondary one. Both overlay can be mixed in a lookup path depending on which routing criterion, local or global, suits for the next hop.

A supernode must have significant processing power and high-bandwidth outgoing links. Preferably, it is a network access point such as gateway or router. The final choice is resolved by an election algorithm or by the responsible ISP. As a result, supernodes are distributed over the underlying network such that each domain is represented by its supernode. Consequently, supernodes become landmarks for IP network domains.

In the naive solution, each supernode maintains a list of all nodes in its group. When a lookup reaches a supernode, the latter determines whether the lookup is destined to a node in a local group or global routing is needed. Ideally, supernodes are endpoints of a tunnel through the secondary overlay that allows transferring a message directly from its current domain to the destination domain. Hence a supernode can forward inter-domain lookups using the secondary overlay for more efficient routing. Also, a regular node can directly forward lookups to its supernode to activate efficient global routing. In any case, a lookup path can mix supernodes and regular nodes, jumping between the overlays.

In contrast to a regular node, a supernode has to maintain additional routing table. Although the location and capacity of supernodes improve the performance, routing in both overlays is still proximity-unaware in the original proposal. An overlay hop can incur many hops in the underlying IP network.

Fig. 7.8 Large hops are at the beginning; then the span is reduced exponentially

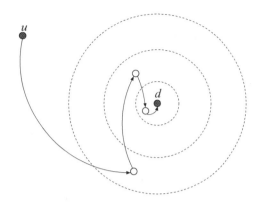

7.3.2.2   eCAN by Xu et al. [66, 73]

eCAN is a hierarchical extension of flat CAN [50, 51]. The design primarily aims at improving the routing performance from $O(\sqrt{N})$ overlay hops to $O(\log N)$. The basic CAN design uses only local neighbors: d nodes, one per dimension in the CAN d-torus. To improve routing, eCAN augments routing tables with long-range links.

The same idea is utilized in many flat DHT designs, leading to pre-hierarchical schemes when each node locally selects additional long-range neighbors [26]. In contrast, eCAN constructs a global overlay (expressway) for each distance scale. An expressway overlay forms own layer $i \geq 2$. Its topology is formed with links of the corresponding span; the higher i the longer span. The bottom layer is a flat CAN DHT overlay, it performs the basic routing function, all upper overlays are auxiliary, and the routing function is incrementally arranged along the vertical dimension.

The eCAN expressway mechanism can be considered as evolution of the Brocade shortcut mechanism to arbitrary number of layers. Instead of the coarse-grained routing scale "local ($i = 1$) vs. global ($i = 2$)", the range $i = 1, 2, \ldots, M$ is used for finer distance granularity. The span scale grows exponentially with i, which agrees with the small-world principle. It leads to paths where the closer the destination the shorter the hop span, see Fig. 7.8, which is analogical to geometrically progressive routing paths of flat DHTs.

As in Brocade, an expressway link preferably has high bandwidth, and more capable nodes take more load being active expressway nodes. In contrast to Brocade, this node responsibility differentiation is implicit; it does not require specific registration in expressways.

Recall that a Brocade shortcut spans long distance in the underlying IP network. In contrast, an eCAN expressway is primarily for long leaps in the overlay ID space. Nevertheless, eCAN employs proximity neighbor selection for long-range links, reducing the routing stretch. Additionally, topology-aware CAN overlay construction [51] allows adopting the node distribution in the ID space to the node location distribution in the underlying IP network.

In the sequel work Xu et al. [68, 69] develop the expressway method further. They diminished strict DHT protocol rules in expressway construction; the distance directly reflects the underlying network topology, leading to better adaptation and lower routing stretch. Registration of high-responsible nodes and publishing proximity-related information in expressways are mandatory.

7.3.2.3   Super-Peer Based Lookup by Zhu et al. [75]

The design is close to the Brocade two-layer ordered vertical architecture. The primary overlay and secondary overlay are for local and global routing, respectively. They are constructed using any flat DHT protocol. Supernodes are elected or selected based on the node capability: network bandwidth, storage capacity, and processing power. The secondary overlay uses its own ID space, and a supernode has two independent routing tables.

Additionally, a supernode acts a centralized server to a set of regular nodes. It maintains an index over the resources available at any regular node of this supernode. Hence, centralized clusters with the star topology appear on the bottom layer, introducing elements of horizontal architecture. When v joins the system, it associates itself with a nearby supernode u and becomes a regular member of the centralized cluster. Also v constructs its routing table for the bottom layer overlay and behaves according to the flat DHT protocol. Participating in the system, v notifies u about all local resources (for any insertion and removal). When v leaves the system, u updates the index appropriately.

In contrast to Brocade, where routing can mix paths from both overlays, this design states that lookup always first tries global routing using the secondary overlay. A regular node forwards a lookup to its supernode. Then the lookup runs in the secondary overlay until the responsible supernode is found. This supernode forwards the lookup to the destination regular node (local one-hop routing based on the local index). If a lookup cannot continue in the secondary overlay then the lookup takes the primary overlay.

The design has no clear prevalence of nested vs. disjointed architectures. The nested architecture property becomes apparent when the secondary overlay topology is formed with long-range inter-domain links in the IP network. The partition onto centralized clusters makes the architecture closer to disjointed. The design balances these two approaches in between two points of view on the routing mechanics: either the top DHT is auxiliary and boosts the global routing performance, or the bottom DHT is auxiliary and preserves the local routing dependability when the top DHT and centralized clusters cannot resolve a lookup.

Note that resource update notifications within each centralized cluster lead to overhead for regular nodes, not only for supernodes. It can be inappropriate for environments with very low-performance nodes and frequent update rates, e.g., mobile and embedded network environments.

Fig. 7.9 At the beginning
hops are fast in the underlying
network. For a distant
destination hops become of
higher routing latency

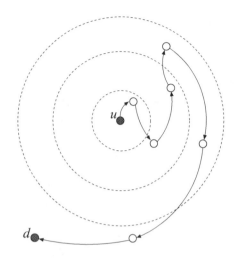

7.3.2.4   Coral by Freedman et al. [7]

The designs supports any number $M > 1$ of layers and applies pure horizontal
architecture. Each layer consists of clusters of nodes with similar RTTs. The node
ID space is the same for all layers. A node belongs to one cluster at each layer—the
extreme property of nested architecture. Chord, Kademlia, or another flat DHT can
be used to implement this cluster-based hierarchy.

The recommended value $M = 3$ has the following reason. There are many fast
clusters with regional coverage (low-level overlays for $i = 3$, a reasonable RTT
threshold is 30 ms), a few clusters with continental coverage (middle-level overlays
for $i = 2$, RTT is up to 100 ms), and one planet-wide cluster (the global overlay for
$i = 1$, RTT is unlimited). A node joins an acceptable cluster, e.g., one in which the
latency to 90 % of nodes is within the cluster diameter. If a node cannot find such a
cluster, it forms its own.

A lookup for k first starts in the cluster on layer $i = 1$, trying to take advantage
of network locality. If the current cluster on layer i does not contain a responsible
node for k, the lookup reaches the closest node u to k in this cluster. Then the lookup
continues on the next layer $i + 1$ in the higher-level cluster that u belongs to. The
process continues until a responsible node is found.

Coral routing guarantees that lookups at the beginning are fast (see Fig. 7.9) since
(a) the small cluster size leads to few overlay hops and (b) proximity-awareness
leads to low routing latency. An additional replication mechanism complements this
routing strategy. Resources (pointers or indexes) are replicated to nodes along paths
to the destination node, and it is likely to resolve many lookups locally.

When a lookup reaches layer i, it first continues on this layer, i.e., within the
cluster. Thus a node needs information which layers incoming lookups must use.
Coral uses cluster IDs and implements a cluster management mechanism: joining
a cluster, merging and splitting clusters. It increases the maintenance overhead
compared to flat DHTs.

The Coral design does not take the heterogeneity of individual node responsibility into account. The hierarchy specifically arranges the routing functionality along vertical layers to achieve higher performance than in flat DHTs. In this case, deploying a few high-responsible nodes is not a radical change; all nodes must afford their capacity non-exclusively. Moreover, if a node becomes an active participant then its nearby nodes also have to be active due to the locality property of Coral. Therefore, the design might be appropriate in environments with domain heterogeneity.

7.3.2.5 HIERAS by Xu et al. [67]

The design is similar to Coral. The key difference in a clustering mechanism: Coral uses ping-pong probes for RTT estimation and Hieras uses distributed binning. Each of M layers consists of several overlays (rings), following the horizontal architecture approach. A ring contains a subset of nearby nodes of the underlying IP network; they take equal responsibilities for the workload within the ring. Values $M = 2, 3$ are recommended for a tradeoff between the routing performance and ring maintenance overhead. The hierarchy can be built on top of a flat DHT; Chord was used for a reference case.

All N nodes are partitioned into several rings on layer $i = 1, 2, \ldots, M$. The top layer $i = M$ has one ring only. A node belongs to exactly one ring on every layer. The lower is the layer, the smaller is the latency between nodes in the ring. Nodes partition themselves into disjointed rings using the distributed binning scheme [51]. It requires landmark nodes, a well-known set of machines spread across the Internet.

A lookup is first performed on the bottom-layer ring where the lookup originator is located. It moves up and eventually reaches the top-layer ring.

Since each node belongs to M rings in sum, it has to maintain M routing tables. There are also ring tables that used to maintain information of different rings. A ring table is stored at the node whose ID is numerically closest to the ring ID as well as duplicated to several other nodes for fault-tolerance. Additional operations at a node are needed, e.g. calculating ring information and requesting ring table when a new node joins the system.

Park et al. [49] proposed P3ON (Proximity based P2P Overlay Network), a two-layer design that is conceptually close to HIERAS and Coral. The global overlay of all nodes is on the top and many local overlays are on the bottom. A local overlay is a Chord ring that connects all nodes in a single AS. Dividing IDs onto the prefix and suffix parts allows node IDs of the same AS to be close in the node ID space (proximity-based ID assignment). Resources are initially stored in the global overlay, then popular resources are replicated in local overlays. As a result, routing first in a local overlay reduces the latency.

Xu and Jin [65] proposed Uinta and SW-Uinta (small-world), designs based on the same two-layer cluster-based nested architecture as in Coral, HIERAS and P3ON. The Uinta design considers underlying network proximity and data semantics in cluster formation on the bottom layer. SW-Uinta changes the deterministic Chord-like neighbor selection strategy of Uinta to the stochastic small-world strategy, which results in reduced maintenance cost and improved routing performance.

7.3.2.6 TOPLUS by Garces-Erice et al. [11]

It is an extreme variant to proximity-aware overlay construction. Underlying network topology is the key factor that influence the overlay hierarchy. A TOPLUS overlay straightforwardly simulates IPv4 subnetwork hierarchy. As a result, a lookup path follows the router-level shortest-distance path, and the routing stretch becomes close to 1.

The node ID space S is the set of all IP addresses. Groups are virtual entities identified with IP network prefixes. The latter are obtained from BGP tables. An IP subnetwork maps to a group in the TOPLUS hierarchy. Proximity close nodes are organized into groups (subnets in ASes). Groups are organized into supergroups (ASes). Supergroups merge to hypergroups (aggregations of ASes). Therefore, a typical case is $M = 3$ layers, similarly to Coral and HIERAS.

The XOR metric, a refinement of longest-prefix matching, supports the above group construction. Assuming close nodes in the IP network have similar IP address prefixes, the XOR distance is small for proximity close nodes.

This hierarchy follows the tree-based model with tree $\mathcal{T}$. Its root corresponds to S. Overlay nodes are leaves of $\mathcal{T}$. A group is a non-leaf vertex and contains all nodes from the descendant groups. For any pair of groups, they are either node-disjointed, or one group is proper nested into the other. The tree is typically irregular and imbalanced: groups on the same layer (even siblings) can be heterogeneous in size and in number of subgroups.

Let m_u be the length of the path in $\mathcal{T}$ from the root to a node u. Then u belongs to m_u groups nested along the path:

$$\mathcal{C}_{u1} \subset \mathcal{C}_{u2} \subset \ldots \subset \mathcal{C}_{um_u} = S.$$

That is, u is contained in an assembly of the telescoping groups. Local links of u is for all $v \in \mathcal{C}_{u1}$, i.e., u knows all participants of the inner-most IP network (closest neighbors). Any $\mathcal{C}_{ui}$, except the root ($i = m_u$), has one or more siblings in $\mathcal{T}$. Node u keeps at least one neighbor from every sibling group of $\mathcal{C}_{ui}$ for $i < m_u$. Therefore, u knows long-range neighbors on all distance scales. Routing is greedy: forwarding a lookup to the neighbor closest to the key.

When u joins the system, u takes the routing table of the closest node to u. First, u notifies all its local neighbors to update their routing tables. Second, u asks each long-range neighbor v for a random node w in v's group and then w replaces v in u's routing table. It provides the diversity property when nodes of the same group likely know different nodes from other groups. Formally, if $u_1, u_2 \in \mathcal{C}$, $u_1 \neq u_2$, u_1 has a long-range link to $v_1 \in \mathcal{C}'$ and u_2 has a long-range link to $v_2 \in \mathcal{C}'$ then $v_1 \neq v_2$ with high probability.

IP network prefixes can reflect proximity inaccurately: nodes with contiguous prefixes might not be adjacent in the IP network. Node capacity does not influence the overlay load balance, and the load of a node directly depends on position of the inner-most IP network in the hierarchy.

7.3.2.7   Canon by Ganesan et al. [9]

Canon provides a generic method for accurate reflection in the overlay topology the real-world hierarchical organization of nodes in the underlying network. Such node organization is due to a global hierarchy of network domains, which can be described with a tree-based model. An example was shown in Fig. 7.3 in Sect. 7.2.1.2. The hierarchy reflection in overlay is primarily targeted to reducing the routing latency because of lower routing stretch. The overlay path length remains the same as in flat DHTs.

The Canon method requires each node to know its own position in the global hierarchy, and any two nodes are able to compute their lowest common ancestor. Note that the Coral and HIERAS designs simulate the global hierarchy using clustering mechanisms with a proximity metric. The TOPLUS design assumes that the global hierarchy is accurately reflected in the IPv4 address space.

Let $\mathscr{T}$ be the tree of global hierarchy with M layers. Overlay nodes are its leaves. Let m_u be the height of node u in $\mathscr{T}$. Non-leaf vertexes are domains consisting of all nodes from lower-level domains. The Canon method ensures that the nodes in any domain form own DHT-based overlay by themselves; it is the case of horizontal architecture. Since a node u belongs exactly to one domain on each of m_u layers, u participate in m_u overlays in total.

The Canon hierarchy construction is bottom up. Given a flat DHT, each set of nodes at a leaf domain forms a DHT-based overlay. At each internal domain, the overlay, containing all nodes in that domain, is constructed by merging all its children overlays. At each merging step, some links are added to routing tables. The top-level (global) overlay is eventually produced; it contains all N nodes.

The key point that links from a lower-level overlay are inherited in higher-level overlays, following the telescoping scheme of (7.6). The number of additional links is moderate. The link addition rules depend on the underlying flat DHT. Some nodes may not form any additional links at all. As a result, the total number of links per node remains comparable with the flat DHT design.

Importantly that the higher layer the longer spans its domain overlays have. Routing tries short-range links first, see Fig. 7.9 above. On each layer, a lookup for k reaches the closest node u in the current domain. Then u is responsible for switching to the next higher layer where the lookup continues. It leads to (a) the path locality property when a path between two nodes of the same domain does not leave this domain and (b) the path convergence property when paths from different nodes of a domain to the same outside destination exit the domain through a common node.

When a new node u joins the system, it must know at least one other existing node in its lowest-level domain ($i = 1$). This knowledge requires an extra mechanism.

Then u joins sequentially the nested overlays on layers $i = 1, 2, \ldots, m_u$ using the DHT protocol.

The Canon method can be applied to transform many flat DHT designs into their hierarchical versions. In particular, [9] describes such transformations for Chord (Crescendo), Symphony (Cacophony), Pastry/Kademlia (Kandy), and CAN (Can-Can).

7.3.2.8 Multi-Ring Hierarchy by Mislove and Druschel [45]

Similarly to Canon, it is a generic method for constructing a hierarchy of DHT overlays (rings). It reflects real-world administrative organization with a tree of rings. The top layer (root in the tree) consists of the only global ring. All nodes participate in the global ring, unless they are connected behind a firewall or a NAT. In contrast to Canon, each overlay may follow own flat DHT protocol, leading to high organizational autonomy.

A node is not required to participate in each of the nested rings on path from the leaf to the root in the tree. The reason is connectivity constraints of the underlying network. Lower-layer rings are connected to their parent ring using gateway nodes, a kind of supernodes. Although the latter feature is from disjointed architecture, the design is still closer to nested architecture (see also Sect. 7.3.1.8 and Fig. 7.7 there). If a node has low capacity or its connectivity in the underlying network is limited then the node participates only in rings on lower layers. Nevertheless, in the extreme case all nodes can become gateways.

Consider the case $M = 2$. The bottom layer consists of independent organizational rings. Each ring has a globally unique ring ID known to all members of the ring. The global ring has ID of all zeros. Any node must know in advance its position in the global hierarchy to join a ring of the given organization. Some nodes become members of more than one ring; they are gateway nodes. A gateway node acts as multiple virtual overlay nodes, one in each ring, but uses the same node ID in each ring. Gateway nodes announce themselves to other members of their rings (which they participate in) by subscribing to an anycast group in each of the rings. A group ID is equal to the associated ring ID.

A lookup carries (in addition) ID of the ring r where a responsible node is located. If a current node u belongs to r then the lookup continues in accordance with the overlay protocol of r. Otherwise, u locates a gateway node to forward the lookup to r. If u is a member of the global ring then u forwards to the anycast group identified with r. Hence, the lookup will be delivered to a gateway node to r, and routing proceeds in r. Otherwise, u anycasts the lookup into the global ring group.

Consider the extension to $M > 2$. A ring ID is a sequence of digits in a configurable base b. Each layer appends an extra digit onto the parent ring ID. Thus, an organization with a given ring ID can dynamically create new rings. In a lookup, if a node u does not belong to the target ring r, then u, as a member of multiple rings, selects the ring with the longest shared prefix. In case of multiple rings with the longest prefix, u uses the ring with shortest ID.

Routing is two phase. First, bottom-up routing is performed until a ring is found whose ID is a prefix of the destination ring. Second, top-down routing continues the lookup towards the destination ring. Similarly to Canon, it provides the path locality property.

The principal maintenance overhead is due to the requirement that nodes must join multiple rings, and thus require additional control traffic for maintaining the routing state in each ring. The design relies on a group anycast mechanism. It requires maintenance of spanning trees consisting of the overlay nodes from group member nodes towards the overlay node that is responsible for the group ID.

7.3.2.9 Diminished Chord by Karger and Ruhl [20]

The nested hierarchical architecture designs above account the global domain hierarchy. Overlay paths become close to underlying network paths, improving the routing properties. However, the designs do not directly allow an individual node to efficiently adopt its participation to its needs. Position of a node in the hierarchy is predetermined, and the node has to provide the responsibility level in accordance with the position.

In service-oriented applications, a group corresponds to a service that collectively provided by the participating nodes. For instance, a group of nodes can be responsible for content for a given topic. A node may individually select which groups it wants to belongs to. In contrast to domain hierarchies, such a group structure can be dynamic when groups frequently appear and disappear. A straightforward way is to implement own DHT overlay for each group, e.g., see content-based hierarchies in Sect. 7.3.1.8. In dynamic environments, it becomes too expensive because of the maintenance overhead.

Diminished Chord applies the group-based model on top of the Chord DHT for grouping any subset of nodes. Nodes of the same group jointly offer a service without forming its own overlay. Instead of additional sum state $O(|\mathscr{C}|\log|\mathscr{C}|)$ for a group $\mathscr{C}$ with own Chord ring, Diminished Chord introduces additional $O(|\mathscr{C}|)$ state uniformly distributed among DHT nodes that are not in $\mathscr{C}$ themselves.

The architecture is two-layer. The bottom layer is the primary Chord ring where all N nodes participate. It provides the basic lookup operation: lookup(k) returns the node d that minimizes $\rho(d,k)$ over all nodes. The top layer is for organizing arbitrary groups. A group $\mathscr{C}$ has a group ID in the Chord space. A node participates in none or many groups according to its individual interests. The possibility of group overlapping makes the design closer to nested architecture.

An additional group lookup service is provided to locate the node responsible for the key in a specified group $\mathscr{C}$. In particular, the group operation lookup($k,\mathscr{C}$) returns

$$d = \arg\min_{u\in\mathscr{C}}\rho(u,k),$$

which is a group-restricted version of (1.1) from Sect. 1.3.

It works by embedding in the primary Chord ring a directory tree for each group $\mathscr{C}$. Vertexes are points in the Chord space, each maps to the closest DHT node, and vertex and node can be used interchangeably. The root corresponds to the group ID. The tree is binary of height $O(\log N)$. All nodes are ordered such that a left descendant precedes a right descendant in terms of the Chord space distance. Edges are links from a child to its parent. An edge $u \rightarrow v$ is either a standard Chord finger (v immediately succeeds $u + 2^i$ for some $i = 0, 1, \ldots, n-1$) or an additional link—a prefinger (v immediately precedes $u + 2^i$ for some $i = 0, 1, \ldots, n-1$).

The key property is that for any node u its right subtree includes the node $v \in \mathscr{C}$ closest to u. The node u either keeps link to v itself or delegate it to the parent w if v is also the closest node to w. The property ensures that a path from a leaf to the root goes through a node that knows the closest node from $\mathscr{C}$ for this leaf node. This scheme requires storing one additional link at some overlay node for every group node, leading to the $O(\log |\mathscr{C}|)$ state overhead.

To resolve lookup$(k, \mathscr{C})$ the responsible node $d' \in N$ is found in the primary ring (Chord lookup, $O(\log N)$ hops). Then the responsible node $d \in \mathscr{C}$ is located by traversing a path in the directory tree from d' towards the root. The traversal takes $O(\log N)$ hops. Therefore, the group lookup has the same routing complexity as basic Chord.

For a given group, the design employs "special" nodes to keep additional information about the group. They have to serve additional group lookups even though the nodes are not members of the group. The number of groups a node participates in is not an accurate metric for the node responsibility, and individual adaptation of the desired participation level becomes complicated. When the number of groups increases the overhead expenses are distributed uniformly among all N nodes, regardless of their heterogeneity.

7.3.2.10 Cyclone by Artigas et al. [1, 2]

It evolves the Canon method (Sect. 7.3.2.7). Similarly, it reflects real-world hierarchical organization of nodes in the underlying network. The Cyclone method aims at better load balancing and scalability in partitioning the system onto autonomous domains when complete knowledge about the global hierarchy is not available at individual nodes. The reference case is Whirl, the Cyclone version of Chord. In general, the method allows different P2P protocols for overlays (clusters) in the hierarchy, including hybrid P2P systems.

The n-bit circular node ID space $S = [0, 2^n)$ with the XOR distance metric is common for all nodes. The partition scheme follows the tree-based model employing suffixes for cluster identification. A node ID consists of two parts: a prefix of $m - p$ bits and suffix of p bits. In a node ID, the suffix identifies the cluster of the node residence, whereas the prefix is an intra-cluster identity. In particular, a p-bit suffix is a hash of a full name of the domain that the cluster represents. Domains may continue subsequent partition taking the next leftwards $1 \leq l_i \leq p$ bits to construct up to 2^{l_i} branches. Such l_i-bit strings label subsequently subdomains at each layer and identify their clusters.

In accordance with the tree-based model, lower-layer clusters are merged to form a network on the next layer, leading to nested cluster structure as in TOPLUS and Canon and telescoping scheme (7.6) for routing tables. The ID prefix–suffix structure makes the process more reusable since any two nodes u and v of a cluster have the same l-bit suffix, hence $u = v \mod 2^l$. In particular in Whirl, if a cluster $\mathscr{C}$ with an l-bit ID contains a node u then u's immediate successor v is at least 2^l away in the ID space, thus the link $u \to v$ is reused in any higher-level cluster of $\mathscr{C}$. As a result, the total number of links per node remains comparable with the flat Chord DHT.

Similar to Canon, Cyclone uses bottom-up routing to take advantage of the network locality. Similarly to HONet and TOPLUS (Sects. 7.3.1.7 and 7.3.2.6), faster inter-cluster routing is possible with link augmentation. A node maintains additional links to nodes from its sibling clusters at different layers. Many of these links are already in its routing table due to the Cyclone link reusability. If a lookup cannot be resolved in the current cluster, the lookup is forwarded directly to the cluster closest to the key.

Inter-cluster links are optional and only a node with enough capacity can additionally maintain them. Such nodes act as supernodes, turning heterogeneity into an advantage for global routing. This optional feature, in fact, is closer to disjointed architecture, when a supernode is a representative of the cluster. Inter-cluster links can be also used to construct link-disjointed Hamilton cycles connecting sibling clusters on each layer. This ability supports multipath routing with improved security and reliability [1].

The Cyclone architecture has the same typical disadvantages as Canon. First, the assumption that a node can participate in overlays on all layer, especially on higher layers, is often violated because of connectivity restrictions of the underlying IP network. Second, there is no clear distinction of the load that different-capacity nodes should take.

7.3.2.11 GTap by Zhang et al. [72]

GTap (Grouped Tapestry) supports organizing various group structures of nodes, similarly to diminished Chord (Sect. 7.3.2.9). A node may belong to several groups simultaneously. The key feature is group-aware routing when (a) a lookup terminates at the node most responsible in a specified group (as in diminished Chord), and (b) a lookup path is constrained within a specified group. Group-aware routing complements basic Tapestry routing.

The Tapestry space is n-digit integers of base b. On the top layer, the global Tapestry overlay consists of all N nodes. If a node u is a member of a group $\mathscr{C}$ then u maintains additional neighbors from $\mathscr{C}$ to form a group Tapestry overlay.

In the global overlay, any node $u = u_{n-1} \cdots u_i \cdots u_0$ for every $i = 0, 1, \ldots, n - 1$ selects $b - 1$ long-range neighbors $v(j) \in N$ such that $v(j)$ resolves the ith digit to j in prefix-matching Tapestry routing, $j = 0, \ldots, u_i - 1, u_i + 1, \ldots, b - 1$. GTap additionally requires that a primary neighbor $v(j)$ is numerically closest to $u_{n-1} \cdots u_{i+1} j u_{i-1} \cdots u_0$. Without this requirement, there are many other candidates for $v(j)$.

The diversity of candidates for $v(j)$ is utilized for constructing group overlays on the bottom layer. In a group-$\mathscr{C}$ overlay, any its node u stores a neighbor $v_{\mathscr{C}}(j) \in \mathscr{C}$ in addition to its primary neighbor $v(j)$. Similarly, for the Tapestry leaf set, u maintains up to $2m_0$ primary leaves from N (global overlay) and up to $2m_0$ leaves from $\mathscr{C}$ (group overlay).

When a lookup is group-unaware, nodes call basic Tapestry routing to find to the closest responsible node. When a lookup is constrained within a group $\mathscr{C}$ (path-constrained routing), only group-$\mathscr{C}$ neighbors are used. When a lookup is constrained with a destination group $\mathscr{C}$ (destination specified routing), then the lookup (a) is routed to a node $v \in \mathscr{C}$ (group discovering) and (b) path-constrained routing within $\mathscr{C}$ completes the lookup.

For the group discovery, GTap uses group rendezvous. A group name $\mathscr{C}$ is hashed to its key $c_{n-1} \cdots c_0$ in the Tapestry space. The node c responsible for this key in the global overlay is $\mathscr{C}$'s rendezvous, and any node may query c for a node from $\mathscr{C}$. To find a rendezvous, u simply initiates a lookup for the group key in the global overlay.

A rendezvous can be found faster if a group-$\mathscr{C}$ node is nearby. GTap hierarchically distributes the rendezvous load using a group-$\mathscr{C}$ group membership rendezvous (GRM) tree. The tree has b^i nodes at levels $i = 0, 1, \ldots, n - 1$, namely the nodes responsible for keys $x_{n-1} \cdots x_{n-i} c_{n-i-1} \cdots c_0$. If $i < n - 1$ such a node has b children at level $i + 1$, namely the nodes responsible for keys $x_{n-1} \cdots x_{n-i} j c_{n-i-2} \cdots c_0$ with $0 \leq j < b$. A path in the group-$\mathscr{C}$ tree from a leaf to the root follows suffix-matching routing in resolving a lookup for the group key.

A GTap node u maintains a finger set of nodes that belong to other groups. When u is queried for a node $v \in \mathscr{C}$, then either u resolves the query if u knows $v \in \mathscr{C}$, or u forwards (a tree hop) the query to its parent in the group-$\mathscr{C}$ tree. In the former case, u is involved in the group-$\mathscr{C}$ tree. In the latter case, routing in the global overlay can be used effectively as follows. Node u finds the largest i such that u is responsible for the key $u_{n-1} \cdots u_{n-i} c_{n-i-1} \cdots c_0$. If $i = 0$ then u is the root of the tree. Otherwise u lookups for the key $u_{n-1} \cdots u_{n-i+1} c_{n-i} \cdots c_0$ to find the responsible node—the parent. Since the primary neighbors are closest to u among all available prefix-matching nodes, the number of hops in the last lookup is small.

When a new node joins, it first joins the global overlay following the Tapestry protocol. Then it joins any existing groups or creates a new one. To join a group $\mathscr{C}$, a node u finds a node $v \in \mathscr{C}$ using the group discovery procedure as above. Then u can find a node w, the current root of the group-$\mathscr{C}$ GRM tree, and complete joining to $\mathscr{C}$.

If u belongs to several groups, then u maintains multiple Tapestry routing tables, and the routing state and maintenance increase proportionally. In addition, u maintains its finger set. Since group names are hashed, the probability that u

is involved in many GRM trees is small and the expected size of the finger set is negligible compared with other routing information. GTap provides $O(\log N)$ routing in the global overlay and $O(\log N_2)$ routing within a group of N_2 nodes. The group discovery takes $O(\log(N))$ hops in the worst case, since the height of a GRM tree is $O(\log N)$.

7.4   Performance Models

Hierarchical architecture allows differentiating nodes and their roles, specifically important for heterogeneous environments. The basic problem is determination of the number of layers and the population size on each layer. On one hand, the more layers the more differentiation. On the other hand, a high hierarchy degree leads can be expensive for maintenance in dynamic environments. In this section we consider basic models for this tradeoff in two-layer architectures. The target problem is the optimal number of supernodes for a given total population of N nodes.

7.4.1   Local State Cost

Consider a disjointed two-layer hierarchical architecture with $N_1 = N$ nodes, forming a flat DHT overlay at the bottom layer. Among them N_2 nodes are selected as supernodes, forming a supernodes overlay at the top layer. The same ID space is used on both layers. Each supernode belongs to two overlays, one for each layer. Since a supernode maintains two routing tables, the total state cost becomes higher compared with the flat DHT case.

Under these assumptions, the basic model for optimal value of N_2 (the optimal number of supernodes for N-nodes network) was introduced in [46]. A supernode keeps routing records of size c_2 bytes for each supernode (fully-connected topology in the worst case). Also a supernode hosts N/N_2 regular nodes on the bottom layer (star topology) and a corresponding record consumes c_1 bytes per node. Then the total storage at each supernode is upper bounded with

$$\text{StateCost} = c_2 N_2 + c_1 \frac{N}{N_2}. \tag{7.7}$$

This cost function is convex with the only minimum $2\sqrt{c_1 c_2 N}$ for $N_2 = \sqrt{\frac{c_1}{c_2} N}$, where N is a free parameter of the total population size. It leads to the requirement $N_2 = \Theta(\sqrt{N})$ in the structured superpeers design, see Sect. 7.3.1.3.

The similar state cost model for a loosely structured two-layer hierarchy appeared in [16] with the aim at the optimal value for N_2, the number of node groups, in dependence on the total population size N and the total resource amount R kept in the system. The cost function is

$$\text{StateCost} = \frac{N}{N_2} + \frac{R}{N_2} + c(N_2 - 1),$$

where R is the total number of resources and c is the number of contacts per foreign group at the bottom layer. Note that this model skips the size coefficients like c_1 and c_2 in (7.7), since they do not affect resultant estimates written in the big-O notation. The cost is minimized at $N_2 = c^{-1}\sqrt{N+R}$. The Kelips design uses fixed c with the optimal tradeoff $N_2 = \Theta(\sqrt{N+R})$. Assuming that $R = O(N)$ (i.e., the total number of resources is proportional to N in the worst case), the design requires $N_2 = \Theta(N)$ groups on the top layer.

When a conventional flat DHT with $O(\log N)$ local state is used, the state cost at a supernode is

$$\text{StateCost} = \log N_2 + \log(N/N_2) = \log N.$$

Hence, the supernode state cost is the same for hierarchical and flat designs. However, the regular node state cost is reduced from $\log N$ in the flat case to $\log(N/N_2)$ in the hierarchical architecture case.

7.4.2 Routing Cost

A routing cost model for the optimal value of N_2 is considered in [40, 41] for the disjointed two-layer CAN-based hierarchy. Let d be the CAN ID space dimension. Starting from a regular node, the number of hops needed to find the destination cluster supernode on the top layer (a CAN overlay of N_2 nodes) is $1 + dN_2^{1/d}$ on average. Then $d(N/N_2)^{1/d}$ hops are needed on average to find the responsible node on the bottom layer (a CAN overlay of N/N_2 nodes). In total, the routing cost in overlay hops is

$$\text{RoutingCost} = f(N_2) = 1 + dN_2^{1/d} + d\left(\frac{N}{N_2}\right)^{1/d}.$$

The first derivate respect to N_2 is

$$f'(N_2) = -\frac{1}{N_2}\left(\frac{N}{N_2}\right)^{1/d} + N_2^{1/d-1}.$$

It is equal to 0 when $N_2 = \sqrt{N}$. This point is a minimum if the second derivate is positive. It is easy to prove that $f''(\sqrt{N}) > 0$ for all $d \in (1, +\infty)$.

The optimal value of N_2 is independent on d. When $N_2 = \sqrt{N}$ the routing cost is minimal for $d = \ln N_2$. It is twice lower than the optimal $d' = \ln N = \ln N_2^2 = 2\ln N_2$ in the basic flat CAN. That is, a regular node has twice lower state. A supernode has the same state as in a flat CAN ($d' = d + d$) due to maintenance of two routing tables, each has d entries.

The hierarchical CAN has the optimal routing cost

$$\text{RoutingCost}_{\text{opt}} = 1 + N^{1/\ln N} \ln N,$$

which is one hop greater then the optimal routing cost in the flat CAN network with N nodes and $d' = \ln N$.

Applying the same technique for a conventional flat DHT with $O(\log N)$ routing, we yield the routing cost (in overlay hops)

$$\text{RoutingCost} = 1 + \log N_2 + \log \frac{N}{N_2} = 1 + \log N,$$

and the routing performance again is almost equal for hierarchical and flat designs. In the Crescendo hierarchical architectures [9] the similar result holds irrespective of the number of layers.

7.4.3   Network Traffic Cost

Consider the model of two-tier hierarchy for optimizing the number of supernodes proposed in [77, 79]. The model describes traffic sent/received by nodes on average; it is similar to the workload model of [25] for Hi3 scalability analysis, see also Chap. 12. Three alternative intra-group connectivity structures are studied: fully-meshed (FuMe), single-connection (SiCo), and DHT, see Sect. 7.3.1.6.

The key model parameter is $\alpha = N_2/N$, the fraction of supernodes in the system. The case $\alpha = 100\%$ corresponds to a flat overlay. Traffic cost is evaluated in terms of the number of transmitted messages and is a function of α. The model focuses on $\alpha \leq 25\%$, i.e., groups on the bottom layer consist of four or more nodes.

The network cost is formed by lookup traffic (λ), maintenance traffic (μ), and republish traffic (ρ). Traffic generation for supernodes and regular nodes is differentiated. The total traffic cost is a sum of all traffic costs over all nodes. Obviously, in the SiCo topology, a centralized overlay network with only one supernode generates the lowest network traffic, because only lookup and ping/pong maintenance messages are exchanged between the supernode and its regular nodes. The network traffic cost typically increases when α grows, mostly caused by maintenance traffic in the top overlay o supernodes. In the FuMe topology, low values of α correspond to high total traffic because of large groups on the bottom layer. In the DHT topology, the extreme cases $\alpha \to 0$ and $\alpha = 100\%$ corresponds to the highest traffic cost, and there exists a tradeoff value for α in between. The exact analytical expressions is not important for our discussion and an interested reader can find them in Table 1 of [79].

On the other hand, the load a supernode can be high for some α. The optimal operation point is achieved for α such that the total network traffic is minimized subject to a non-overload constraint for every supernode. To find the minimal necessary α, a load factor is defined for every supernode s as the ratio between the traffic b_s for s at a specific time, and the bandwidth limit B_s, i.e., $L_s(\alpha) = b_s/B_s$. A supernode is overloaded if its load factor exceeds 100%.

The model of [15] also accounts the traffic costs in three-layer hierarchy with N_1 regular nodes, N_2 unit leaders (level 2), and N_3 slice leaders (layer 3), see the OneHop design in Sect. 7.3.1.4. The most bandwidth consuming nodes are slice leaders and the model aims at minimization of bandwidth utilization for them. The optimum is $N_2 = O(\sqrt{N})$ and $N_3 = O(\sqrt{N})$ with the constants dependent on membership rates and messages sizes.

7.5 Summary

This chapter discussed hierarchical architectures applicable in structured P2P designs. Based on the basic arrangement models the nodes can be structured onto groups. Recursive application allows further structuring within groups. It is a crucial step in constructing the global network hierarchy.

We considered Kleinberg's models of decentralized network hierarchy and showed that for a P2P system its global group structure is arranged along the vertical and horizontal dimensions. We derived three conceptual models of hierarchy to cover cluster-based, tree-based, and group-based constructions. All they frequently appear in various forms in existing hierarchical P2P designs.

A hierarchical architecture defines how node groups are formed in the P2P system and then inter-connected along the vertical and horizontal dimensions. We described and clarified the existing classification of hierarchical P2P architectures: vertical, horizontal, and their subclasses. Based on the conceptual models we analyzed these architectures and formulated certain design principles that affect design choices in a hierarchical P2P architecture. The principles concretize the means of hierarchical decomposition of P2P nodes onto inter-connected groups.

Finally, we overviewed particular hierarchical P2P designs. We selected from the existing literature the most representative designs based on their popularity and on our personal point of view. Most of the designs are for HDHTs, since they represent "pure structured" P2P systems. The overview shows various design solutions to hierarchical architectures. It can be considered as state-of-the-art summarized using the general framework.

References

1. Artigas, M.S., Lopez, P.G., Skarmeta, A.F.G.: A novel methodology for constructing secure multipath overlays. IEEE Internet Comput. **9**(6), 50–57 (2005). doi: http://dx.doi.org/10.1109/MIC.2005.117
2. Artigas, M.S., Lopez, P.G., Ahullo, J.P., Skarmeta, A.F.G.: Cyclone: a novel design schema for hierarchical DHTs. In: IEEE P2P '05: Proceedings of 5th International Conference on Peer-to-Peer Computing, pp. 49–56. IEEE Computer Society (2005). doi: http://dx.doi.org/10.1109/P2P.2005.5

3. Chiang, M., Low, S.H., Calderbank, A.R., Doyle, J.C.: Layering as optimization decomposition: a mathematical theory of network architectures. Proc. IEEE **95**, 255–312 (2007). doi: http://dx.doi.org/10.1109/JPROC.2006.887322
4. Dabek, F., Kaashoek, M.F., Karger, D., Morris, R., Stoica, I.: Wide-area cooperative storage with CFS. In: Proceedings of 18th ACM Symposium Operating Systems Principles (SOSP '01), pp. 202–215. ACM, New York (2001). doi: http://doi.acm.org/10.1145/502034.502054
5. Duchon, P., Hanusse, N., Lebhar, E., Schabanel, N.: Towards small world emergence. In: Proceedings of 18th Annual ACM Symposium on Parallelism in Algorithms and Architectures (SPAA '06), pp. 225–232. ACM, New York (2006). doi: http://doi.acm.org/10.1145/1148109. 1148145
6. Fonseca, P., Rodrigues, R., Gupta, A., Liskov, B.: Full-information lookups for peer-to-peer overlays. IEEE Trans. Parallel Distrib. Syst. **20**(9), 1339–1351 (2009)
7. Freedman, M.J., Mazières, D.: Sloppy hashing and self-organizing clusters. In: IPTPS '03: Proceedings of 2nd International Workshop on Peer-to-Peer Systems. Lecture Notes in Computer Science, vol. 2735, pp. 45–55. Springer, Berlin (2003)
8. Ganesan, P., Manku, G.S.: Optimal routing in Chord. In: SODA '04: Proceedings of 15th Annual ACM-SIAM Symposium Discrete Algorithms, pp. 176–185. Society for Industrial and Applied Mathematics (2004)
9. Ganesan, P., Gummadi, K., Garcia-Molina, H.: Canon in G major: designing DHTs with hierarchical structure. In: ICDCS '04: Proceedings of 24th International Conference Distributed Computing Systems, pp. 263–272. IEEE Computer Society (2004)
10. Garcés-Erice, L., Biersack, E., Felber, P.A., Ross, K.W., Urvoy-Keller, G.: Hierarchical peer-to-peer systems. In: Euro-Par 2003: Proceedings of ACM/IFIP International Conference Parallel and Distributed Computing, pp. 643–657 (2003)
11. Garcés-Erice, L., Ross, K.W., Biersack, E.W., Felber, P., Urvoy-Keller, G.: Topology-centric look-up service. In: Proceedings of 5th International Conference Group Communications and Charges (NGC 2003), Workshop on Networked Group Communication. Lecture Notes in Computer Science, vol. 2816, pp. 58–69. Springer, Berlin (2003)
12. Godfrey, P.B., Stoica, I.: Heterogeneity and load balance in distributed hash tables. In: Proceedings of IEEE INFOCOM'05, pp. 596–606. IEEE (2005). doi:10.1109/INFCOM.2005. 1497926
13. Guisheng, Y., Jie, S., Xianghui, W.: Hierarchical small-world P2P networks. In: ICICSE '08: Proceedings of International Conference Internet Computing in Science and Engineering, pp. 452–458. IEEE Computer Society (2008). doi: http://dx.doi.org/10.1109/ICICSE.2008.94
14. Gummadi, K., Gummadi, R., Gribble, S., Ratnasamy, S., Shenker, S., Stoica, I.: The impact of DHT routing geometry on resilience and proximity. In: Proceedings of ACM SIGCOMM'03, USENIX Association. Berkeley, CA, USA, pp. 381–394. ACM, New York (2003). doi: http://doi.acm.org/10.1145/863955.863998
15. Gupta, A., Liskov, B., Rodrigues, R.: Efficient routing for peer-to-peer overlays. In: Proceedings of 1st Symposium on Networked Systems Design and Implementation (NSDI '04), USENIX Association. Berkeley, CA, USA (2004). URL: citeseer.ist.psu.edu/gupta04efficient. html
16. Gupta, I., Birman, K., Linga, P., Demers, A., van Renesse, R.: Kelips: building an efficient and stable P2P DHT through increased memory and background overhead. In: IPTPS '03: Proceedings of 2nd International Workshop on Peer-to-Peer Systems. Lecture Notes in Computer Science, vol. 2735, pp. 160–169. Springer, Berlin (2003)
17. Heristyo, A., Masuyama, H., Kasahara, S., Takahashi, Y.: User-search time analysis for hierarchical peer-to-peer overlay networks with time-dependent user-population process. In: Proceedings of 4th International Conference Queueing Theory and Network Applications (QTNA'09), pp. 5:1–5:4. ACM, New York (2009). doi: http://doi.acm.org/10.1145/1626553. 1626558
18. Hu, J., Li, M., Zheng, W., Wang, D., Ning, N., Dong, H.: Smartboa: constructing P2P overlay network in the heterogeneous internet using irregular routing tables. In: IPTPS

'04: Proceedings of 3rd International Workshop on Peer-to-Peer Systems. Lecture Notes in Computer Science, vol. 3279, pp. 278–287. Springer, Berlin (2004)

19. Joung, Y.J., Wang, J.C.: Chord2: a two-layer chord for reducing maintenance overhead via heterogeneity. Comput. Commun. **51**(3), 712–731 (2007)

20. Karger, D.R., Ruhl, M.: Diminished Chord: a protocol for heterogeneous subgroup formation in peer-to-peer networks. In: IPTPS '04: Proceedings of 3rd International Workshop on Peer-to-Peer Systems. Lecture Notes in Computer Science, vol. 3279, pp. 288–297. Springer, Berlin (2004)

21. Karrels, D.R., Peterson, G.L., Mullins, B.E.: Structured P2P technologies for distributed command and control. Peer-to-Peer Netw. Appl. **2**(4), pp. 311–333, Springer US (2009). doi:10.1007/s12083-009-0033-y

22. Kempe, D., Kleinberg, J., Demers, A.: Spatial gossip and resource location protocols. J. ACM **51**, 943–967 (2004). doi: http://doi.acm.org/10.1145/1039488.1039491

23. Kleinberg, J.M.: The small-world phenomenon: an algorithm perspective. In: Proceedings of 32nd Annual ACM Symposium Theory of Computing (STOC '00), pp. 163–170. ACM, New York (2000). doi: http://doi.acm.org/10.1145/335305.335325

24. Kleinberg, J.M.: Complex networks and decentralized search algorithms. In: Proceedings of International Congress of Mathematicians (ICM 2006). European Mathematical Society (2006)

25. Korzun, D., Gurtov, A.: On scalability properties of the Hi3 control plane. Comput. Commun. **29**(17), 3591–3601 (2006)

26. Korzun, D., Gurtov, A.: Survey on hierarchical routing schemes in "flat" distributed hash tables. Peer-to-Peer Netw. Appl. **4**, 346–375 (2011). doi: http://dx.doi.org/10.1007/s12083-010-0093-z

27. Krishnamurthy, B., Wang, J., Xie, Y.: Early measurements of a cluster-based architecture for P2P systems. In: IMW '01: Proceedings of 1st ACM SIGCOMM Workshop on Internet Measurement, pp. 105–109. ACM. New York (2001). doi: http://doi.acm.org/10.1145/505202.505216

28. Le, L., Kuo, G.S.: Hierarchical and breathing peer-to-peer SIP system. In: ICC 2007: Proceedings of IEEE International Conference Communications, pp. 1887–1892. IEEE (2007)

29. Lee, J.W., Schulzrinne, H., Kellerer, W., Despotovic, Z.: mDHT: multicast-augmented DHT architecture for high availability and immunity to churn. In: CCNC'09: Proceedings of 6th IEEE Conference Consumer Communications and Networking Conference, pp. 760–764. IEEE (2009)

30. Leong, B., Liskov, B., Demaine, E.: Epichord: parallelizing the Chord lookup algorithm with reactive routing state management. In: ICON 2004: Proceedings of 12th International Conference on Networks, pp. 270–276 (2004)

31. Li, J., Stribling, J., Morris, R., Kaashoek, M.F.: Bandwidth-efficient management of DHT routing tables. In: Proceedings of the 2nd Symposium on Networked Systems Design and Implementation (NSDI '05), pp. 99–114 (2005)

32. Li, J., Stribling, J., Morris, R., Kaashoek, M.F., Gil, T.M.: A performance vs. cost framework for evaluating DHT design tradeoffs under churn. In: Proceedings of IEEE INFOCOM'05, vol. 1, pp. 225–236. IEEE (2005). doi: 10.1109/INFCOM.2005.1497894

33. Li, X., Wu, J.: Hierarchical P2P systems in a small world. In: LACCEI'2004: Proceedings of 2nd Latin American and Caribbean Conference for Engineering and Technology (2004). URL: http://www.laccei.org/proceedings2004/FinalPapers/IT_046.pdf

34. Lian, J., Naik, K., Agnew, G.B.: A framework for evaluating the performance of cluster algorithms for hierarchical networks. IEEE/ACM Trans. Netw. **15**, 1478–1489 (2007). doi: http://dx.doi.org/10.1109/TNET.2007.896499

35. Lloret, J., Palau, C., Boronat, F., Tomas, J.: Improving networks using group-based topologies. Comput. Commun. **31**, 3438–3450 (2008). doi:10.1016/j.comcom.2008.05.030

36. Loguinov, D., Kumar, A., Rai, V., Ganesh, S.: Graph-theoretic analysis of structured peer-to-peer systems: routing distances and fault resilience. IEEE/ACM Trans. Netw. **13**(5), 1107–1120 (2005)

37. Lua, E.K., Crowcroft, J., Pias, M., Sharma, R., Lim, S.: A survey and comparison of peer-to-peer overlay network schemes. IEEE Commun. Surv. Tutor. **7**(2), 72–93 (2005)
38. Manku, G.S., Bawa, M., Raghavan, P.: Symphony: distributed hashing in a small world. In: USITS'03: Proceedings of 4th USENIX Symposium on Internet Technologies and Systems, pp. 127–140. USENIX Association (2003)
39. Martinez-Yelmo, I., Bikfalvi, A., Guerrero, C., Rumin, R.C., Mauthe, A.: Enabling global multimedia distributed services based on hierarchical DHT overlay networks. Int. J. Internet Protoc. Tech. (IJIPT) **3**(4), 234–244 (2008). doi: http://dx.doi.org/10.1504/IJIPT.2008.023772
40. Martinez-Yelmo, I., Cuevas, R., Guerrero, C., Mauthe, A.: Routing performance in a hierarchical DHT-based overlay network. In: PDP 2008: Proceedings of 16th Euromicro Conference Parallel, Distributed and Network-Based Processing, pp. 508–515. IEEE Computer Society (2008). doi: http://dx.doi.org/10.1109/PDP.2008.79
41. Martinez-Yelmo, I., Guerrero, C., Rumín, R.C., Mauthe, A.: A hierarchical P2PSIP architecture to support skype-like services. In: PDP 2009: Proceedings of 17th Euromicro International Conference Parallel, Distributed and Network-Based Processing, pp. 316–322. IEEE Computer Society (2009). doi: http://doi.ieeecomputersociety.org/10.1109/PDP.2009.27
42. Mesaros, V.A., Carton, B., Roy, P.V.: S-Chord: Using symmetry to improve lookup efficiency in Chord. Proceedings of the International Conference on Parallel and Distributed Processing Techniques and Applications, PDPTA '03, Las Vegas, Nevada, USA, Vol. 4, CSREA Press, (2003)
43. Meshkova, E., Riihijärvi, J., Petrova, M., Mähönen, P.: A survey on resource discovery mechanisms, peer-to-peer and service discovery frameworks. Comput. Netw. **52**(11), 2097–2128 (2008). doi: http://dx.doi.org/10.1016/j.comnet.2008.03.006
44. Min, S.H., Holliday, J., Cho, D.S.: Optimal super-peer selection for large-scale P2P system. In: ICHIT '06: Proceedings of 2006 International Conference Hybrid Information Technology, pp. 588–593. IEEE Computer Society (2006). doi: http://dx.doi.org/10.1109/ICHIT.2006.188
45. Mislove, A., Druschel, P.: Providing administrative control and autonomy in structured peer-to-peer overlays. In: IPTPS '04: Proceedings of 3rd International Workshop on Peer-to-Peer Systems. Lecture Notes in Computer Science, vol. 3279, pp. 162–172. Springer, Berlin (2004)
46. Mizrak, A.T., Cheng, Y., Kumar, V., Savage, S.: Structured superpeers: Leveraging heterogeneity to provide constant-time lookup. In: WIAPP 2003: Proceedings of 3rd IEEE Workshop on Internet Applications, pp. 104–111 (2003)
47. Monnerat, L.R., Amorim, C.L.: Peer-to-peer single hop distributed hash tables. In: Proceedings of IEEE Globecom'09 (2009)
48. Ou, Z., Harjula, E., Koskela, T., Ylianttila, M.: GTPP: General truncated pyramid peer-to-peer architecture over structured DHT networks. Mob. Netw. Appl. **15**, 729–749 (2010). doi:10.1007/s11036-009-0193-2
49. Park, K., Pack, S., Kwon, T.: Proximity based peer-to-peer overlay networks (P3ON) with load distribution. In: Proceedings of International Conference Information Networking (ICOIN 2007). Towards Ubiquitous Networking and Services. Revised Selected Papers, pp. 234–243. Springer, Berlin (2008). doi: http://dx.doi.org/10.1007/978-3-540-89524-4_24
50. Ratnasamy, S., Handley, P.F.M., Karp, R., Shenker, S.: A scalable content-addressable network. In: Proceedings of ACM SIGCOMM'01, pp. 161–172. ACM, New York (2001)
51. Ratnasamy, S., Handley, M., Karp, R., Shenker, S.: Topologically-aware overlay construction and server selection. In: Proceedings of IEEE INFOCOM'02 (2002)
52. Ren, X.J., Gu, Z.M.: SA-Chord: A novel P2P system based on self-adaptive joining. In: Proceedings of 6th International Conference Grid and Cooperative Computing (GCC 2007), pp. 75–81. IEEE Computer Society (2007). doi: http://dx.doi.org/10.1109/GCC.2007.109
53. Risson, J., Moors, T.: Survey of research towards robust peer-to-peer networks: search methods. Comput. Netw. **50**(17), 3485–3521 (2006). doi: http://dx.doi.org/10.1016/j.comnet.2006.02.001
54. Risson, J., Qazi, S., Moors, T., Harwood, A.: A dependable global location service using rendezvous on hierarchic distributed hash tables. In: ICN/ICONS/MCL '06: Proceedings of International Conference Networking, International Conference Systems and International

Conference Mobile Communications and Learning Technologies. IEEE Computer Society (2006). doi:10.1109/ICNICONSMCL.2006.11

55. Risson, J., Harwood, A., Moors, T.: Stable high-capacity one-hop distributed hash tables. In: ISCC '06: Proceedings of 11th IEEE Symposium on Computers and Communications, pp. 687–694. IEEE Computer Society (2006). doi: http://dx.doi.org/10.1109/ISCC.2006.152

56. Rufino, J., Alves, A., Exposto, J., Pina, A.: A cluster oriented model for dynamically balanced DHTs. In: IPDPS'04: Proceedings of 18th International Symposium on Parallel and Distributed Processing. IEEE Computer Society (2004)

57. Singh, A., Liu, L.: A hybrid topology architecture for P2P systems. In: ICCCN 2004: Proceedings of 13th International Conference on Computer Communications and Networks, pp. 475–480 (2004)

58. Stoica, I., Morris, R., Liben-Nowell, D., Karger, D., Kaashoek, M.F., Dabek, F., Balakrishnan, H.: Chord: a scalable peer-to-peer lookup service for internet applications. IEEE/ACM Trans. Netw. **11**(1), 17–32 (2003)

59. Surana, S., Godfrey, B., Lakshminarayanan, K., Karp, R., Stoica, I.: Load balancing in dynamic structured peer-to-peer systems. Perform. Eval. **63**(3), 217–240 (2006). doi: http://dx.doi.org/10.1016/j.peva.2005.01.003

60. Tang, C., Xu, Z., Dwarkadas, S.: Peer-to-peer information retrieval using self-organizing semantic overlay networks. In: Proceedings of ACM SIGCOMM'03, pp. 175–186. ACM, New York (2003). doi: http://doi.acm.org/10.1145/863955.863976

61. Tanta-ngai, H., McAllister, M.: A peer-to-peer expressway over Chord. Math. Comput. Model. **44**(7-8), 659–677 (2006). doi: http://dx.doi.org/10.1016/j.mcm.2006.02.003

62. Tian, R., Xiong, Y., Zhang, Q., Li, B., Zhao, B.Y., Li, X.: Hybrid overlay structure based on random walks. In: IPTPS '05: Proceedings of 4th International Workshop on Peer-to-Peer Systems. Lecture Notes in Computer Science, vol. 3640, pp. 152–162. Springer, Berlin (2005)

63. Viswanath, B., Post, A., Gummadi, K.P., Mislove, A.: An analysis of social network-based sybil defenses. In: Proceedings of ACM SIGCOMM 2010 Conference Applications, Technologies, Architectures, and Protocols for Computer Communication, pp. 363–374. ACM, New York (2010). doi: http://doi.acm.org/10.1145/1851182.1851226

64. Wan, Y., Asaka, T., Takahashi, T.: A hybrid P2P overlay network for non-strictly hierarchically categorized contents. In: CCGRID '08: Proceedings of 8th IEEE International Symposium on Cluster Computing and the Grid, pp. 41–48. IEEE Computer Society (2008). doi: http://dx.doi.org/10.1109/CCGRID.2008.10

65. Xu, J., Jin, H.: A structured P2P network based on the small world phenomenon. J. Supercomput. **48**, 264–285 (2009). doi:10.1007/s11227-008-0219-8

66. Xu, Z., Zhang, Z.: Building low-maintenance Expressways for P2P systems. Techical Report HPL-2002-41, HP Labs, Palo Alto (2002)

67. Xu, Z., Min, R., Hu, Y.: HIERAS: A DHT based hierarchical P2P routing algorithm. In: ICPP 2003: Proceedings of 32nd International Conference Parallel Processing, pp. 187–194. IEEE Computer Society (2003)

68. Xu, Z., Tang, C., Zhang, Z.: Building topology-aware overlays using global soft-state. In: Proceedings of 23rd International Conference Distributed Computing Systems (ICDCS'03), pp. 500–508. IEEE Computer Society (2003). doi: http://dx.doi.org/10.1109/ICDCS.2003.1203500

69. Xu, Z., Mahalingam, M., Karlsson, M.: Turning heterogeneity into an advantage in overlay routing. In: Proceedings of IEEE INFOCOM'03, pp. 1499–1509 (2003)

70. Yang, B., Garcia-Molina, H.: Designing a super-peer network. In: ICDE'03: Proceedings of 19th International Conference on Data Engineering, pp. 49–60 (2003). doi: http://doi.ieeecomputersociety.org/10.1109/ICDE.2003.1260781

71. Zhang, X.M., Wang, Y.J., Li, Z.: Research of routing algorithm in hierarchy-adaptive P2P systems. In: ISPA 2007: Proceedings of 5th International Symposium Parallel and Distributed Processing and Applications. Lecture Notes in Computer Science, vol. 4742, pp. 728–739. Springer, Berlin (2007)

72. Zhang, Y., Li, D., Chen, L., Lu, X.: Flexible routing in grouped DHTs. In: IEEE P2P '08: Proceedings of 8th International Conference Peer-to-Peer Computing, pp. 109–118. IEEE Computer Society (2008). doi: http://dx.doi.org/10.1109/P2P.2008.43

73. Zhang, Z., Shi, S.M., Zhu, J.: Self-balanced P2P Expressway: When Marxism meets Confucian. Techical Report MSR-TR-2002-72, Microsoft Research Asia (2002)

74. Zhao, B.Y., Duan, Y., Huang, L., Joseph, A.D., Kubiatowicz, J.D.: Brocade: Landmark routing on overlay networks. In: IPTPS '02: Proceedings of 1st International Workshop on Peer-to-Peer Systems. Lecture Notes in Computer Science, vol. 2429, pp. 34–44. Springer, Berlin (2002)

75. Zhu, Y., Wang, H., Hu, Y.: A super-peer based lookup in structured peer-to-peer systems. In: Proceedings of ISCA 16th International Conference Parallel and Distributed Computing Systems (PDCS 2003), pp. 465–470 (2003)

76. Zoels, S., Eichhorn, M., Tarlano, A., Kellerer, W.: Content-based hierarchies in DHT-based peer-to-peer systems. In: SAINT Workshops 2006: Proceedings of International Symposium Applications and the Internet Workshops, pp. 105–108. IEEE Computer Society (2006). doi: http://dx.doi.org/10.1109/SAINT-W.2006.12

77. Zoels, S., Despotovic, Z., Kellerer, W.: Cost-based analysis of hierarchical DHT design. In: IEEE P2P '06: Proceedings of 6th International Conference Peer-to-Peer Computing, pp. 233–239. IEEE Computer Society (2006). doi: http://dx.doi.org/10.1109/P2P.2006.13

78. Zoels, S., Despotovic, Z., Kellerer, W.: Load balancing in a hierarchical DHT-based P2P system. In: COLCOM '07: Proceedings of 2007 International Conference Collaborative Computing: Networking, Applications and Worksharing, pp. 353–361. IEEE Computer Society (2007). doi: http://dx.doi.org/10.1109/COLCOM.2007.4553855

79. Zoels, S., Despotovic, Z., Kellerer, W.: On hierarchical DHT systems — an analytical approach for optimal designs. Comput. Commun. 31(3), 576–590 (2008). doi: http://dx.doi.org/10.1016/j.comcom.2007.08.033

Chapter 8
Cyclic Routing

Abstract Distributed Hash Tables (DHT) provide a lookup service in peer-to-peer overlay networks. The known problem is that lookups function poorly when no direct IP connectivity is available to some nodes (e.g., located behind a NAT or firewall) or in the presence of overloaded or malicious nodes. In this chapter, we describe a method for DHT-based routing called Cyclic Routing (CR). It generalizes existing single-hop look-ahead approach (also known as "Know thy neighbor's neighbor") and supports multipath routing. The method provides a systematic way for collecting stable and efficient overlay paths. Cyclic routing has the same theoretical dependability and efficiency upper bounds as basic DHT routing but it is more resilient when IP connectivity is limited or when the overlay suffers from overloaded nodes. The CR method was implemented in the CR-Chord protocol and its experimental evaluation showed improvement in the lookup availability.

8.1 Introduction

A Distributed Hash Table (DHT) provides a lookup of identifiers and routing to the node storing a corresponding value.[1] Each node keeps contacts (node IDs and IP addresses in the local routing table) to some other nodes (neighbors), forming an overlay on the underlying network. For routing a packet, nodes sequentially forward it using communication primitives of the underlying network.

The IP-level reachability applies certain restrictions on DHT functioning. First, a pure DHT assumes a node may access an IP address of another node. In an untrusted

---

[1]©2009 IEEE. Reprinted, with permission, from D. Korzun, B. Nechaev, A. Gurtov, Cyclic Routing: Generalizing Look-ahead in Peer-to-Peer Networks, in Proceedings of the 7th IEEE International Conference on Computer Systems and Applications, May 2009. Some material is also adapted in Chaps. 3 and 11.

D. Korzun and A. Gurtov, *Structured Peer-to-Peer Systems: Fundamentals of Hierarchical Organization, Routing, Scaling, and Security*, DOI 10.1007/978-1-4614-5483-0_8, © Springer Science+Business Media New York 2013

environment, however, a node can prefer not to give its IP address to untrusted nodes to avoid direct attacks. Only trustworthy nodes may contact it directly via IP; communication with other nodes uses overlay node IDs.

Second, IP addresses are less stable than node IDs. For instance, when a node is mobile it can change its IP address frequently. Consequently, many routing table entries can become invalid, and high maintenance cost is required to provide routing tables up to date. In contrast, IP address changes should not necessarily lead to changes in P2P overlay paths.

Third, some nodes have no global IP addresses or a network can exhibit non-transitivity [4]. A node u cannot communicate directly with a node v, and an intermediate node w is needed, e.g., when u and v are separated by a NAT.

For a DHT it means that a node cannot add neighbors freely. Consequently, some routing improvements based on enhancing node knowledge about the overlay become less attractive (e.g., proximity neighbor selection [1] or routing tables of variable size [10]). Iterative routing becomes troublesome since a source node cannot always contact directly each node in a route. As a result, DHT routing suffers from IP-level reachability restrictions.

In this chapter, we propose a new method for overlay routing on top of a DHT. We call the method *cyclic routing* (CR) since it uses cyclic paths (cycles). It generalizes existing ideas of single-hop lookahead also known as "Know thy neighbor's neighbor" [11] and of redundant multipath routing [18].

Each node maintains a collection of cycles additionally to its routing table. A cycle is a path that starts from the source node to its neighbor, then runs through the overlay and returns to the source. Only node IDs are stored to identify a cycle. Cycles present global knowledge about paths in the overlay. In routing, a node can alternatively use either cycles or the underlying DHT. If there is a cycle with a node close to the destination, the packet is sent along the cycle. Otherwise, the underlying DHT selects the next hop. For illustration, we use the Chord DHT [15].

In fact, cyclic routing provides a systematic way for collecting stable and efficient overlay paths. It does not require knowledge of additional IP addresses nor conflict with routing improvements that enhance node knowledge about the overlay using IP addresses. In contrast, our method complements these proposals when they fail because of the IP-level reachability restrictions. When IP-level reachability is restricted, CR is more resilient but preserves provable dependability and efficiency upper bounds of basic DHT routing. CR also intends to provide more resilient routing in the presence of malicious nodes.

The rest of the chapter is organized as follows. Section 8.2 summarizes P2P routing and discusses existing proposals that allow nodes to learn more about the global network. Section 8.3 introduces the cyclic routing method. We analyze its properties and discuss the key design principles. Section 8.4 presents evaluation of lookup success rate of basic Chord and CR-Chord under churn (more detailed analysis of CR-Chord can be found in Chap. 11). In Sect. 8.5, we discuss additional capabilities provided by cyclic routing, including multi-path and secure routing.

8.2   Problem Domain

Consider a DHT-based overlay of N nodes. Let node IDs be assigned from an identifier space with a distance metric ρ. In a pure DHT, distance metric calculation is based on node IDs only. If $\rho(u,d) > \rho(v,d)$ then v is closer to d or v is in between u and d. Let n_{ud} be the number of nodes in between u and d.

Each node s maintains a local routing table T_s of entries (u, IP_u), where u is a neighbor and IP_u is its IP address. In a dynamic environment, nodes collaborate and adopt their routing tables to an up-to-date state. The number of neighbors $|T_s|$ is the node degree. Typically $|T_s| \ll N$, meaning that routing uses *local knowledge* about the overlay network.

In basic DHT routing, s forwards packets to u via the underlying IP network. A packet to a destination d goes sequentially through nodes whose IDs are progressively closer to d according to the distance metric. If $v \in T_s$ then s can forward a packet to v forming the one-hop path $s \rightarrow v$. Consequently, a multi-hop P2P path $s \rightarrow^+ d$ is constructed.

In the ideal case, DHT routing is dependable (a packet eventually reaches its destination) and efficient (a typical upper bound is $O(\log N)$ hops). In real environments, however, P2P paths can be long, inefficient for the underlying IP network, faulty, and insecure. Recently, a few methods have been proposed to improve efficiency and security of DHT routing.

Lookahead. Many DHTs use greedy routing when nodes forward a packet to the neighbor closest to d. More efficient DHT routing techniques (e.g. [6]) do exist. Nevertheless, in previous proposals a node has little idea on the remaining route. An alternate approach exploits knowledge of neighbor's neighbor (one level of lookahead) [11] producing shorter paths compared to greedy routing [12].

Flexible routing table maintenance. To reduce the route latency, s selects neighbors using knowledge of the underlying IP network. In particular, neighbors are chosen based on a proximity metric (proximity neighbor selection); selection of next hops depends not only on the ID distance but also on the geographical closeness (proximity route selection) [1, 5]. Moreover, nodes can increase the number of routing table entries. In this case, s varies $|T_s|$ independently on other nodes. When there is enough local resources, s inserts a new neighbor. When a neighbor is likely to have failed or to be malicious, it is removed. Proximity information, random sampling, and behavioral statistics are actively used [9, 10, 17].

Multipath routing. Having many neighbors, s forwards a packet via several alternate paths (using several neighbors), either in parallel or sequentially. This redundancy allows routing around failed or malicious nodes [2, 9, 18].

These methods use *additional knowledge* about the network. Nevertheless, in the first method, only a small step is done to analyze the remaining route; the last two require IP addresses, facing the problems of IP reachability restrictions.

Below we propose a routing method that generalizes lookahead, supports flexible routing tables and multipath routing, but works with node IDs only. It complements the above IP-based proposals when they fail or become inefficient.

8.3 The Cyclic Routing Algorithm

Cycles are natural for bidirectional P2P communications. Let a node s send a packet to d; the path is $s \to^+ d$. Then d replies to s; the path is $d \to^+ s$. These paths form the cycle $s \to^+ d \to^+ s$. Taking intermediate nodes, $c = \left(s \to v_1 \to^+ \cdots \to^+ v_{l-1} \to^+ s \right)$, where $(v_1, \text{IP}_{v_1}) \in T_s$ (direct IP contact). Nodes $v_2, \ldots, v_{l-1}$ may represent some but not all the nodes visited. It leads to the notation "$\to^+$" instead of "$\to$" in c.

Such cycles present possible routes in the overlay, generalizing the one-level lookahead [11, 12]. Given a cycle c, a node s may assume that a packet sent to v_1 would go through $v_2, \ldots, v_{l-1}$. That is, s looks ahead of its neighbors.

A cycle uses only node IDs, except v_1 whose IP address is already in the routing table. Therefore, even if a node u cannot use IP_v, the node v can serve as an intermediate node being presented in a cycle stored at u. This is a key to deal with IP-level reachability restrictions.

Let a node u maintain (additionally to T_u) a collection of q cycles, $\mathscr{C}_u = \{c_j\}_{j=1}^q$, where $c_j = (u \to v_{j1} \to^+ \cdots \to^+ v_{j,l-1} \to^+ u)$. We call $\mathscr{C}_u$ a network *cyclic structure* known to u. It represents additional knowledge about the network.

The idea of cyclic routing (CR) is to find the next hop analyzing the remaining route. The latter is constructed at a current node using the cyclic structure. When no appropriate cycle exists, the underlying DHT is used, see Algorithm 1.

8.3.1 Cycle Transitions

Let a source node s select a cycle $c = (s \to^+ u \to^+ d \to^+ s)$. In Fig. 8.1a, every intermediate node u uses the same cycle. The nodes, however, may be less

---

Algorithm 1 Cyclic Routing (CR)

Require: Packet p (traveling from s to d) arrives to $u \neq d$.
    The node u maintains T_u and $\mathscr{C}_u$.

Find $c \in \mathscr{C}_u$ such that
    $c = (u \to v_1 \to^+ \cdots \to^+ v_i = d' \to^+ \cdots \to^+ v_{l-1} \to^+ u)$ where d' is close to d;
if c is found **then**
    Let v_1 be the next-hop node v;
else
    Find the next-hop node $v \in T_u$ according
    to the underlying DHT;
end if
Forward p to v;

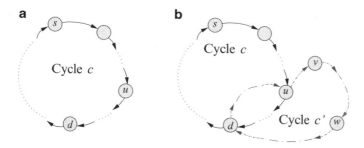

Fig. 8.1 Cyclic routing $s \rightarrow^+ d$: (**a**) one-cycle routing; (**b**) two-cycle routing

coordinated. In Fig. 8.1b, u selects a cycle $c' = \left(u \rightarrow v \rightarrow^+ w \rightarrow^+ d \rightarrow^+ u\right)$ such that its path $u \rightarrow^+ d$ does not coincide with the path $u \rightarrow^+ d$ of c.

In fact, the remaining route in CR is approximate. A packet follows a cycle, then transits to another, later it changes the cycle again and so forth. One cycle is enough to route a packet, but there are transitions due to lack of coordination between nodes, absence of an appropriate cycle, or failures when a cycle does not reflect correctly the current network topology.

A node can fix a cycle when sending a packet. In this case, cycle node IDs are piggybacked to the data packet. The next-hop node tries to follow this cycle when possible.

8.3.2 Dependable Routing and Path Length Upper Bounds

In this section we consider two problems, (1) the CR dependability since cycle transitions make it questionable and (2) the CR path length compared with basic DHT routing.

We follow Castro et al. [3] in definition of the routing dependability. P2P routing is dependable if a packet sent from s to d is always delivered to d in a stable and correct network. In general, Algorithm 1 is not dependable. In Fig. 8.1b, if w selects a cycle to reach d via u, then looping happens.

A dependable variant of the CR method exploits the same idea as in many DHTs—progressive routing. Any node forwards a packet progressively closer to the destination according to a distance metric ρ.

Theorem 8.1. *Algorithm 1 is dependable if any next-hop node v satisfies*

$$\rho(v,d) < \rho(u,d). \tag{8.1}$$

The number of hops is at most $\rho(s,d)/\delta$ where $\delta = \min_{u,v}(\rho(u,d) - \rho(v,d))$.

Proof. In (8.1) states that v is closer to the destination, regardless of whether a cycle or the underlying DHT is used. Therefore, the distance to the destination decreases monotonically, and after a finite number of hops (not more $\rho(s,d)/\delta$) the packet arrives to d.   □

Some DHTs provide more efficient paths of length $O(\log N)$ using geometrical progressive routing when at every hop the distance is reduced by a constant $k > 1$. For instance it happens in greedy routing when a node selects the next hop as close as possible to the destination.

Theorem 8.2. *Algorithm 1 is dependable if for some constant $k > 1$ any next-hop node v satisfies*

$$\rho(v,d) \leq \frac{\rho(u,d)}{k}. \tag{8.2}$$

There are $O(\log_k N)$ hops if n_{uv} is proportional to $\rho(u,v)$.

Proof. Inequality (8.2) is a particular instance of (8.1), thus the dependability is due to Theorem 8.1.

Let the route be $s \to v_1 \to v_2 \to \cdots \to v_{l-1} \to d$, where l is the hop number. Initially $n_{sd} < N$. Then $\rho(v_1,d) \leq \rho(s,d)/k$, and the proportional reduction is $n_{v_1 d} < N/k$. At the next hop, $\rho(v_2,d) \leq \rho(v_1,d)/k \leq \rho(s,d)/k^2$, and $n_{v_2 d} < N/k^2$. Finally, it yields $N/k^{l-1} = 1$, and $l = O(\log_k N)$.   □

As an example consider the Chord DHT. The distance $\rho(u,w)$ is length (the number of all possible IDs) of the ring arc between u and w clockwise. A neighbor of u is a finger or successor. The ith finger is the closest node v such that $\rho(u,v) > 2^{i-1}$ ($i = 1,2,\ldots,n$; 2^n is the ID space size). A successor is one of the closest nodes to u. When v is a successor, then $\rho(u,d) - \rho(u,v) \geq 1$ (case of local routing, $\delta \geq 1$ in Theorem 8.1). When the next-hop node v is a finger, then $\rho(u,v) \geq \rho(u,d)/2$ (case of global routing, $k = 2$ in Theorem 8.2).

According to Theorems 8.1 and 8.2, CR may follow the same rules of progressive routing as the underlying DHT. To satisfy these rules, only the next-hop nodes are used. It is enough to provide the same dependability and not to degrade the performance upper bounds. We discuss more efficient routing in Sect. 8.3.4.

8.3.3   Constructing Cycles

In (8.1) and (8.2) do not provide a direct way for constructing appropriate cycles. The dependence on d means that there should be a cycle at a node u for any destination d. In this section, we consider the problem of constructing a cycle appropriate for many destinations.

DHT Lookups: A consequence of bidirectional P2P communications. The method caches paths discovered in lookups.

Let a lookup packet follow the path $s \to^+ d$ collecting some visited node IDs (v_1, ..., v_{i-1}). They are included into the acknowledgment that d sends to s. Similarly, some node IDs (v_i, ..., v_{l-1}) are collected in the backward trip $d \to^+ s$. According to the theorems above, the cycle

$$s \to v_1 \to^+ \cdots \to^+ v_{i-1} \to^+ v_i \to^+ \cdots \to^+ v_{l-1} \to^+ s$$

can be used later for dependable routing to any d' such that $\rho(v_1, d') < \rho(s, d')$.

Lookahead: Let u maintain $\mathscr{C}_u$ such that for any $c \in \mathscr{C}_u$

$$c = \left(u \to v_1 \to v_2 \to \cdots \to v_{l-1} \to^+ u \right). \tag{8.3}$$

Nodes $u, v_1, \ldots v_{l-1}$ are consecutive neighbors ($v_1 \in T_u$, $v_2 \in T_{v_1}$, ...) and $l - 2$ is the level of lookahead (different for different nodes). NoN-greedy routing [11, 12] is a particular case of (8.3) when $l = 3$ and u maintains all cycles $u \to v_1 \to v_2 \to^+ u$ (for all neighbor's neighbors).

To construct a cycle in form (8.3) u asks its neighbor v. It replies with some neighbors w. Then u constructs cycles $u \to v \to w \to^+ u$. Also v can reply with cycles $v \to w_1 \to w_2 \to \ldots \to w_{l-1} \to^+ v$. In this case, u constructs cycles $u \to v \to w_1 \to w_2 \to \ldots \to w_{l-1} \to^+ u$.

The lookahead construction can be easily combined with DHT lookups; a lookup packet collects only the first n nodes of the route $s \to^+ d \to^+ s$.

These methods have the following properties.

Using existing overlay paths: This property preserves the routing dependability (in terms of Theorems 8.1 and 8.2). Moreover, it provides a kind of synchronization among nodes, since the nodes of the route are likely to use the same path that has been already in use.

Incrementality: CR can start from an overlay without any knowledge of cycles. Either passive or proactive cycle construction is possible. The former piggybacks on existing lookup packets running in the overlay. The latter introduces extra packets for cycle construction.

No substitution of local routing tables: A direct IP contact is more efficient than a multi-hop path. If v provides its IP to u, then u puts (v, IP_v) into T_u. However, when node IPs are not accessible the routing table cannot be updated, and the cyclic structure is a way to keep the additional information.

Overhead control: Cycle nodes are collected in lookup packets; it causes some overhead in terms of packet size. As we discuss in Sect. 8.3.5, CR needs only short cycles bounded with $O(\log N)$ hops. Moreover, it is not required to keep all intermediate nodes; some of them may be skipped. For instance, in the lookahead method only the first $l - 1$ nodes are collected and a node can vary l.

8.3.4 Finding the Most Efficient Cycle

According to Theorems 8.1 and 8.2, the selection of an appropriate cycle depends only on the distance between the current node, next-hop node and destination. Algorithm 1, however, supports more efficient routing. It relates to the definition of "d' is close to d" and depends on routing criteria, e.g., within few hops, with low latency or with high success probability.

The simplest case is exact matching $d' = d$ when an efficient cycle contains the destination. Another way defines a neighborhood with $H > 0$; in an efficient cycle, d' is a nearby node, i.e., $\rho(d', d) \leq H$.

Let u find such a cycle that

$$c = (u \to v_1 \to^+ \cdots \to^+ v_{l-1} \to u) \text{ and}$$
$$\rho(d', d) \text{ is minimum among all } c \in \mathscr{C}_u, d' = v_i \in c. \tag{8.4}$$

The same is in NoN-greedy routing [11] for $l = 3$ in (8.3), improving the efficiency from $O(\log N)$ to $O(\log N / \log \log N)$ hops in skip-graphs and small-world graphs.

Then u has to decide either using c or the underlying DHT. Basically, c is preferable when $\rho(d', d) \ll \rho(v, d)$. Some other parameters can also be important.

Assume that the underlying DHT provides geometrically progressive routing. If u forwards to v, the number of hops is estimated as $L_{\text{DHT}} = \log_k n_{ud} + 1$ (similarly to the proof of Theorem 8.2). If u uses c then routing takes either $L_{\text{CR}} = i$ hops when $\rho(d', d) \leq H$ (i.e., $i - 1$ hops to reach $d' = v_i$ and then one hop to d), or $L_{\text{CR}} = i + \log_k n_{d'd}$ hops when $\rho(d', d) > H$. Accordingly, c is preferable if $L_{\text{CR}} < L_{\text{DHT}}$.

Some DHTs estimate the round-trip time, $\text{RTT}(u, v)$ and $\text{RTT}(u, v_1)$. Taking the latency into account, c is preferable if $L_{\text{CR}} \cdot \text{RTT}(u, v_1) < L_{\text{DHT}} \cdot \text{RTT}(u, v)$. It is similar to the proximity route selection [5].

Some P2P systems rank nodes according to their behavior, $0 \leq r_v(u) \leq 1$. For instance, $r_v(u)$ is the probability that v successfully serves lookups sent from u. Then a cycle rank is $r_c = \prod_{j=1}^i r_{v_j}$, and c is preferable if $r_c > r_v r_{\min}^{L_{\text{DHT}}-2}$, where $r_{\min}$ is the default rank for unknown nodes.

Applying the criteria above, we can modify (8.4) in different ways. For instance, minimize i or $i \cdot \text{RTT}(u, v_1)$ (or maximize $\prod_{j=1}^i r_{v_j}(u)$) subject to $\rho(v_i, d) \leq H$.

8.3.5 Maintaining Cycles

Each node u maintains the cyclic structure $\mathscr{C}_u$ as a cache, inserting new cycles and removing invalid cycles.

New cycles: Constructing methods are described in Sect. 8.3.3. According to Sect. 8.3.4 a node inserts only efficient cycles. The same efficiency criteria can be applied except that there is no destination d fixed. An efficient cycle is short ($O(\log N)$ hops) and its first node is IP-close ($\text{RTT}(u, v_1)$ is small, as in the proximity neighbor selection [5]). Path-specific metrics can also be used, e.g., the round trip time along the cycle.

Memory cost: Given the typical bound $|T_u| = O(\log N)$. Preferably, each neighbor should be first in several cycles since finding a good cycle needs a comprehensive set of candidates. We assume that the number of such cycles is $O(\log N)$ because of the same reason as in geometrically progressive routing. An efficient cycle is short and consists of $O(\log N)$ nodes. As a result, a reasonable bound for IDs a node maintains is $O(\log^3 N)$.

Note, however, that a node may vary the cyclic structure size according to available resources and independently on other nodes. CR works even if there are nodes that do not maintain any cyclic structure.

Cycle rank: A node ranks cycles according to the routing history. Low-ranked cycles are removed or replaced. It takes into account the cycle usage rate, cycle success probability, cycle round trip time, and cycle stability. Note that these metrics can be estimated passively using regular lookups.

Invalid cycles: In a DHT, nodes ping neighbors for availability and stability. Similarly, cycle checking uses lookups, either passively or proactively. As was shown above, a node maintains $O(\log^2 N)$ cycles compared to $O(\log N)$ neighbors. Nevertheless, when a node checks a cycle then the neighbor is tested too. While pinging a node does not provide a new neighbor, checking a cycle can produce a new one.

In dynamic environments, a cycle is less stable than a node. Let p be the probability that a node is alive. A cycle of $n > 1$ nodes is alive with the lower probability p^n. However, cyclic structures gradually accumulate stable cycles since most of the nodes in real P2P systems are long-lived [16].

8.4 Simulation Results

CR-Chord was implemented as an extension of the MIT Chord simulator and simulation experiments were performed to compare Chord vs. CR-Chord in the presence of malicious DHT nodes and under churn. The results showed that CR-Chord has better lookup availability. In this section, we use a part of the data produced with the Chord and CR-Chord simulation in [7] and described further in Chap. 11 in detail. Compared to it, we provide further analysis of the basic CR properties. Note that [7] does not take into account the IP-level reachability restrictions; and produces more pessimistic estimates for the CR method.

8.4.1   Chord and CR-Chord Implementations

Both Chord and CR-Chord use flexible routing table maintenance when a node keeps n mandatory fingers and also n_{af} additional fingers to know more about the network.

CR-Chord implements cyclic routing when an appropriate cycle (if any) is piggybacked into the packet. Only short cycles are stored for reuse; the number of hops is at most $2\log_2 N$, where a node approximates $\log_2 N$ as $\min(f,n)$, f is the number of fingers and n is the number of ID bits.

CR-Chord supports a kind of multipath routing when m_d duplicate packets are sent using fingers closest to the destination. This mechanism is used for constructing cycles (see the DHT lookups method in Sect. 8.3.3).

8.4.2   Simulation Setup

Chord DHT of size $N = 1,000$ nodes is simulated. The Chord ID space is of $n = 24$ (2^{24} node IDs). Other parameters are $n_{af} = 12$ and $m_d = 3$. Each simulation experiment is executed ten times for averaging.

We assume that malicious nodes ignore all data lookups for which they are responsible. That is, data at such nodes are lost. They process correctly all other requests trying to hide the malicious activity.

Network construction consists of three steps. First, a Chord DHT of G good nodes is constructed; nodes are joining randomly. Second, $D = 100N = 10^5$ documents are distributed uniformly. Third, $M = fN$ malicious nodes join the network at the same time, where $G + M = N$ and f is the fraction of malicious nodes (varied from 5 % up to 50 %).

There are 100 documents per a node on the average but about $100M$ documents are lost. Good nodes initiate randomly $L = 0.01N^2 = 10^4$ requests in total with rate $1.0\,\mathrm{s}^{-1}$ (requests per second). The standard Chord stabilization procedure is executed every 30 s.

Churn consists of node joins and leaves. To preserve the ratio of good and malicious nodes, any node join is accompanied with a node leave and vice versa; those happen with rates $R = 0, 0.02, 0.2\,\mathrm{s}^{-1}$. When the lookup rate is $1.0\,\mathrm{s}^{-1}$ as above, a churn join/leave pair happens every 5 lookups for $R = 0.2\,\mathrm{s}^{-1}$ and every 50 lookups for $R = 0.2\,\mathrm{s}^{-1}$.

8.4.3   CR-Chord vs. Chord

Let $S_{chord}(f,R)$ and $F_{chord}(f,R) = L - S_{chord}(f,R)$ be counters of successful and failed lookups for Chord. Similarly, $S_{cr}(f,R)$ and $F_{cr}(f,R) = L - S_{cr}(f,R)$ are for CR-Chord. Table 8.1 presents the estimates of S and F when varying the fraction

Table 8.1 Lookup success improvement and lookup failure reduction ($N = 10^3$, $L = 0.01N^2 = 10^4$).

$M = fN$, %	10%		20%		30%		40%		50%		Average	
R, s^{-1}	0	0.2	0	0.2	0	0.2	0	0.2	0	0.2	0	0.2
S_{chord}	6724.9	5141.0	4353.8	3110.1	2673.9	1900.4	1501.1	994.7	808.2	534.9	3212.4	2336.2
$\Delta_c(S_{\text{chord}})$, %	23.6		28.6		28.9		33.7		33.8		29.7	
S_{cr}	7958.4	7557.9	5985.4	5262.2	4197.9	3526.2	2617.3	2042.3	1518.5	1154.9	4455.5	3908.7
$\Delta_c(S_{\text{cr}})$, %	5.0		12.1		16.0		22.0		23.9		15.8	
$\dfrac{S_{\text{cr}} - S_{\text{chord}}}{S_{\text{chord}}}$, %	18.3	47.0	37.5	69.2	57.0	85.6	74.4	105.3	87.9	115.9	55.0	84.6
F_{chord}	3275.1	4859.0	5646.2	6889.9	7326.1	8099.6	8498.9	9005.3	9191.8	9465.1	6787.6	7663.8
$\Delta_c(F_{\text{chord}})$, %	48.4		22.0		10.6		6.0		3.0		18.0	
F_{cr}	2041.6	2442.1	4014.6	4737.8	5802.1	6473.8	7382.7	7957.7	8481.5	8845.1	5544.5	6091.3
$\Delta_c(F_{\text{cr}})$, %	19.6		18.0		11.6		7.8		4.3		12.3	
$\dfrac{F_{\text{chord}} - F_{\text{cr}}}{F_{\text{chord}}}$, %	37.7	49.7	28.9	37.0	20.8	20.1	13.1	11.6	7.7	6.6	21.6	23.8

Fig. 8.2 Lookup failure rates for different churn rates. CR-Chord always outperforms Chord. The CR efficiency is lower for larger f and R

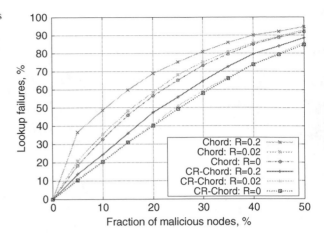

of malicious nodes. The relative lookup success improvement and relative lookup failure reduction,

$$\Delta(S) = \frac{S_{cr} - S_{chord}}{S_{chord}} \quad \text{and} \quad \Delta(F) = \frac{F_{chord} - F_{cr}}{F_{chord}},$$

are also given. In these relative estimates, Chord is a base.

Two cases—with churn ($R = 0.2$) and without churn ($R = 0$)—are considered. The effect of churn is estimated with the relative variation

$$\Delta_c(x) = \frac{x(f,0) - x(f,0.2)}{x(f,0)} \quad \text{for } x = S, F,$$

where $R = 0$ is a basement.

Both in Chord and CR-Chord, the lookup success (failure) rate decreases (increases) with more malicious nodes and churn. CR-Chord always has better values for these rates. Interestingly that CR-Chord under churn ($R = 0.2$) outperforms Chord in stable networks ($R = 0$). Figure 8.2 is from [7] and graphically shows the estimates of lookup failure rate.

When the fraction of malicious nodes is small (5–10 %) the CR-Chord lookup failure rate is almost twice lower. When the fraction of malicious nodes is large (40–50 %) the CR-Chord lookup success rate is about twice higher.

Churn affects the CR-Chord lookup availability less. In particular, $\Delta_c(S_{chord}) > \Delta_c(S_{cr})$ for all fractions of malicious nodes, and $\Delta_c(S_{chord})/\Delta_c(S_{cr}) \approx 2$ on the average. In terms of the lookup failure rate, $\Delta_c(F_{chord})$ becomes slightly lower than $\Delta_c(F_{cr})$ only for large f. On average, $\Delta_c(F_{chord})/\Delta_c(F_{cr}) \approx 1.5$.

In fact, CR accelerates the routing mechanism of underlying DHT. The effect is achieved with a small set of cycles per node (Fig. 8.3). The length of cycles is short (Fig. 8.4).

Fig. 8.3 The average number of cycles per node ($N = 10^3$, $R = 0$). There is a lack of cycles constructed in very malicious networks

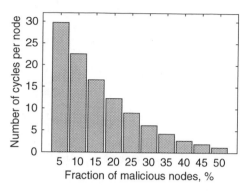

Fig. 8.4 The average cycle length (in hops) for different network sizes ($R = 0$, $L = 0.01N$). In a larger network, cycles are longer. Cycles become shorter in malicious networks

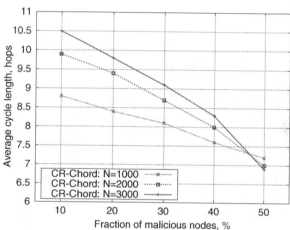

The CR efficiency is reduced for large f and R. The reason is that underlying DHT lookups do not allow constructing good cycles (although good paths do exist). As a result, there is a lack of cycles constructed at a node (see Fig. 8.3), they are too short and hence less global (see Fig. 8.4), and the difference between Chord and CR-Chord becomes small.

As was shown in [7], known methods like flexible routing table maintenance (parameter n_{af}) multipath routing (parameter m_d) become inefficient for finding good paths when f or R are large. From this point of view, the lookahead method seems more promising for constructing cycles.

8.5 Discussion on Advanced Capabilities

The basic CR design is an additional routing mechanism to deal primarily with the IP-level reachability restrictions. It can be combined with known methods such as flexible routing table maintenance and multipath routing. The simulation showed

Fig. 8.5 In a lookup path
$s \rightarrow^+ d$, CR acts primarily as
global routing $s \rightarrow^+ d'$; local
routing $d' \rightarrow^+ d$ takes
typically one hop

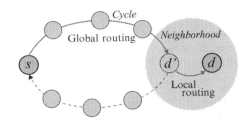

that this combination also works efficiently in networks with malicious nodes
and churn. This section briefly discusses some advanced capabilities that CR can
provide.

8.5.1   Global and Local Routing

In many DHTs, routing to a destination neighborhood is an important point. Let us
mention two reasons.

First, DHT routing is usually key-based, not nodeID-based, performing routing
to a node responsible for a given key. Since a node is responsible for all keys close
to its ID, let a message arrive to a node close to the key. Moreover, DHT replication
techniques distribute data among nearby nodes.

Second, a node typically knows its neighborhood well since many nearby nodes
are in the routing table. As an example, the Chord DHT keeps several successors
and predecessors. Consequently, one hop is needed when a message is at a nearby
node.

Therefore, we distinguish two parts of P2P routing, *global* and *local*. In global
routing, a message is delivered close to the destination. In local routing, the
destination is at a nearby node. By design, CR aims at global routing (Fig. 8.5).
When an appropriate cycle is found, then global routing happens, crossing a large
part of the overlay and arriving to a nearby node d'.

Cyclic routing can also act as local when gaps appear in routing tables because
of IP-level reachability restrictions, e.g., a nearby node is located behind a NAT.
Moreover, cycles provide some information to prioritize routing. For instance, a
neighbor is good if cycle success rate is high for those cycles that this neighbor
starts.

8.5.2   Cyclic vs. Acyclic Paths

In principle, a node may maintain paths to a set of destinations instead of cycles, as
in link-state routing protocols [8] or MPLS networks [13]. In link-state routing, a
node keeps a graph of the whole network. Global graph analysis calculates entries

(d, IP_v), where v is the-next hop node of the best path to d. In MPLS networks, communication between s and d is preceded with construction of a path $s \to^+ d$, where all nodes know to which node to forward a message. These approaches are essentially global and do not scale well.

On the other hand, basic DHT routing uses a local view of the network when a node knows a limited set of neighbors. This approach scales well but in some cases more knowledge about the network is needed.

The CR design allows a tradeoff between the local and global cases. When a node does not use cyclic structure, only the local view is presented. With cyclic structure a node knows more, up to the extreme case when all cycles are known.

A cycle $s \to^+ d \to^+ s$ contains more information than a non-cyclic path since both forward and backward paths are presented. First, it introduces explicit support for bidirectional communication. Second, there is no fixed destination in a cycle compared with a source-destination path, and the same cycle can be used for routing to many destinations.

8.5.3 Multi-Path Routing

Multi-path routing is a way to deal with problems such as path congestion, node failures, high path latency, and the presence of malicious nodes [2, 14, 18].

In DHTs, when s experiences a problem with the best next-hop, another one is used. This mechanism is enhanced in CR. In fact, CR provides a way to route messages around overloaded, failed or malicious nodes.

First, a node can duplicate a message using both the most efficient cycle and DHT next-hop node (see Sect. 8.3.4). Basically, only a source node makes this duplication (for example, in CR-Chord [7]). However, other strategies are possible.

Second, there can be several appropriate cycles in $\mathscr{C}_s$ for a given destination d. They provide a precomputed pool of alternate paths $s \to^+ d$. That is, s sends duplicates through several cycles like in the neighbor-set anycast [2] or the constrained multicast [18]). Moreover, s can sort alternate cycles according to their efficiency, as in the sorted backup paths technique [18].

The CR method also supports disjointed paths [14]. A node s treats alternate cycles to d as semi-disjointed if their subpaths $s \to^+ d$ do not share common intermediate nodes. The prefix "semi" appears since s does not necessarily know all nodes of these cycles. To date, although disjointed paths are very valuable for multi-path and secure routing, most of P2P systems do not support them.

In multicast communication, a node s sends a message to a set D of destinations. In the CR method, s forms a pool $\mathscr{M}_s(D) \subseteq \mathscr{C}_s$ of appropriate cycles that covers all destinations. The message is sent using these cycles. Although in the worst case the size of $\mathscr{M}_s(D)$ can be equal to $|D|$, several destinations can be reached with one cycle. Hence, the duplication is reduced.

8.5.4 Security

The CR efficiency for networks with malicious nodes is shown in Sect. 8.4 and in our simulations described in Chap. 11 and [7].

Recall that the CR method reduces the use of IP addresses. Even if a node restricts access to its IP address, other nodes still can gain knowing that the node appears as an intermediary in some lookups. Consequently, a DHT equipped with CR is less exposed to IP-level attacks, such as Denial-of-Service (DoS).

The CR-method supports bidirectional communications. The security level is higher when both directions are independent, and traffic goes through disjoint paths. Similarly to semi-disjointed cycles, a source node selects a cycle with independent forward and backward directions.

In CR method, a node has more possibilities to detect incorrect or suspicious routes. For instance, let a node u appear at s in a large fraction of cycles. This fact signals that u becomes dangerous for s. At least, u is a bottleneck; in a worse case, u can disrupt communications of s with other nodes. Therefore, the cyclic structure helps s to detect such a risky situation in advance.

Assume a node s detects that a node u is malicious. In a basic DHT, s can only remove u from its routing table. In the CR method, s also removes all cycles containing u as well as s can decrease the priority of those neighbors that routed messages to u (they are first-hop nodes in the cycles with u).

8.6 Summary

Recent progress in DHT routing made possible many practical P2P-based applications in the Internet. However, they work poorly when no direct IP connectivity is available to some nodes (e.g., located behind a NAT or firewall) and in the presence of malicious or overloaded nodes.

This chapter introduced the design of Cyclic Routing (CR), a method that improves routing over basic DHT algorithms. Experimental analysis is provided using data from simulations [7] for the Chord DHT enhanced with CR.

In simulations, malicious nodes dropped lookup packets but there were no restrictions on IP addressing. The results suggest that CR improves the Chord lookup availability. For instance, when the number of malicious nodes is small (5–10%) CR-Chord has almost twice lower lookup failure rate. When the number of malicious nodes is large (40–50%) CR-Chord has about twice higher lookup success rate.

The CR method does not substitute known methods of flexible routing table maintenance and multipath routing but complements them when the IP-level reachability fails. The CR method allows a node to have a broader view on the global network. Since CR only uses node IDs, it is applicable even when a routing table entry insertion is not possible, e.g., when the node's IP address is not globally routable or frequently changes.

Cyclic routing generalizes the idea of lookahead approach to DHT routing, maintaining a trade-off between the lookahead level and available resources. Therefore, CR takes a place between two extreme cases: (1) DHT routing using only neighbors (local) and (2) shortest-path routing (global).

Using additional knowledge of the global network, cycles enable multiple-path packet delivery and routing around malicious or overloaded nodes. Instead of selecting a path based on neighbors only, CR exploits paths already discovered in the overlay. From this point of view, cyclic structures at nodes accumulate stable efficient paths in the overlay improving its routing.

References

1. Castro, M., Drushel, P., Hu, Y., Rowstron, A.: Exploiting network proximity in peer-to-peer networks. Technical Report MSR-TR-2002-82, Microsoft Research (2002)
2. Castro, M., Drushel, P., Ganesh, A., Rowstron, A., Wallach, D.S.: Secure routing for structured peer-to-peer overlay networks. In: Proceedings of 5th USENIX Symposium on Operating System Design and Implementation (OSDI 2002), pp. 299–314. ACM, Boston (2002)
3. Castro, M., Costa, M., Rowstron, A.: Performance and dependability of structured peer-to-peer overlays. Technical Report MSR-TR-2003-94, Microsoft Research (2003)
4. Freedman, M.J., Lakshminarayanan, K., Rhea, S., Stoica, I.: Non-transitive connectivity and DHTs. In: Proceedings of the 2nd USENIX Workshop on Real, Large Distributed Systems (WORLDS'05), pp. 55–60, USENIX Association San Francisco (2005)
5. Gummadi, K., Gummadi, R., Gribble, S., Ratnasamy, S., Shenker, S., Stoica, I.: The impact of DHT routing geometry on resilience and proximity. In: Proceedings of ACM SIGCOMM'03, pp. 381–394. ACM, New York (2003). doi: http://doi.acm.org/10.1145/863955.863998
6. Kaashoek, M.F., Karger, D.R.: Koorde: A simple degree-optimal distributed hash table. In: IPTPS '03: Proceedings of 2nd International Workshop on Peer-to-Peer Systems. Lecture Notes in Computer Science, vol. 2735, pp. 98–107. Springer, Berlin (2003)
7. Korzun, D., Nechaev, B., Gurtov, A.: CR-Chord: Improving lookup availability in the presence of malicious DHT nodes. HIIT Technical Report 2008-2, Helsinki Institute for Information Technology HIIT (2008). URL: http://www.hiit.fi/nrg-publications
8. Kurose, J.F., Ross, K.: Computer Networking: A Top-Down Approach Featuring the Internet. Addison-Wesley Longman Publishing Co. Inc., Boston (2002)
9. Leong, B., Liskov, B., Demaine, E.: Epichord: parallelizing the Chord lookup algorithm with reactive routing state management. In: ICON 2004: Proceedings of 12th International Conference on Networks, IEEE, USA pp. 270–276 (2004)
10. Li, J., Stribling, J., Morris, R., Kaashoek, M.F.: Bandwidth-efficient management of DHT routing tables. In: Proceedings of the 2nd Symposium on Networked Systems Design and Implementation (NSDI '05), USENIX Association. USA pp. 99–114 (2005)
11. Manku, G.S., Naor, M., Wieder, U.: Know thy neighbor's neighbor: the power of lookahead in randomized P2P networks. In: STOC '04: Proceedings of 36th Annual ACM Symposium on Theory of Computing, pp. 54–63. ACM, New York (2004). doi: http://doi.acm.org/10.1145/1007352.1007368
12. Naor, M., Wieder, U.: Know thy neighbor's neighbor: better routing for skip-graphs and small worlds. In: IPTPS '04: Proceedings of 3rd International Workshop on Peer-to-Peer Systems. Lecture Notes in Computer Science, vol. 3279. Springer, Berlin (2004)
13. Sharma, V., Hellstrand, F.: Framework for multi-protocol label switching (MPLS)-based recovery. RFC 3469, IETF (2003). URL: http://www.ietf.org/rfc/rfc3469.txt

14. Srivatsa, M., Liu, L.: Vulnerabilities and security threats in structured overlay networks: a quantitative analysis. In: ACSAC '04: Proceedings of 20th Annual Computer Security Applications Conference, pp. 252–261. IEEE Computer Society, USA (2004). doi: http://dx. doi.org/10.1109/CSAC.2004.50
15. Stoica, I., Morris, R., Liben-Nowell, D., Karger, D., Kaashoek, M.F., Dabek, F., Balakrishnan, H.: Chord: a scalable peer-to-peer lookup service for internet applications. IEEE/ACM Trans. Netw. **11**(1), 17–32 (2003)
16. Stutzbach, D., Rejaie, R.: Understanding churn in peer-to-peer networks. In: IMC '06: Proceedings of 6th ACM SIGCOMM Conference on Internet Measurement, pp. 189–202. ACM, New York (2006). doi: http://doi.acm.org/10.1145/1177080.1177105
17. Zhang, H., Goel, A., Govindan, R.: Incrementally improving lookup latency in distributed hash table systems. In: Proceedings of 2003 ACM SIGMETRICS International Conference Measurement and Modeling of Computer Systems, pp. 114–125. ACM, New York (2003). doi: http://doi.acm.org/10.1145/781027.781042
18. Zhao, B.Y., Huang, L., Stribling, J., Joseph, A.D., Kubiatowicz, J.D.: Exploiting routing redundancy via structured peer-to-peer overlays. In: ICNP '03: Proceedings of 11th IEEE International Conference on Network Protocols, pp. 246–257 (2003)

Chapter 9
Diophantine Routing

Abstract An important problem in any structured P2P overlay network is what routes are available between nodes. Understanding the structure of routes helps to solve problems related to routing performance, security, and scalability. In this chapter, we propose a theoretical approach for describing routes and their structures. It is based on recent results in the linear Diophantine analysis. A route aggregates several P2P paths that packets follow. A commutative context-free grammar describes the forwarding behavior of P2P nodes. Derivations in the grammar correspond to P2P routes. Initial and final strings of a derivation define packet sources and destinations, respectively. Based on that we construct a linear Diophantine equation system, where any solution counts forwarding actions in a route representing certain integral properties. Therefore, P2P paths and their composition into routes are described by a linear Diophantine system—a Diophantine model of P2P routes. Its finite basis of the solution set defines the structure of available routes.

9.1 Introduction

In computer networks, a classical problem is routing a packet from the source to the destination host. Among multiple available routes, the routing algorithm selects an optimal one according to some metric such as a hop count, the delay or bandwidth. The algorithm is executed independently by network nodes based on a limited view of the network, often only based on neighbor links. This is exactly the P2P network case and the corresponding problem of local knowledge. Such algorithms require moderate memory and computational resources but produce suboptimal routes. Traditional algorithms do not support routing among multiple parallel paths for reliability and load balancing.

Although a lot of routing algorithms for P2P networks have been proposed, we agree with Gummadi et al. [7] that the P2P routing literature very seldom provides general insight across algorithms. We believe that such an analysis of routing mechanics can be based on an appropriate theory, general for a wide variety of P2P

D. Korzun and A. Gurtov, *Structured Peer-to-Peer Systems: Fundamentals of Hierarchical Organization, Routing, Scaling, and Security*, DOI 10.1007/978-1-4614-5483-0_9, © Springer Science+Business Media New York 2013

networks. In this chapter, we focus on the concept of a route, a key element in P2P routing problem. We propose a generic approach for describing routes; it uses recent progress in the linear Diophantine analysis. Although we also discuss some possible applications, our primary intention is to introduce the theoretical background for P2P routing in general. This chapter extends our previous work [10].

We consider a mathematical Diophantine model of P2P routes, where a route aggregates several P2P paths that packets follow. Such aggregation is due to multipath routing (packet duplications and retransmissions) as well as multiple sources and destinations. The model is based on abstract parallel process algebra [6], and we use a commutative context-free grammar to describe forwarding behavior of P2P nodes. A derivation in the grammar corresponds to a P2P route such that derivation initial and final strings define packet sources and destinations, respectively.

Given a grammar, we construct a linear Diophantine equation system [11, 13]. Any of its solutions counts forwarding actions in a route. The basis of a linear Diophantine system (Hilbert basis) describes all solutions in a finite way [4, 17, 20, 21]; we use this fact to define a routing structure. Manipulation with parameters of packet sources and destinations specifies which routes the model targets.

The model extends well-known models based on network topology graphs and discrete network flows [3, 8, 20]. Since the latter are very popular in routing models, we believe that our model has potential for various applications. We also hope that the complexity problem, typical for large-scale discrete models, can be overcome in our approach. There are efficient algorithms for solving such linear Diophantine systems as appear in our model [12, 13].

The rest of the chapter is organized as follows. Section 9.2 formulates the mathematical background for commutative context-free grammars and non-negative linear Diophantine equations. In Sect. 9.3, we show how a grammar can be used to describe routes in a P2P overlay network. A routing model combining a grammar and a Diophantine equation system are developed in Sect. 9.4. Section 9.5 presents a discussion of possible model applications to optimize routing. It also compares our approach to related work. Details in Sects. 9.3–9.5 are explained using a simple network common for all examples. All examples take only cyclic routes since they relate to the primary class of linear Diophantine equations that can be solved efficiently with the syntactic algorithm available via the Web-SynDic system (http://websyndic.cs.karelia.ru).

9.2 Mathematical Background

In this section we summarize our mathematical background developed in [4, 10, 11, 13]. Later in Sects. 9.3 and 9.4 the model of P2P routes will be formulated in terms of string transformations, which follow a commutative context-free grammar and represent grammar derivations. The grammar assigns a linear Diophantine equation system, where non-negative integer solutions correspond to grammar derivations.

9.2.1 Commutative Context-Free Grammars

Let $\mathbb{Z}_+$ be the set of non-negative integers. Given a finite alphabet Π, a commutative string over Π is $\{\pi^{a_\pi}\}_{\pi\in\Pi}$, where $a_\pi \in \mathbb{Z}_+$ is the number of occurrences of $\pi \in \Pi$ in the string. In other words, the order of symbols is ignored and a string is a multiset of symbols. For a string α, denote its exponents

$$\#_\pi[\alpha] = a_\pi \in \mathbb{Z}_+ \text{ for } \pi \in \Pi \quad \text{and} \quad \#[\alpha] = a \in \mathbb{Z}_+^{|\Pi|}.$$

The reversion is $\star[a] = \{\pi^{a_\pi}\}_{\pi\in\Pi}$.

All strings over Π (including the empty string ε) form the free commutative monoid Π^* in respect to concatenation. Given $\alpha', \alpha'' \in \Pi^*$, both $\alpha'\alpha''$ and $\alpha' + \alpha''$ denote the concatenation of α' and α'', which corresponds to the sum of exponents. The reverse operation $\alpha' - \alpha''$ is defined if the exponents remain non-negative. If so we write $\alpha'' \leq \alpha'$.

A commutative context-free grammar (CCF-grammar) without a start symbol is a 3-tuple $G = (N, \Sigma, P)$, where N and Σ are finite disjointed sets, called the nonterminal and terminal alphabets, respectively, P is a finite subset of $N \times \Sigma^* N^*$, called the set of rules. A rule is written $u \to \tau\rho$, where $u \in N$, $\tau \in \Sigma^*$, $\rho \in N^*$.

A CCF-grammar provides a mechanism to transform strings $\varkappa\alpha \in \Sigma^* N^*$. Let

$$\varkappa', \varkappa'' \in \Sigma^*, \quad \alpha = \star[(b_v^-)_{v\in N}], \quad \beta = \star[(b_v^+)_{v\in N}].$$

We say that $\varkappa'\alpha$ directly derives $\varkappa''\beta$, written $\varkappa'\alpha \Rightarrow \varkappa''\beta$, if $b_u^- > 0$, $\varkappa'' = \varkappa'\tau$, and $\beta = \alpha - u + \rho$ for some $(u \to \tau\rho) \in P$. In other words, one instance of u in α is substituted with $\tau\rho$. A finite sequence of direct derivations $\varkappa'\alpha \Rightarrow \cdots \Rightarrow \varkappa''\beta$ is called a derivation and denoted

$$\varkappa'\alpha \Rightarrow^* \varkappa''\beta.$$

The derivation length k is the number of direct derivations; k may be written in the exponent, $\varkappa'\alpha \Rightarrow^k \varkappa''\beta$. By definition, $\varkappa'\alpha \Rightarrow^0 \varkappa''\beta$ iff $\varkappa'\alpha = \varkappa''\beta$. If $k > 0$ we write $\varkappa'\alpha \Rightarrow^+ \varkappa''\beta$. Let $\#_r[\varkappa'\alpha \Rightarrow^* \varkappa''\beta]$ denote the number of applications of the rule r in the derivation. Then $\#[\varkappa'\alpha \Rightarrow^* \varkappa''\beta]$ is a non-negative integer vector of rule applications.

Let $\varkappa', \varkappa'' \in \Sigma^*$ and $\alpha, \beta \in \Upsilon^*$. A derivation $\varkappa'\alpha \Rightarrow^+ \varkappa''\beta$ is (1) cyclic if $\alpha \leq \beta$, (2) a cycle if $\alpha = \beta$ and $\varkappa' = \varkappa''$, (3) simple if it does not contain proper cycles. Clearly, $\varkappa' \leq \varkappa''$ holds always since terminals do not disappear in a derivation.

9.2.2 Nonnegative Linear Diophantine Equations

Let $\mathbb{Z}$ be the set of integers. A nonnegative linear Diophantine equation (NLDE) system consists of n equations in m unknowns,

$$Ax = b, \quad \text{where } A \in \mathbb{Z}^{n\times m}, \ b \in \mathbb{Z}^n, \ x \in \mathbb{Z}_+^m, \tag{9.1}$$

where A is called the coefficient matrix, b is called the constant term, and x is the column of unknowns. An NLDE system is homogeneous (a homNLDE system) when $b = \mathbb{O}$.

A solution to (9.1) is *irreducible* if it is not a sum of two non-zero solutions to the same system. For a homNLDE system, the set $\mathscr{H}$ of all its irreducible solutions is called *the Hilbert basis*. For (9.1), a pair $(\mathscr{N}, \mathscr{H})$ is a basis if $\mathscr{N}$ is the set of all irreducible solutions to (9.1) and $\mathscr{H}$ is the Hilbert basis of the homNLDE system. Such a basis is unique and finite. Given the basis $(\mathscr{N}, \mathscr{H})$ of (9.1), the general solution is

$$x = x' + \sum_{h \in \mathscr{H}} c_h h \ \text{ for some } x' \in \mathscr{N},\ c_h \in \mathbb{Z}_+.$$

For further reading on basic facts about NLDE and NLDE systems please refer to [1, 8, 17, 20, 21].

Moving terms with negative coefficients in each equation to another side, we rewrite (9.1) as

$$A'x + b^- = A''x + b^+.$$

The subclass of NLDE systems associated with CCF-grammars (ANLDE systems) was introduced in [4, 11]. It restricts A'' to a special form. In particular, a homANLDE system associates with a CCF-grammar $G = (N, \Sigma, P)$ as follows. Non-terminals (N) and terminals (Σ) correspond to equations, grammar rules (P) correspond to unknowns. Let

$$P_u = \{r \in P \mid r = (u \to \tau_r \rho_r),\ \tau_r = \star[(a_{\sigma r})_{\sigma \in \Sigma}],\ \rho_r = \star[(a_{vr})_{v \in N}]\}.$$

A homANLDE system consists of $n = |N| + |\Sigma|$ equations,

$$\begin{cases} \displaystyle\sum_{r \in P} a_{ur} x_r = \sum_{r \in P_u} x_r, & u \in N, \\ \displaystyle\sum_{r \in P} a_{\sigma r} x_r = 0, & \sigma \in \Sigma. \end{cases} \quad (9.2)$$

Let $t = |\Sigma|$ and $m = |P|$. Unknowns are interpreted as the number of grammar rule applications in cycles, i.e., $x = \#[\alpha \Rightarrow^+ \alpha]$. Note that only the first N equations in (9.2) are essential; the last t equations can be eliminated with setting $x_r = 0$ for all r such that $a_{\sigma r} > 0$ for some $\sigma \in \Sigma$. Rules assigned with eliminated unknowns are not applied in the cycle derivation.

Let $b^-, b^+ \in \mathbb{Z}_+^{n-t}$ and $b \in \mathbb{Z}_+^t$. In addition to G, consider non-terminal strings $\alpha = \star[b^-]$ and $\beta = \star[b^+]$ and a terminal string $\varkappa = \star[b]$, where $\alpha, \beta \in N^*$ and $\varkappa \in \Sigma^*$. An ANLDE system associates with the triple $(G; \alpha, \varkappa\beta)$ as

$$\begin{cases} \displaystyle\sum_{r \in P} a_{ur} x_r + b_u^- = \sum_{r \in P_u} x_r + b_u^+, & u \in N, \\ \displaystyle\sum_{r \in P} a_{\sigma r} x_r = b_\sigma, & \sigma \in \Sigma, \end{cases} \quad (9.3)$$

Unknowns are interpreted similarly to the homANLDE case, i.e., $x = \#[\alpha \Rightarrow^* \varkappa\beta]$.

Assume $\min(b_u^+, b_u^-) = 0 \; \forall u \in N$; otherwise take $d_u = \min(b_u^+, b_u^-)$ and reassign $b_u^+ := b_u^+ - d_u$, $b_u^- := b_u^- - d_u$. The following theorem relates grammar derivations and ANLDE system solutions [13].

Theorem 9.1 (Solution to an ANLDE system). *Let* $(\mathscr{H}, \mathscr{N})$ *be the basis of (9.3).*

1. *(ANLDE system): x is a solution to (9.3) if and only if*

$$x = x^{\alpha, \varkappa' \beta'} + x^{\varkappa'' \beta''} + x^{\varepsilon}, \quad \text{where} \tag{9.4}$$

- $\varkappa' \varkappa'' = \varkappa$ *and* $\beta' \beta'' = \beta$;
- $x^{\alpha, \varkappa' \beta'} = \#[\alpha \Rightarrow^* \varkappa' \beta']$, *a simple non-cyclic derivation;*
- $x^{\varkappa'' \beta''} = \#[\alpha' \Rightarrow^* \varkappa'' \alpha' \beta'']$ *for a some* $\alpha' \in N^*$, *a simple cyclic derivation but not a cycle;*
- $x^{\varepsilon} = \#[\alpha'' \Rightarrow^+ \alpha'']$ *for a some* $\alpha'' \in N^*$, *a cycle.*

2. *(Basis): $x \in \mathscr{N}$ if and only if $x^{\varepsilon} = \mathbb{O}$ in (9.4).*
3. *(homANLDE system): x is a solution to (9.2) if and only if $x = x^{\varepsilon}$.*
4. *(Hilbert basis): $x \in \mathscr{H}$ if and only if $x = x^{\varepsilon}$, where x^{ε} is defined by a simple cycle.*

The idea behind Theorem 9.1 is that any derivation $\alpha \Rightarrow^* \varkappa \beta$ corresponds to a solution to (9.3). We satisfy (9.4) by taking[1]

$$\varkappa \beta = \varkappa' \beta', \quad \alpha' = \varkappa'' = \beta'' = \varepsilon, \quad x^{\varkappa'' \beta''} = x^{\varepsilon} = \mathbb{O}$$

Nevertheless, there can be solutions that correspond to a collection of derivations:

$$\alpha \Rightarrow^* \varkappa' \beta', \quad \alpha' \Rightarrow^* \varkappa'' \alpha' \beta'', \quad \alpha'' \Rightarrow^+ \alpha''.$$

They can be combined into the derivation

$$\alpha \alpha' \alpha'' \Rightarrow^* \varkappa \alpha' \alpha'' \beta.$$

It is the same as the initial derivation $\alpha \Rightarrow^* \varkappa \beta$ except that a cyclic part $\alpha' \alpha''$ appears on both sides. Note that (9.3) associates with the triple $(G; \alpha \alpha' \alpha'', \varkappa \alpha' \alpha'' \beta)$.

This relation between solutions and derivations leads to efficient (polynomial and pseudo-polynomial) algorithms for solving homANLDE systems and for solving some classes of inhomogeneous ANLDE systems [12].

---

[1] For brevity we ignore the case when $\alpha \Rightarrow^* \varkappa \beta$ is not simple. It also satisfies (9.4) but requires extracting all simple cycles and moving them to x^{ε}, so leading to $x^{\varepsilon} \neq 0$.

9.3   Routing Grammar

In this section we introduce basic model assumptions on routing in P2P overlay networks and define forwarding options that a node implements. At each step the current node selects candidates for the next hop. Then, having observed that candidates can be represented as a string over N, we construct a CCF-grammar, where each rule models a forwarding option.

9.3.1   Routing and Forwarding

An overlay topology can be represented with a digraph $G_T = (N, E)$, where the arc set E consists of all outgoing links for all nodes; any entry $(v, IP_v) \in T_u$ corresponds to the arc $(u, v) \in E$. Such a graph model for the routing problem is well-known, e.g., see [3, 8], as well as actively applied for structured P2P networks, e.g., see [7, 9, 15, 16, 19].

Let a node s (source) send a packet to a node d (destination). The packet follows in G_T either a path $s \Rightarrow^* d$ (the destination is reached by the only path), or a path $s \Rightarrow^* v$ (the destination is not reached since the packet completed the path at $v \neq d$), or a collection of paths as above (the packet is duplicated; its copies follow by their own paths).

Based on this observation, we call *an atomic route* the collection of all paths that a packet and its copies have followed starting from a given source node s. Such a route is denoted

$$s \Rightarrow^* d_1^{b_1^+} \cdots d_k^{b_k^+},$$

where d_i are all nodes at which the packets completed paths, and nonnegative integer b_i^+ shows[2] how many packets completed their paths at d_i.

Let source nodes $s_1, \ldots, s_l$ send $b_1^-, \ldots, b_l^-$ packets,[3] respectively. Combining their atomic routes yields *the aggregated route*:

$$s_1^{b_1^-} \cdots s_l^{b_l^-} \Rightarrow^* d_1^{b_1^+} \cdots d_k^{b_k^+}. \tag{9.5}$$

The left- and right-hand sides in (9.5) can be interpreted as commutative strings $\{u^{b_u^-}\}_{u \in N}$ and $\{u^{b_u^+}\}_{u \in N}$ over N, when the order of symbols is ignored.

The definition of a P2P route is at the overlay scale. Now consider the routing process at the node scale. Let a packet targeted to a node d be received at a node u. When $u \neq d$ then u selects one or more next-hop nodes among all nodes in T_u then several options are available for u.

---

[2] The sign "+" in b_i^+ means that packets arrive to d_i.

[3] The sign "−" in b_i^- means that packets leave s_i.

Base forwarding: Exactly one node v in T_u is selected for the next hop.

Retransmissions: The node u having sent the packet to v waits for an acknowledgment. If it has not been received in a predefined time, then u retransmits the packet. A positive integer parameter a_v defines the number of attempts.

Sequential forwarding: Multiple alternative directions are used in retransmissions. There are several candidates $v_1, v_2, \ldots, v_k$ in T_u for the next hop. Initially, u forwards the packet to v_1. If no success is achieved in the predefined time, then u tries sequentially through v_2, v_3, and so on up to v_k. For each next hop v_i, the number of attempts is $a_i = a(v_i)$, $i = 1, 2, \ldots, k$.

Parallel forwarding: Similar to the previous case, but u forwards the packet simultaneously to $v_1, v_2, \ldots,$ and v_k as in multicast communication.

Path completion: A packet has reached a node u and is not forwarded further. In Sparticular, this happens when (1) u is a destination node ($u = d$), (2) u discards the packet, e.g., because of overload, or (3) some alternative directions are not used.

9.3.2 Forwarding Options as Grammar Rules

Taking an arbitrary node $u \in N$, consider the following formal representation of a forwarding option:

$$u \to v_1^{a_1} v_2^{a_2} \cdots v_k^{a_k} \tag{9.6}$$

Let us call (9.6) *a forwarding rule of u*.
    A particular case is when $k = 1$ in (9.6); the rule is reduced to

$$u \to v^a,$$

which models the base forwarding option. That is, u selects only one next-hop v and forwards to it up to $a = a_v$ times.
    When $k = 0$ in (9.6), the rule is reduced to

$$u \to \varepsilon,$$

modeling a path completion at u.
    When $k > 1$, rule (9.6) can be interpreted as sequential or parallel forwarding. Nodes $v_1, v_2, \ldots, v_k$ are candidates for the next hop, and there are $a_i = a_{v_i}$ transmission attempts for each. To distinguish between sequential and parallel forwarding, we introduce terminals σ in (9.6):

$$u \to \sigma v_1^{a_1} v_2^{a_2} \cdots v_k^{a_k} \tag{9.7}$$

Let us take the alphabet $\Sigma = \{\sigma_{\text{par}}, \sigma_{\text{seq}}\}$, where σ_{par} is designated for parallel forwarding and σ_{seq} is for sequential forwarding. Depending on $\sigma \in \Sigma$, rule (9.7) represents either parallel or sequential forwarding. When no preceding σ, then the difference between the options is ignored.

In general, terminals are used to classify rules according to a given finite set of behavioral forwarding types. The same idea is applicable for the path completion. Another role of terminals is for modeling communication cost c between u and its next-hop nodes. Let the cost be measured in discrete units of $\{0, 1, \ldots, \bar{c}\}$, where $c = 0$ and $c = \bar{c}$ are the cheapest and most expensive cases, respectively. For instance, c reflects the latency in the ternary scale: "small," "medium," and "big."

The following extensions of (9.7) include the cost into the grammar. One way is to introduce a terminal σ_{cst}, designated for cost counting on the right-hand side:

$$u \to \sigma_{\mathrm{cst}}^c v_1^{a_1} v_2^{a_2} \cdots v_k^{a_k} \tag{9.8}$$

for an appropriate $c \in \{0, 1, \ldots, \bar{c}\}$.

An alternative is employing a separate terminal σ_c for each value of the discrete cost c:

$$u \to \sigma_c v_1^{a_1} v_2^{a_2} \cdots v_k^{a_k}. \tag{9.9}$$

Generalizing (9.6)–(9.9), routing at a node u to a destination d is modeled with the forwarding rule $r = r_u(d)$:

$$u \to \tau_r v_{1r}^{a_{1r}} v_{2r}^{a_{2r}} \cdots v_{kr}^{a_{kr}}, \tag{9.10}$$

where nodes $v_{1r}, v_{2r}, \ldots, v_{kr}$ are candidates for the next hop, $k = k(r)$, positive integers $a_{1r}, a_{2r}, \ldots, a_{kr}$ define the number of transmission attempts, string $\tau_r \in \Sigma^*$ represents behavioral and cost attributes.

Let P_u be the set of all forwarding rules at u. Note that for the same destination, there may be several different forwarding rules. If the difference is essential, one can use extra terminals to classify the rules in P_u. On the other hand, many different destinations are often produced on the same right-hand side because of the P2P locality property: the same neighbor is used for many destinations. Hence the size of P_u is typically less than the overlay size N.

Consider the right-hand side of (9.10). Extend its exponents $(a_{ir})_{i=1}^k$ to a vector $a_r \in \mathbb{Z}_+^{n-t}$ by adding zero entries. Using a string $\tau_r \star [a_r] = \tau_r \rho_r \in \Sigma^* N^*$ rewrite (9.10)

$$u \to \tau_r \star [a_r]. \tag{9.11}$$

Ignoring the order of symbols requires a discussion on appropriateness of (9.11) in modeling sequential forwarding. Obviously, the order of next hops that u advances in forwarding is ignored. Instead, the model reflects only that u implements sequential forwarding. Section 9.4 will show that this simplification leads to analysis of all possible paths a route consists of, regardless of whether a parallel or sequential forwarding option is used. Nevertheless, the difference is still preserved in integral metrics, such as how many times each option was used in a route.

Summarizing what was stated above, we assign with a P2P overlay network the CCF-grammar $G = (N, \Sigma, P)$, called *the routing grammar*. It models how nodes forward messages to the next hop.

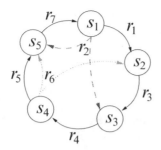

Fig. 9.1 An example network of five nodes. Clockwise links r_1, r_3, r_4, r_5, r_7 form a ring; r_2 represents parallel forwarding (*dashed arcs*), when s_1 sends a message simultaneously to s_3 and s_5; r_6 represents sequential forwarding (*dotted arcs*), when s_4 first tries to forward to s_2, then to s_5. Parameters of transmission attempts are equal to one

Example 9.1. Consider the network illustrated in Fig. 9.1. It is an instance of a basic Chord ring (successors only) with an addition of parallel forwarding at s_1 and sequential forwarding at s_4. The routing grammar is defined with $N = \{s_1, \ldots, s_5\}$, $\Sigma = \{\sigma_{par}, \sigma_{seq}\}$, and $P = \{r_1, \ldots, r_7\}$,

$$r_1, r_2 : s_1 \rightarrow s_2 \mid \sigma_{par} s_3 s_5 \qquad r_3 : s_2 \rightarrow s_3$$
$$r_5, r_6 : s_4 \rightarrow s_5 \mid \sigma_{seq} s_2 s_5 \qquad r_4 : s_3 \rightarrow s_4$$
$$r_7 : s_5 \rightarrow s_1$$

9.4   A Diophantine Model of Routes

From (9.5) we see that routing in a P2P overlay can be treated as a transformation of a source string into a destination string. In this section, the transformation is implemented as a derivation in the routing grammar. A model of a given set of routes is formulated in terms of an ANLDE system; its solutions map to routes and vice versa. Routes can be finitely generated by basic ones, which correspond to basis solutions to the ANLDE system. Consequently, the basis solutions represent a structure of routes.

9.4.1   Routes and Grammar Derivations

Let each node $u \in N$ initiate $b_u^- \in \mathbb{Z}_+$ packets, $\alpha = \star[b^-]$. They are routed through the overlay. The forwarding process at nodes is modeled with a routing grammar G as described in Sect. 9.3. In the issue, each node $v \in N$ receives $b_v^+ \in \mathbb{Z}_+$ packets,

$\beta = \star[b^+]$. A routing attribute marked with $\sigma \in \Sigma$ has been applied $b_\sigma \in \mathbb{Z}_+$ times,[4] $\varkappa = \star[b]$. The resultant route corresponds directly to a derivation $\alpha \Rightarrow^* \varkappa\beta$ in G.

The terms *cyclic*, *cycle*, and *simple* for derivations are directly applied for routes. To include b^-, b^+, b explicitly in the notation, we will denote such a route

$$b^- \xrightarrow{b} b^+ \tag{9.12}$$

Therefore, we can conclude that derivations in a routing grammar describe all routes (9.12) in a P2P overlay for various b^-, b, b^+.

Example 9.2. Consider the network in Example 9.1. Let $b^- = (1,0,0,0,0)$, where for simplicity we omit using transpose notation and treat row and column vectors interchangeably since the context eliminates the confusion. Starting from s_1, a packet can run clockwise through every node and finally returns back. The derivation is a simple cycle:

$$s_1 \Rightarrow s_2 \Rightarrow s_3 \Rightarrow s_4 \Rightarrow s_5 \Rightarrow s_1.$$

The route is $(1,0,0,0,0) \xrightarrow{(0,0)} (1,0,0,0,0)$. For the same source, another route $(1,0,0,0,0) \xrightarrow{(1,0)} (2,0,0,0,0)$ is possible. It consists of two simple cyclic paths and corresponds to the derivation

$$s_1 \Rightarrow \sigma_{\mathrm{par}} s_3 s_5 = \left(\sigma_{\mathrm{par}} \begin{matrix} s_3 \Rightarrow s_4 \Rightarrow s_5 \Rightarrow s_1 \\ s_5 \Rightarrow s_1 \end{matrix} \right) = \sigma_{\mathrm{par}} s_1^2,$$

where the duplication at s_1 is due to parallel forwarding.

9.4.2   Routes and ANLDE System Solutions

Considering routes as derivations we partially fix the order of grammar rule applications. However, integral properties of routing do not depend on this order. Treating routes with the same number of grammar rule applications as equivalent, we can formulate the model of routes in terms of an ANLDE system and its solutions.

Below we introduce several instances of our model; each defines certain restrictions to the route parameters b^-, b^+, and b in (9.12). We omit proofs from the scenarios; they can be derived directly from Theorem 9.1, see page 249.

Instance 1: The model is ANLDE system (9.3) and describes all routes $b^- \xrightarrow{b} b^+$ for fixed b^-, b^+, and b.

---

[4]For instance, $b_{\sigma_{\mathrm{par}}}$ is the total number of parallel forwarding applications, see (9.7); $b_{\sigma_{\mathrm{cost}}}$ is the integral cost of a route, see (9.8).

Theorem 9.2. *Any solution to (9.3) maps to a P2P route*

$$b^- + d \xrightarrow{b} b^+ + d \text{ for some } d \in \mathbb{Z}^{n-t} \text{ such that } b^- + d, b^+ + d \in \mathbb{Z}_+^{n-t}$$

and vice versa. A solution is in the basis if and only if the route is simple.

Each u-equation in (9.3) states the balance between packet arrivals and departures taking into account that u initiates b_u^- and finally receives b_u^+ packets. Each σ-equation in (9.3) states the exact equality for routing attribute σ.

Theorem 9.2 relates to Theorem 9.1 as follows. A route $b^- \xrightarrow{b} b^+$ corresponds to a derivation $\alpha \Rightarrow^* \varkappa\beta$, where $b^- = \#[\alpha]$, $b^+ = \#[\beta]$, and $b = \#[\varkappa]$. Such a route leads to a solution to (9.3) taking $d = \mathbb{O}$ (in Theorem 9.1, $\alpha'' = \varepsilon$). Since only the difference $b^- - b^+$ affects (9.3), there can be solutions that do not correspond to $b^- \xrightarrow{b} b^+$ but to $b^- + d \xrightarrow{b} b^+ + d$ (in Theorem 9.1, to combined derivation $\alpha\alpha'\alpha'' \Rightarrow^* \varkappa\alpha'\alpha''\beta$).

The relation to Theorem 9.1 brings an interesting interpretation of a route $b^- + d \xrightarrow{b} b^+ + d$. By changing the order of rule applications in the derivation this route can be transformed into a composition of tree derivations:

$$\alpha \Rightarrow^* \varkappa'\beta', \quad \alpha' \Rightarrow^* \alpha'\varkappa''\beta'', \quad \alpha'' \Rightarrow^+ \alpha'',$$

$$\text{where } \#[\alpha\alpha'\alpha''] = b^- + d, \quad \#[\alpha'\alpha''\beta'\beta''] = b^+ + d,$$
$$\#[\varkappa'\varkappa''] = b, \quad \text{and} \quad \forall u \in N \; \min(\#_u[\alpha], \#_u[\beta']) = 0.$$

The first part of sources, α, does not consist of destinations; it initiates packets and feeds the destinations in β', $\alpha \Rightarrow^* \varkappa'\beta'$. The second part, α', receives packets that it initiated as well as feeds destinations in β'', $\alpha' \Rightarrow^* \alpha'\varkappa''\beta''$. The last part, α'', receives all packets that it initiated, $\alpha'' \Rightarrow^+ \alpha''$.

Example 9.3. In Example 9.1, let s_1 send one and then receive two packets applying parallel forwarding once, i.e., $s_1 \Rightarrow^* \sigma_{\text{par}}s_1^2$. All such routes are described by the ANLDE system

$$x_7 + 1 = x_1 + x_2 + 2, \; x_1 + x_6 = x_3,$$
$$x_2 + x_3 = x_4, \qquad\qquad x_4 = x_5 + x_6,$$
$$x_2 + x_5 + x_6 = x_7, \qquad x_2 = 1.$$

Here $b^- = (1,0,0,0,0)$, $b^+ = (2,0,0,0,0)$, and $b = (1,0)$. The basis solution $x = (0,1,0,1,1,0,2)$ defines a simple route

$$s_1 \Rightarrow \sigma_{\text{par}}s_3s_5 \Rightarrow \sigma_{\text{par}}s_4s_5 \Rightarrow \sigma_{\text{par}}s_5^2 \Rightarrow \sigma_{\text{par}}s_1s_5 \Rightarrow \sigma_{\text{par}}s_1^2.$$

It is in the form $\alpha\alpha'\alpha'' \Rightarrow^* \alpha'\alpha''\varkappa'\varkappa''\beta'\beta''$, where $\alpha = \alpha'' = \varkappa' = \beta' = \varepsilon$, $\alpha' = \beta'' = s_1$, $\varkappa' = \sigma_{\text{par}}$.

Instance 2: The model of cycles. Any node receives as many packets as it has initiated ($b^- = b^+ = d$).

Case 2A: No routing attributes are applied ($b = \mathbb{O}$). The model is homANLDE system (9.2) and describes all possible cycles $d \xrightarrow{\mathbb{O}} d$ for arbitrary $d \in \mathbb{Z}_+^N$.

Theorem 9.3. *Any solution of (9.2) maps to a cycle and vice versa. A solution is in the Hilbert basis if and only if the cycle is simple.*

Case 2B: No restrictions to routing attributes. The model describes all routes $d \xrightarrow{b} d$ for arbitrary $b \in \mathbb{Z}_+^t$ and $d = b^- = b^+ \in \mathbb{Z}_+^N$.

This case is reduced to the previous one. Eliminate all terminals from the grammar ($\Sigma = \varnothing$). Then the homANLDE system (9.2) contains only equations for non-terminals, and Theorem 9.3 is still valid.

Case 2C: Routing attributes b are fixed. The model consists of all non-terminal equations of (9.2) and all terminal equations of (9.3). The case is hence reduced to Theorem 9.2. That is, any solution maps to a cyclic route $d \xrightarrow{b} d$ for some $d \in \mathbb{Z}_+^N$ and vice versa.

Example 9.4. Consider the routing grammar of Example 9.1 but ignore the terminals (in rules r_2, r_6). The homANLDE system is

$$x_7 = x_1 + x_2, \ x_4 = x_5 + x_6,$$
$$x_1 + x_6 = x_3, \ x_2 + x_5 + x_6 = x_7.$$
$$x_2 + x_3 = x_4,$$

It has the only basis solution $x = (1, 0, 1, 1, 1, 0, 1)$, which maps to a cycle $s_1 \Rightarrow s_2 \Rightarrow s_3 \Rightarrow s_4 \Rightarrow s_5 \Rightarrow s_1$.

9.4.3   Path Structure

Any solution of an ANLDE system can be represented by the basis. Since in the Diophantine model a route corresponds to a solution, the basis defines *basic routes*; any route is a combination of basic ones. On the other hand, a route composes several paths. Therefore, the model provides a finite structure of P2P paths. The following example explains the idea.

Example 9.5. Consider a network in Example 9.1. Let all nodes except s_1 use the path completion option. The model reflects this by adding to the routing grammar the rules $s_i \rightarrow \varepsilon$ for $i = 2, \ldots, 5$, which correspond to unknowns z_i.

Any cyclic route $u^p \Rightarrow^+ \varkappa u^p$ shows how p packets initiated at u run through the network; all p packets return, but extra copies complete their paths at $v \neq u$. There is no restriction to routing attributes $\varkappa$. Figure 9.2 shows schematically the structure of a basic route when $u = s_1$. Such routes are described with the following homANLDE system.

Fig. 9.2 The path structure of a basic route $s_1 \Rightarrow^+ \varkappa s_1$ in Example 9.5. There is a cycle that contains s_1. Cycle nodes can forward packet copies out of the cycle

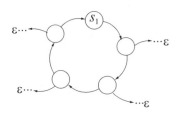

Table 9.1 Basis solutions for routes $\alpha \Rightarrow^+ \varkappa\alpha$ in Example 9.5

Grammar rules	$s_1 \to s_2$	$s_1 \to \sigma_{par} s_3 s_5$	$s_2 \to s_3$	$s_3 \to s_4$	$s_4 \to s_5$	$s_4 \to \sigma_{seq} s_2 s_5$	$s_5 \to s_1$	$s_2 \to \varepsilon$	$s_3 \to \varepsilon$	$s_4 \to \varepsilon$	$s_5 \to \varepsilon$
	x_1	x_2	x_3	x_4	x_5	x_6	x_7	z_2	z_3	z_4	z_5
I. No duplication											
1	1	0	1	1	1	0	1	0	0	0	0
II. Parallel forwarding option											
2	1	1	0	1	1	0	2	1	0	0	0
3	0	1	0	0	0	0	1	0	1	0	0
4	0	1	0	1	0	0	1	0	0	1	0
5	0	1	0	1	1	0	1	0	0	0	1
III. Sequential forwarding option											
6	1	0	1	1	0	1	1	1	0	0	0
7	1	0	2	1	0	1	1	0	1	0	0
8	1	0	2	2	0	1	1	0	0	1	0
9	0	0	1	1	0	1	0	0	0	0	1
IV. Parallel and sequential forwarding options											
10	1	1	0	1	0	1	2	2	0	0	0
11	0	1	0	1	0	1	1	1	0	0	1

$$x_7 = x_1 + x_2, \qquad x_4 = x_5 + x_6 + z_4,$$
$$x_1 + x_6 = x_3 + z_2, \quad x_2 + x_5 + x_6 = x_7 + z_5.$$
$$x_2 + x_3 = x_4 + z_3,$$

Its Hilbert basis is shown in Table 9.1; all basic routes are listed in Table 9.2.

Routes are placed into groups I, ..., IV depending on parallel and sequential option usage. Group I consists simply of one route that reflects clockwise traversing of all nodes.

Each route in group II uses the parallel forwarding option at s_1 once. In route 2, s_1 uses both rules it has for forwarding. One packet reaches s_2 and completes the path. Another packet is duplicated at s_4 and both copies returns to s_1. In routes 3, 4, and 5, s_1 starts with the parallel forwarding option. One copy returns to s_1, another one completes the path at s_3, s_4, and s_5, respectively.

Similarly, each route in group III duplicates a packet using the sequential forwarding option at s_4. One copy runs a cycle, another one completes its path at s_2, s_3, s_4, or s_5.

Table 9.2 Basic routes $\alpha \Rightarrow^+ \varkappa\alpha$ in Example 9.5

#	Route (derivation)
I. No duplication	
1	$s_1 \xrightarrow{x_1} s_2 \xrightarrow{x_3} s_3 \xrightarrow{x_4} s_4 \xrightarrow{x_5} s_5 \xrightarrow{x_7} s_1$
II. Parallel forwarding option	
2	$s_1 \xrightarrow{x_1} s_2 \xrightarrow{z_2} \varepsilon,$
	$s_1 \xrightarrow{x_2} \sigma_{\mathrm{par}} s_3 s_5 \xrightarrow{x_4} \sigma_{\mathrm{par}} s_4 s_5 \xrightarrow{x_5} \sigma_{\mathrm{par}} s_5^2 \xrightarrow{2\times x_7} \sigma_{\mathrm{par}} s_1^2$
3	$s_1 \xrightarrow{x_2} \sigma_{\mathrm{par}} s_3 s_5 \xrightarrow{z_3} \sigma_{\mathrm{par}} s_5 \xrightarrow{x_7} \sigma_{\mathrm{par}} s_1$
4	$s_1 \xrightarrow{x_2} \sigma_{\mathrm{par}} s_3 s_5 \xrightarrow{x_4} \sigma_{\mathrm{par}} s_4 s_5 \xrightarrow{z_4} \sigma_{\mathrm{par}} s_5 \xrightarrow{x_7} \sigma_{\mathrm{par}} s_1$
5	$s_1 \xrightarrow{x_2} \sigma_{\mathrm{par}} s_3 s_5 \xrightarrow{x_4} \sigma_{\mathrm{par}} s_4 s_5 \xrightarrow{x_5} \sigma_{\mathrm{par}} s_5^2 \xrightarrow{z_5} \sigma_{\mathrm{par}} s_5 \xrightarrow{x_7} \sigma_{\mathrm{par}} s_1$
III. Sequential forwarding option	
6	$s_1 \xrightarrow{x_1} s_2 \xrightarrow{x_3} s_3 \xrightarrow{x_4} s_4 \xrightarrow{x_6} \sigma_{\mathrm{seq}} s_2 s_5 \xrightarrow{z_2} \sigma_{\mathrm{seq}} s_5 \xrightarrow{x_7} \sigma_{\mathrm{seq}} s_1$
7	$s_1 \xrightarrow{x_1} s_2 \xrightarrow{x_3} s_3 \xrightarrow{x_4} s_4 \xrightarrow{x_6} \sigma_{\mathrm{seq}} s_2 s_5 \xrightarrow{x_3} \sigma_{\mathrm{seq}} s_3 s_5 \xrightarrow{z_3} \sigma_{\mathrm{seq}} s_5 \xrightarrow{x_7} \sigma_{\mathrm{seq}} s_1$
8	$s_1 \xrightarrow{x_1} s_2 \xrightarrow{x_3} s_3 \xrightarrow{x_4} s_4 \xrightarrow{x_6} \sigma_{\mathrm{seq}} s_2 s_5 \xrightarrow{x_3} \sigma_{\mathrm{seq}} s_3 s_5 \xrightarrow{x_4} \sigma_{\mathrm{seq}} s_4 s_5 \xrightarrow{z_4} \sigma_{\mathrm{seq}} s_5 \xrightarrow{x_7} \sigma_{\mathrm{seq}} s_1$
9	$s_2 \xrightarrow{x_3} s_3 \xrightarrow{x_4} s_4 \xrightarrow{x_6} \sigma_{\mathrm{seq}} s_2 s_5 \xrightarrow{z_5} \sigma_{\mathrm{seq}} s_2$
IV. Parallel and sequential forwarding options	
10	$s_1 \xrightarrow{x_1} s_2 \xrightarrow{z_2} \varepsilon,$
	$s_1 \xrightarrow{x_2} \sigma_{\mathrm{par}} s_3 s_5 \xrightarrow{x_4} \sigma_{\mathrm{par}} s_4 s_5 \xrightarrow{x_6} \sigma_{\mathrm{par}} \sigma_{\mathrm{seq}} s_2 s_5^2 \xrightarrow{z_2} \sigma_{\mathrm{par}} \sigma_{\mathrm{seq}} s_5^2 \xrightarrow{2\times x_7} \sigma_{\mathrm{par}} \sigma_{\mathrm{seq}} s_1^2$
11	$s_1 \xrightarrow{x_2} \sigma_{\mathrm{par}} s_3 s_5 \xrightarrow{x_4} \sigma_{\mathrm{par}} s_4 s_5 \xrightarrow{x_6} \sigma_{\mathrm{par}} \sigma_{\mathrm{seq}} s_2 s_5^2 \xrightarrow{z_2} \sigma_{\mathrm{par}} \sigma_{\mathrm{seq}} s_5^2 \xrightarrow{z_5} \sigma_{\mathrm{par}} \sigma_{\mathrm{seq}} s_5 \xrightarrow{x_7} \sigma_{\mathrm{par}} \sigma_{\mathrm{seq}} s_1$

Group IV shows paths when both forwarding options are used. In route 10, two packets complete their paths at s_2, and the other two packets follow a cycle. In route 11, two packets complete their paths at s_2 and s_5, respectively; one packet follows a cycle.

9.5 Applications and Discussion

Understanding the structure of P2P routes helps to solve problems of routing performance, security, and scalability. In this section, we discuss how some problems in recent P2P research can be approached with the Diophantine model.

9.5.1 *Workload and Utilization*

According to Theorems 9.2 and 9.3, ANLDE systems (9.3) and (9.2) are formal structural models [14]. They define dependence analytically between the initial workload (packet sources), P2P overlay resources that routing consumes (nodes and links), and the final distribution (packet destinations).

Let us focus on estimating P2P resource consumption, which is described typically in relative terms like node and link utilization. A close characteristic is congestion, which happens when the consumption exceeds a certain bound.

Xu et al. addressed the issue of the P2P routing workload and congestion [22]. They considered a uniform all-to-all communication load, when for each pair of nodes $u \neq v$, a unit of traffic is imposed. The idea behind is that such a workload leads to uniform utilization and congestion in a uniform network. It gives a way to test a network for uniformity.

In the Diophantine model, a unit of traffic is a packet. Let $b_u^- = b_u^- = N - 1 \; \forall u \in N$, i.e., each node initiates and then receives $N - 1$ packets. Clearly, when u sends a packet to v and v sends a packet to u, these two paths form a cycle $u \Rightarrow^+ v \Rightarrow^+ u$. Therefore, the case is reduced to Theorem 9.3, and the Hilbert basis describes the structure of possible P2P routes for a uniform all-to-all communication load. Note that (9.5) generalizes the notion of routing applied in [19]. Given an overlay topology digraph $G_T = (N, E)$, they defined routing as the set of $N(N - 1)$ paths specified for every (ordered) pair of vertices of G_T.

Example 9.6. Consider the network in Example 9.5. Assume all-to-all communications and cycles $u \Rightarrow^+ v \Rightarrow^+ u$ for all $u \neq v$ without restrictions to routing attributes. It leads to the same homANLDE system as in Example 9.5; the path structure is defined by basis solutions shown in Table 9.1. To construct a uniform workload model, one should define for all $u \neq v$ which cycle $u \Rightarrow^+ v \Rightarrow^+ u$ is used when u sends a packet to v and v sends a packet to u. These details depend on a given routing protocol.

Compared to [19, 22] the Diophantine model allows more general workload scenarios than the uniform all-to-all communication. For instance, given $u \in N$, $b_u^- = 1$, and $b_v^- = 0 \; \forall v \neq u$, one scenario is how the unit activity of u loads the P2P overlay. Taking $b^- = (1, 1, \ldots, 1)$ provides a scenario when all nodes start activity simultaneously.

Metrics of resource consumption are easily derived from the model. Given a basis solution x, its component x_r counts how many times the rule $r = (u \to \tau_r \rho_r)$ was applied in the route. That is, $L_u = \sum_{r \in P_u} x_r$ is a usage metric of a node u; $L_{uv} = \sum_{r \in P_u, v \in \rho_r} x_r$ is a usage metric of an outgoing link (u, v). Routing attributes τ_r provide other useful metrics.

Example 9.7. As in Example 9.6 assume all-to-all communications but let each basic route be used once. Since cycles $u \Rightarrow^+ v \Rightarrow^+ u$ are not distributed uniformly among pairs (u, v) this assumption does not necessarily satisfy the uniform communication but preserves the simplicity in this example. Table 9.3 shows the link and node usage metrics. They are computed summing over all basis solutions. Node s_5 is most utilized. It receives 15 packets; 3 of them complete paths; all others go to s_1 making link (s_5, s_1) the most congested. Another congested link is (s_3, s_4), which is used 11 times.

Table 9.3 Link and node
usage in the network from
Example 9.5

Usage	s_1	s_2	s_3	s_4	s_5
s_1	12/6/0/0	6	6		6
s_2		12/0/0/5	7		
s_3			13/0/0/2	11	
s_4		6		11/0/6/2	9
s_5	12				15/0/0/3

A non-diagonal element ($u \neq v$) contains link usage L_{uv}; when
a link is not used the element is empty. A diagonal element
is $L_u/L_u^{\mathrm{par}}/L_u^{\mathrm{seq}}/L_u^{\mathrm{fin}}$, where L_u is node usage, L_u^{par} is parallel
option usage, L_u^{seq} is sequential option usage, and L_u^{fin} is
completion option usage

9.5.2 Connectivity

Loguinov et al. [16] and Gummadi et al. [7] studied the problem of P2P connectivity.
It relates to resilience against node or link failure, i.e., the number and location of
failures a P2P network can tolerate without becoming disconnected. Two routes are
node (link) disjointed if they consist of different nodes (links). For practical purpose
this definition can be weakened introducing a parameter k for the number of joint
nodes (links). Routes are disjointed when $k = 0$; they are almost disjointed when
$k > 0$ and small.

Consider a pair of nodes, $u \neq v$. They are connected if there are paths $u \Rightarrow^+ v$ and
$v \Rightarrow^+ u$. Again, Theorem 9.3 is applicable in this case, and the Hilbert basis defines
the connectivity structure. The number of disjointed cycles shows the resilience.
When a failure happens it breaks only those cycles that contain the failed node/link.
Cycles that disjointed with the broken ones are not affected, thus preserving the
connectivity. In terms of the Diophantine model, two basis solutions x and x' define
disjointed cycles when either $\forall u \in N$ and $\forall r, p \in P_u \; x_r x'_p = 0$ (node disjointed
cycles), or $\forall r \in P \, x_r x'_r = 0$ (link disjointed cycles). That is, analysis of basis solutions
provides conclusions on resilience.

Similar cycle-based analysis can also focus on a given node u or link (u, v).
The more cycles use the node or link, the less resilient the network is. Note that
the importance of cycles for connectivity maintenance in structured P2P network
designs was emphasized in several works, see [15, 18] and references therein.

Example 9.8. Consider again the network and route structure of Example 9.5.
Let us now ignore routing attributes. Then there is a cycle $u \Rightarrow^+ v \Rightarrow^+ u$ for any
$u, v \in \{s_1, s_2, s_3, s_4, s_5\}$ concluding that the network is connected. For instance, when
$u = s_2$ and $v = s_4$ there exists cycle $s_2 \Rightarrow^+ s_4 \Rightarrow^+ s_2$ (row 9 in Tables 9.1 and 9.2).
There are no node disjointed cycles. There are only two link disjointed cycles
(rows 3 and 9). It means that the network has low resilience to node/link removal.
The congestion metrics of Example 9.7 sort nodes and links in the order of their
importance to the resilience. More congested nodes/links are less resilient.

More sophisticated characterization of connectivity can be obtained based, for
instance, on cyclic routes $u \Rightarrow^+ \varkappa u^k$, when several (up to k) alternate paths are

used. It means that the Diophantine model defines a multipath routing structure between nodes u and v. This multipath characterization correlates with work of Gummadi et al. [7] and Artigas and Skarmeta [2]. They focused on the importance of alternative paths between sources and destinations for routing dependability, performance, and security. In contrast to the Diophantine model, their approach does not define a generic way to model multipath routing.

9.5.3 Performance

Loguinov et al. [16] and Xu et al. [22] considered the problem of P2P routing performance. They relate it to the routing diameter (maximum distance between any two nodes), which gives the worst-case routing performance. Again, we can use cycles to identify the longest paths.

For instance, consider one-to-all routes for a given u, where $b_u^- = N - 1$ and $b_u^+ = 0$ while $b_v^- = 0$ and $b_v^+ = 1$ $\forall v \neq u$. The longest path $u \Rightarrow^+ v$ over all basic routes defines the worst case. Considering all such u, we move to the case of all-to-all communication with cycles $u \Rightarrow^+ v \Rightarrow^+ u$.

Recall that a route consists of several paths because of packet duplications and drops, typical in practice. In contrast to [16, 22], our approach allows more freedom in selecting performance metrics other than the primary path length (counting only links of $u \Rightarrow^+ v$).

One option is additionally counting the secondary activity induced by packet duplications and drops, that is, considering some other (secondary) paths of the route. For instance, even if a primary path is short, it generates a lot of packet copies that reduces the overall network performance.

Another case happens in request-response communications when u sends a request packet to v and then receives the response. That is, both forward $u \Rightarrow^+ v$ and backward path $v \Rightarrow^+ u$ are important. The optimality criterion here is not

$$|u \Rightarrow^+ v| \to \min$$

but

$$|u \Rightarrow^+ v| + |v \Rightarrow^+ u| = |u \Rightarrow^+ u| \to \min.$$

Example 9.9. Consider Example 9.5 again. Ignore routing attributes and assume for simplicity equal performance in all forwarding actions. It means that the performance of a cycle can be computed summing all components of the basis solution.

According to Table 9.1, the worst-case performance happens when cycles 8 and 10 are used, each taking 8 forwarding actions in all its paths. When the packet completion option is not taken into account, the worst-case cycle is 8 consuming six hops. Table 9.4 shows the performance of basic cycles. The high average values show that the network performance is very much affected by packet duplication.

Table 9.4 Performance of basic cycles $s_i \Rightarrow^+ s_i$ in Example 9.5

Cycle	1	2	3	4	5	6	7	8	9	10	11	Average
#Actions, in total	5	7	3	4	5	6	7	8	4	8	6	5.7
#Completions, $s_i \to \varepsilon$	4	6	2	3	4	5	6	7	3	6	4	4.5

9.5.4 Comparisons

Some comparison remarks were mentioned above when discussing applications of the Diophantine model. Now let us compare the theoretical base of the linear Diophantine approach to the other ones.

The description of routes using a CCF-grammar enhances traditional network graph models. The latter use the topology graph; its analysis can be targeted to path availability (connectivity and network diameter [9, 16]), to path overlap and convergence (congestion [22] and fault resilience [16]), or to disjointed paths (dependable and secure routing [2, 5]).

Such a graph model corresponds to a routing grammar consisting of rules $u \to v$ ($u, v \in N$) (only one path per packet is considered, no retransmissions and multicast duplications). It produces a Diophantine system that is based on the graph incidence matrix and describes network flow circulation [3, 8]:

$$\sum_{v:\, r=(v \to u)} x_r + b_u^- = \sum_{v:\, r=(u \to v)} x_r + b_u^+, \quad u \in N$$

Allowing a packet to be duplicated because of retransmission (rules $u \to v^{a_{uv}}$ in a routing grammar) produces a Diophantine system that describes generalized flows [3]:

$$\sum_{v:\, r=(v \to u)} a_{uv} x_r + b_u^- = \sum_{v:\, r=(u \to v)} x_r + b_u^+, \quad u \in N$$

At the same time, the Diophantine model allows more general rules that can aggregate several next-hop candidates as well as take routing attributes into account, see rules (9.10). It produces the enhanced class of Diophantine systems defined by (9.3). Non-terminal equations is a generalization of the network flow balance at nodes allowing hyper-arcs, when a packet is forwarded to several next-hop nodes. Terminal equations constrain routing attributes.

Esparza [6] introduced a communication-free Petri net, an abstract model in process algebras, which is conceptually similar to the Diophantine model. However, Esparza's model is not related to concrete application. The Diophantine model is for the case of P2P routing, and we explicitly define how to use and interpret non-terminals, terminals and grammar rules. For instance, terminals in the model allows capturing sequential behavior, cost attributes, and some other routing details, while Petri nets can target only pure parallel processes.

In contrast to [6], we ground on the notion of NLDE basis, the specific structure of ANLDE system, and the relation between NLDE and CCF-grammars. It allows clear tracing from P2P routes to grammar derivations and then to ANLDE system solutions. Esparza's model says a little about solutions and nothing about their form and finite structure.

Moreover, to be used in practice, any model needs efficient computational support. In case of the Diophantine model, efficient algorithms for solving some classes of ANLDE systems are available, see the discussion in the next section.

9.5.5 Computational Complexity

Solving Diophantine equations is computationally demanding [20]. It also concerns ANLDE systems (9.3), a subclass of NLDE systems, since the uniform word problem for CCF-grammars[5] is NP-complete [6]. Finding the NLDE basis is an overNP problem, since the number of basis solutions can depend exponentially on NLDE system dimensions and values of coefficients.

On the other hand, according to [12], some ANLDE systems can be solved efficiently by polynomial algorithms, when finding a non-zero solution, and by pseudo-polynomial algorithms, when finding the basis. In the latter case, the complexity is pseudo-polynomial because of using the number of basis solutions as a parameter.

Obviously, if a model describes all possible routes, then the computation cost may be subject to exponential growth, even if the pseudo-polynomial algorithms are used. For instance such a case happens when the Diophantine model is equivalent to the network topology graph. However, even in this case the computational cost may be reduced based on the order in which the basis solutions are constructed.

For instance, the syntactic algorithm actually produces solutions with short length first ($|x| = \sum_r x_r$). Due to practical interpretation of the length (e.g., a performance metric), optimal routes are among short routes. Thus it is enough to limit the model with a certain amount of short basic routes, and no more routes are computed. A similar idea is applied in the theory of k-optimization when constraints define all possible solutions but only k best solutions are under search.

In general, our model intends to describe not all routes but a small subset of them. This follows the principle that there are a lot of bad routes but very few good ones. Such a restriction can be done by carefully selecting coefficients in (9.2) and (9.3) as well as adding extra constraints. For instance, inequation $\sum_r x_r \leq L$ explicitly constraints the length of routes to search.

Another interesting feature relates to the case when routes have to be computed frequently. When the network changes, it triggers a modification of the model, for instance, adding and removing an equation or changing a coefficient value. Actually, the syntactic algorithms are incremental in many cases. That is, solutions to the initial system can be used in construction of solutions to the modified one.

---

[5]This problem is of deciding, given a CCF-grammar and strings α, β, if $\alpha \Rightarrow^* \beta$.

We saw the importance of cycles when discussing possible applications of the Diophantine model. In this particular case, the model requires solving a homANLDE system, the primary class of ANLDE systems allowing efficient computations. In [12], a polynomial algorithm is constructed for finding a non-zero solution as well as a pseudo-polynomial algorithm for finding the Hilbert basis.[6]

These algorithms also work well in practice, and a reader can experiment with them using the Web-SynDic system available at http://websyndic.cs.karelia.ru. Note that homANLDE systems are not the only class with efficient solutions computation. Moreover, the research for finding more classes of ANLDE systems is going on as well as development of new algorithms for solving.

9.6   Summary

This chapter has considered the problem of modeling P2P routes and proposed a generic discrete approach to describe them. We contribute a linear Diophantine model that restricts the analysis to routes that have given properties. Particularly, it supports various scenarios of message sending and receiving as well as of forwarding behavior at nodes.

We defined a P2P route as all paths that given messages follow. The model is formulated as a linear Diophantine equation system, solutions to which correspond to routes. Since the basis of such a system is unique and finite, the model defines a certain structure of P2P paths.

To construct the model we used a relation between NLDE systems and formal grammars. A forwarding process at nodes can be described by a CCF-grammar, a routing grammar for a P2P overlay, where any derivation simulates a P2P route. Then an ANLDE system associates with the routing grammar forming the Diophantine model of routes.

We discussed several possible applications of the model. Having a finite path structure, one can compute metrics related for instance to utilization (load to nodes and links), connectivity (availability of alternate paths), and performance (number of hops). Detailed analysis of model applications is subject to our current research. The model can be supported with efficient algorithms that allow polynomial and pseudo-polynomial computations (not discribed in this chapter).

---

[6]Theory says that there is a polynomial algorithm for finding a non-zero solution of an arbitrary homNLDE system, not necessarily a homANLDE system. Unfortunately, to the best of our knowledge, we know no implementation appropriate for practical large-scale applications.

References

1. Aardal, K., Weismantel, R., Wolsey, L.A.: Non-standard approaches to integer programming. Discrete Appl. Math. **123**(1–3), 5–74 (2002). doi: http://dx.doi.org/10.1016/S0166-218X(01)00337-7

2. Artigas, M.S., Lopez, P.G., Skarmeta, A.F.G.: A novel methodology for constructing secure multipath overlays. IEEE Internet Comput. **9**(6), 50–57 (2005). doi: http://dx.doi.org/10.1109/MIC.2005.117

3. Bertsekas, D.P.: Network Optimization: Continuous and Discrete Models. Athena Scientific, USA (1998)

4. Bogoyavlenskiy, Y.A., Korzun, D.G.: On solutions of a system of linear Diophantine equations associated with a context-free grammar. Trans. Petrozavodsk State University Appl. Math. Comp. Sci. **6**, 79–94 (1998) (in Russian)

5. Castro, M., Costa, M., Rowstron, A.: Performance and dependability of structured peer-to-peer overlays. Technical Report MSR-TR-2003-94, Microsoft Research (2003)

6. Esparza, J.: Petri nets, commutative context-free grammars, and basic parallel processes. Fundam. Inform. **30**, 23–41 (1997)

7. Gummadi, K., Gummadi, R., Gribble, S., Ratnasamy, S., Shenker, S., Stoica, I.: The impact of DHT routing geometry on resilience and proximity. In: Proceedings of ACM SIGCOMM'03, pp. 381–394. ACM, New York (2003). doi: http://doi.acm.org/10.1145/863955.863998

8. Hu, T.C.: Integer Programming and Network Flows. Addison-Wesley, Reading (1969)

9. Kleinberg, J.M.: The small-world phenomenon: an algorithm perspective. In: Proceedings 32nd Annual ACM Symposium Theory of Computing (STOC '00), pp. 163–170. ACM, New York (2000). doi: http://doi.acm.org/10.1145/335305.335325

10. Korzun, D., Gurtov, A.: A Diophantine model of routes in structured P2P overlays. ACM SIGMETRICS Perform. Eval. Rev. **35**(4), 52–61 (2008)

11. Korzun, D.G.: On an interrelation of formal grammars and systems of linear Diophantine equations. Bull. Young Sci. **3**, 50–56 (2000) (in Russian)

12. Korzun, D.G.: Grammar-based algorithms for solving certain classes of nonnegative linear Diophantine systems. In: Proceedings of the Annual Finnish Data Processing Week at Petrozavodsk State University (FDPW 2000) on Advances in Methods of Modern Information Technology, vol. 3, pp. 52–67. Petrozavodsk State University, Petrozavodsk (2001)

13. Korzun, D.G.: Syntactic methods in solving linear Diophantine equations. In: Proceedings of the Annual Finnish Data Processing Week at Petrozavodsk State University (FDPW 2004) on Advances in Methods of Modern Information Technology, vol. 6, pp. 151–156. Petrozavodsk State University, Petrozavodsk (2005)

14. Lam, S.F., Chan, K.H.: Computer Capacity Planning. Academic, San Diego (1987)

15. Li, X., Misra, J., Greg Plaxton, C.: Maintaining the Ranch topology. J. Parallel Distrib. Comput. **70**(11), 1142–1158 (2010). doi: http://dx.doi.org/10.1016/j.jpdc.2010.06.004

16. Loguinov, D., Kumar, A., Rai, V., Ganesh, S.: Graph-theoretic analysis of structured peer-to-peer systems: Routing distances and fault resilience. IEEE/ACM Trans. Netw. **13**(5), 1107–1120 (2005)

17. Pottier, L.: Minimal solutions of linear Diophantine systems: Bounds and algorithms. In: Proceedings 4th International Conference Rewriting Techniques and Applications (RTA '91), pp. 162–173. Springer, Berlin (1991)

18. Risson, J., Robinson, K., Moors, T.: Fault tolerant active rings for structured peer-to-peer overlays. In: Proceedings IEEE Conference Local Computer Networks (LCN '05), pp. 18–25. IEEE Computer Society, USA (2005). doi: http://dx.doi.org/10.1109/LCN.2005.69

19. Sánchez-Artigas, M., García López, P.: Echo: A peer-to-peer clustering framework for improving communication in DHTs. J. Parallel Distrib. Comput. **70**, 126–143 (2010). doi: http://dx.doi.org/10.1016/j.jpdc.2009.06.002

20. Schrijver, A.: Theory of Linear and Integer Programming. Wiley, New York (1986)

21. Sebö, A.: Hilbert Bases, Caratheodory's Theorem and Combinatorial Optimization. In: Proceedings of 1st Integer Programming and Combinatorial Optimization Conference, pp. 431–455. University of Waterloo Press, Canada (1990)
22. Xu, J., Kumar, A., Yu, X.: On the fundamental tradeoffs between routing table size and network diameter in peer-to-peer networks. IEEE J. Sel. Areas Commun. **22**(1), 151–163 (2004)

Chapter 10
Structural Ranking

Abstract Open P2P networks are subject for selfish and incorrect behavior of nodes and even intentional attacks. A maintenance protocol must preserve topology structure not only according with the rules of the efficient graph for routing, but also accounting the cooperation level of nodes. Although the actions taken by intermediate nodes in the P2P network are hidden from the source node, the latter must have some guarantees on their correctness. This problem leads to incentive mechanisms to encourage cooperation of nodes. Such mechanisms provide each node with estimates of others behavior and the node makes own decisions when it selects its actions. In Chap. 6 we consider local ranking models when a node u decides its actions for node v based solely on directly observed past behavior of v. In this chapter, we focus on structural ranking models when network topology structure essentially information for computing ranks of nodes, as it is assumed in such well-known graph-based algorithms as PageRank and EigenTrust. We study models with partial knowledge and distributed computations when each node maintains some knowledge about the global network topology. The knowledge is a topology subgraph that aggregates, in form of cycles, direct and indirect observations of past behavior of nodes.

10.1 Introduction

The efficiency of a structured P2P system is due to maintenance of rigorous network topology, like it happens in conventional DHT designs with geometrically progressive routing. There is, however, an inherent tension between system (collective) welfare and individual rationality that threatens the viability of P2P systems. Such a system can consist of rational participants with diverse and selfish interests. Each node operates as an independent entity and does not always follows properly all rules of the P2P protocol. For instance, a stingy or overloaded node ignores some incoming requests, malicious node falsifies its responses and cheat others.

D. Korzun and A. Gurtov, *Structured Peer-to-Peer Systems: Fundamentals of Hierarchical Organization, Routing, Scaling, and Security*, DOI 10.1007/978-1-4614-5483-0_10, © Springer Science+Business Media New York 2013

To preserve the efficiency, a P2P maintenance protocol must account the cooperation level of nodes. A reputation system is often employed to detect malicious nodes or reward well-behaving ones [10, 18, 20, 31]. It provides information for local-level decision making when a node u selects proper neighbors for its routing table among generally selfish nodes. The selection eliminates from the cooperation overloaded or malicious nodes as well as free-riders that do not contribute sufficiently in return. Consequently, the overall system performance is preserved from the degradation. In terms of network topology, the presence of a link $u \to v$ reflects the cooperation quality between these two nodes.

If all nodes maintain their neighbors based on their past cooperation then a source node has some guarantees on performance and security, though the actions taken by intermediate nodes in the P2P network are hidden from the source node [2, 11, 14, 15]. The problem is to impel nodes for this maintenance, leading to an incentive mechanism that encourages the cooperation of nodes. In reciprocity-based schemes [15, 18, 24, 32], nodes maintain histories of past behavior of other nodes and use this information in their local decision making processes. When a node v become low-contributing then other nodes u detect this behavior and reduce their cooperation with v: links $u \to v$ eventually disappear, so changing the global network topology. Consequently, if v is still interested in participation in the system then v has to increase its contribution to the system.

We discussed incentive mechanisms that are based local ranking models in Chap. 6. This case corresponds to direct reciprocity: a node u decides its actions for node v based solely on directly observed past behavior of v. Nevertheless, indirect observation can provide useful extension of the local knowledge. For instance, if v is a neighbor of u and v has many well-behaving neighbors then v is a good neighbor for u for routing; similar idea is used in NoN-routing and CR-routing, see Chap. 8. In P2P resource sharing systems, bilateral exchange $u \leftrightarrow v$ is not always possible, since for two arbitrary nodes u and v the former may have no resources that the latter needs. Thus exchange cycles [1, 4, 27, 29, 32, 36] like $u \to w \to v \to u$ should be formed, when u has resources interested for w, w has resources interested for v, and finally v has resources interested for u.

In contrast to local ranking, structural ranking models assume that network topology structure essentially determines ranks of nodes. PageRank [35] and EigenTrust [21] are most known ranking algorithms [13, 17, 25] that follow this assumption and computes node ranks by analyzing the graph. However, analysis of the global graph is unrealistic in large-scale high-dynamic networks, and models with partial knowledge is needed. Moreover, complex analysis is not always as helpful as it was expected, and simple average aggregation algorithm performs better [28]. In this chapter, we study models with partial knowledge and distributed computation of this knowledge. They enhance local ranking models, where only neighbors are observed, with additional knowledge about the connectivity beyond the neighbors. The knowledge is a topology subgraph that aggregates direct and indirect observations of past behavior of nodes.

Section 10.2 introduces important selection problems where structural ranking can be used to arrange available candidates. It includes route and neighbor selection, which we also considered in Chap. 3, as well as neighbor differentiation for resource sharing, which we also considered in Chap. 6. Then we overview two known structural ranking algorithms—PageRank and EigenTrust—both require the knowledge of global network topology graph for computing rank values, though the computations can be distributed.

Section 10.3 shows that the computational efficiency and robustness to churn of structural ranking algorithms can be improved by letting nodes to maintain only partial knowledge on the global topology. We apply the same idea as in CR-routing; local cyclic structure is a cyclic subgraph, which in turn can be analyzed by a structural ranking algorithm to compute ranks. Assuming that the cyclic structure accounts the most important topology properties observed for the activity of a given node, these ranks are adequate for differentiation of nodes, though the ranks can differ from the globally computed ranks.

10.2 P2P Ranking

In this section we first introduce some important P2P design problems whose solutions need P2P ranking. They belong to the class of selection problems when ranks are used to arrange available candidates. As in the web information retrieval [26], we assume that the P2P overlay link structure is an essential factor for such rankings. Accordingly we then describe the known PageRank and EigenTrust algorithms, which we modify later in this chapter for the needs of P2P selection problems.

10.2.1 P2P Selection Problems

Consider a P2P overlay network of N nodes that shares R resources. (Recall that N and R are the index sets of all nodes and resources in the system, respectively.) Let $N_u \subset N$ be the set of neighbors of a node $u \in N$ and u consumes external resources $k \in R_u^+$ and provides local resources $k \in R_u^-$; let $R_u = R_u^+ \cup R_u^-$.

To implement the lookup operation, any P2P routing protocol has to solve the route selection problem [5]. Given a lookup for resource $k \in R$, a node u should find a next-hop node $v \in N_u$. As we discussed in Part II, the selection can be pure local. For instance, in progressive routing a distance metric arranges neighbors based on their closeness to the resource. In resource exchange systems, u computes its own ranking to differentiate already known neighbors based on their exchange history. Therefore, u arranges its neighborhood N_u in dependence on k.

Local strategies do not take into account more than one-hop relations between nodes in the system. The local knowledge can be enhanced by looking ahead the neighbors. Instead of enlarging the neighbor set N_u, the lookahead approach allows u to know some paths beyond its neighbors. In l-lookahead, u knows up to $l + 1$ subsequent nodes on a path. For instance, NoN-routing [30, 33] is progressive routing with 1-lookahead, see also Chap. 3. Cyclic routing (CR) [23,34] provides a more flexible mechanism to collect and analyze observable cycles kept in cyclic structure $\mathscr{C}_u$; it allows estimating the rest of routing paths, see also Chap. 8. Therefore, u arranges $\mathscr{C}_u$ in dependence on k.

In these non-local cases, the route selection problem needs arrangements of available paths, though some of paths can go through the same neighbor. Clearly, such paths capture the link structure of the overlay network. For instance, cyclic structure $\mathscr{C}_u$ represents a locally known subgraph of the topology graph.

Further, to cope with dynamics, P2P nodes perform the maintenance control to keep the overlay correctly connected. Some $v \in N_u$ can be incorrect at current time instance, which leads to another important problem—neighbor selection. A node u has to detect incorrect neighbors in N_u and to find candidates to insert into N_u. The first part (detection) of the problem can be based on arrangements in the neighborhood N_u. The second part, however, needs arrangements in the potentially unobservable set $N \setminus N_u$. Non-local approaches like CR can be applied here: nodes on the paths beyond the neighbors are candidates.

Resource exchange is a source of further complications for P2P selection problems. A node u has to arrange local and external resources in R_u to control its provision and consumption. Moreover, such prioritization also concerns transit resources, which u additionally serves to earn good reputation in the system.

It is known that when a node interacts with small percentage of other nodes the efficiency of bilateral exchange or direct reciprocity is low [4, 16]. Multilateral exchange provides better efficiency, but the complexity is to find a valid exchange path of indirect reciprocity [1, 9, 27]. This efficiency/complexity gap between bilateral and multilateral P2P exchange has motivated the extensions to the bilateral approach that with lower complexity achieve similar efficiency [27, 29, 32, 36].

Menasché et al. [32] formulated a model that reduces the multilateral case to bilateral. The reduction is due to barter when nodes obtain replicate external resources locally to serve other nodes. It is similar to the local ranking model from Chap. 6 since transit resources form a kind of barter. In contrast, their model is not local since there should be distributed brokers that dynamically search appropriate exchange cycles and transform the multilateral network into bilateral one.

Liu et al. [29] presented a bilateral exchange model where node consumption and provision are maintained asynchronously over time and nodes. In terms of the local ranking model, its parameters $a_{vk}(u)$ and $b_{vk}(u)$ become u's long-term debt and credit. They must be also updated indirectly, when v operates as a mediator for resource k. The model is not local since a distributed credit transfer mechanism is used for searching exchange paths in the network.

Piatek et al. [36] proposed a reputation protocol that allows at most one level of indirection, i.e., node's neighbors are the only providers of information beyond node's direct observations. In terms of the local ranking model, u maintains $a(u)$ and $b(u)$ using in addition recommendations from all $v \in N_u$. Then the BitTorrent TFT model is applied for differentiating the neighbors. A general mechanism for at most l levels of indirection is analyzed in [27].

10.2.2 PageRank Algorithm

PageRank [6, 35] is widely regarded as the best method for the static ranking of web pages, see also [13, 17, 25]. The basic idea behind PageRank for any directed graph is that a link from a node to another states an endorsement of the latter node, indicating the quality. PageRank takes advantage of the global link structure to order nodes according to their perceived quality.

Consider the following random walk in a graph of N nodes. At each step either with probability α a node u selects a link $u \to v$ to follow uniformly among available $v \in T_u$ or with $1 - \alpha$ jumps to a random node among $N - 1$. Jumping prevents permanent confinement in a strongly connected component of the graph. The parameter α is the *damping factor* denoting the probability of following the link structure (usually $\alpha = 0.85$). The probability the random walk is in node u is

$$p_u = \alpha \sum_{v:u \in N_v} \frac{p_v}{N_v} + \frac{1 - \alpha}{N - 1}. \tag{10.1}$$

The sum is over all v such that $u \in N_v$. If the walk is at such a node v then $1/N_v$ is the uniform probability of selecting $v \to u$ and thus the walk returns to u. The PageRank value for node u is defined as this probability. For example, if $p_u = 0.1$ then the random walk visits u every tenth step on average.

The rank p_u is divided among u's forward links $u \to v$ for $v \in N_u$ evenly to contribute to the ranks p_v of the nodes they point to. A node u has high p_u if the sum of the ranks of its ingoing links $v \to u$ is high. Either u has many ingoing links or u has a few highly ranked ingoing links.

Instead a random jump in (10.1), a *personalization vector* π of damping factors can be used. Each $\pi_u \geq 0$ represents the likelihood of jumping to node u. This non-uniform modification is called *personalized PageRank*. Another non-uniform modification is *weighted PageRank* where forward links have probabilities relative to the link weights. Let $b_{uv} > 0$ be the weight of a link $u \to v$ and p_{uv} be the probability of selection of $u \to v$, then

$$p_{uv} = b_{uv} / \sum_{w \in N_u} b_{uw}.$$

PageRank with these two modifications can be computed iteratively starting from some initial values $p_u^{(0)}$:

$$p_u^{(i+1)} = \alpha \sum_{v:u \in N_v} p_{vu} p_v^{(i)} + (1 - \alpha) \pi_u. \tag{10.2}$$

There is a close connection between p_u and the number and length of closed walks $u \to^+ u$, which correspond to cycles in the graph. Consider the nodes and links of graph as the states and transitions of Markov chains, respectively.

For Markov chains that are irreducible and have only positive recurrent states, the expected time a walk first returns to u is $1/p_u$ [38]. If $\alpha < 1$ and $\pi_u > 0 \forall u$, then any transition between two states has non-zero probability, and the irreducibility follows trivially. Since a PageRank random jump occurs at probability $1 - \alpha$ taking a node u with probability π_u, the probability to reach u at each step is at least $(1 - \alpha)\pi_u$. Consequently, the expected time to reach u from any node is finite and thus any state u is positive recurrent.

Similarly to web pages, a link $u \to v$ of P2P topology can be considered as indication of the quality of v for u. Computing PageRank in the global topology graph, we yield the following interpretation. If p_u is high then u is a neighbor of many other nodes or a neighbor of a few highly ranked nodes. This interpretation is useful for route and neighbor selection problems; each node u prefers those neighbors that have high rank.

10.2.3   EigenTrust Algorithm

BitTorrent-like systems [8] provides local incentives, which are based on private history (direct reciprocity). Mechanisms with purely private history do not scale [16]: coverage will be so small that cheaters are indistinguishable from good participants, e.g., whitewashers among newcomers. Private history is suitable for symmetric resource interest, i.e., any pair of nodes has resources to exchange which both nodes are interested in. In the asymmetric case, it may happen that u has resources for v, v has resources for w, w has resources for u, but exchange $u \leftrightarrow v$ fails because v has no resource u is interested in. On the other hand, exchange cycle $u \to v \to w \to u$ with indirect reciprocity can solve the problem. Scaling to higher turnover and mitigating asymmetry of interest requires shared history, which is, however, is prone to false reports and collusion.

The EigenTrust algorithm [21] is designed for reputation management in P2P resource sharing systems. It computes a trust score (rank p_v) that indicates how likely a node $v \in N$ is to be malicious (or stingy). The system evolution is divided into rounds $i = 1, 2, \ldots$ in which nodes interact by making queries and consuming resources. At the end of round i, the record of correct and incorrect consumptions is used to calculate the trust values $p_u(i + 1)$ for the next round.

A node u rates each transaction with its neighbor $v \in N_u$. In the original EigenTrust, a transaction is rated as positive (1) or negative (-1). Then, based on these rates, a local trust value d_{uv} is defined, e.g., as the sum of the ratings of the individual transactions that node u has performed from node v. The original EigenTrust computes d_{uv} as the difference between the sum of positive transactions and the sum of negative transactions. Another example is the local ranking model from Chap. 6, where this local value can be defined as surplus $d_{uv} = a_v(u) - b_v(u)$. The local trust values are normalized; let D be the normalized matrix.

The idea behind EigenTrust is to aggregate the local trust values in a distributed manner, as it happens in many reputation systems, e.g., see [3, 7].

EigenTrust is based on the notion of transitive trust: a node u will have a high opinion of those nodes who have provided it resources well. Moreover, node u is likely to trust the opinions of those nodes, since nodes who are generous about the resources they provide are also likely to be correct in reporting their local trust values.

EigenTrust global trust value is defined iteratively,

$$p_u^{(j+1)} = (1 - \varepsilon) \sum_{v \in N} D_{vu} p_v^{(j)} + \varepsilon t_u, \tag{10.3}$$

where $0 \leq \varepsilon < 1$ is a constant and $t = (t_v)_{v \in N}$ is a probability distribution over pre-trusted nodes. Pre-trusted nodes are essential, as they guarantee convergence and break up malicious collectives. In the simplest case, $t_u = 1/N$. The iterations $j = 0, 1, \ldots$ start with some initial vector $p^{(0)}$ and stop when $|p_u^{(j+1)} - p_u^{(j)}|$ is made small enough.

The definition is similar to PageRank iterations (10.2) with damping factor $\alpha = 1 - \varepsilon$ and personalized vector $\pi = t$. Consequently, the same interpretation and theoretical analysis, based on Markov chains and random walks, is also applicable for EigenTrust. If a random surfer was searching for reputable nodes it can crawl the network using the following rule: at each node u, it will crawl to node v with probability D_{uv}. After crawling for a while in this manner, the surfer is more likely to be at reputable nodes than unreputable ones. The stationary distribution of the Markov chain is the global trust vector $p = (p_u)_{u \in N}$.

Now we can define the distributed EigenTrust algorithm.

1. For round $i = 1$, initialize the trust based on a priory trust distribution, $p_u(1) = t_u$ for all u.
2. Until the end of the current round i, each node u makes its queries q. From the set of neighbors appropriate for serving q, a node v is selected with probability proportional to $p_v(i)$.
3. At the end of round i, each node u sends its local estimates $D_{uv} p_u(i)$ to all nodes v from which u consumed resources. In parallel, u collects $D_{vu} p_v(i)$ from nodes v to which u provides resources.
4. If $d_{vu} p_v(i)$ has not been received then set $D_{vu} = t_v$. Compute global trust value $p(i+1)$ for the next round using iterations (10.3).

EigenTrust is an uttermost form of shared history (indirect reciprocity), accounting all paths exited between participants and assigning each node a unique global trust value, analogous to the PageRank measure for web pages. In fact, the structure of indirect reciprocity cycles is analyzed and nodes that participate in many such cycles become of high rank. Global knowledge is exploited: all nodes in the network cooperate to compute and store the global trust vector. The original EigenTrust can be combined with additional trust metrics to limit the effects that misbehaving nodes can cause to the global network [11].

The distributed EigenTrust computation is questionable in open large-scale high-dynamic systems. The correctness essentially depends on the assumption that there are pre-trusted nodes in the network. Although EigenTrust has been proved very effective against several natural attacks from malicious coalitions, it performs poorly on some sophisticated attacks organized with different kinds of malicious peers [11]. Furthermore, EigenTrust is vulnerable to false negatives and false positives [27].

Lian et al. [27] provided a gradual solution to extend private history with shared history, leading to EigenTrust in the extreme case. The solution requires accounting all paths from a local node to other nodes up to a fixed length bound.

In contrast to the "all-paths" approach of EigenTrust, Anagnostakis and Greenwald [1] suggested to intentionally construct dedicated cycles for appropriate *l*-exchange transfers. The system gives them higher priority than for non-exchange transfers. The construction and then cycle maintenance introduce essential overhead.

Menasche et al. [32] proposed a decomposition of any indirect reciprocity cycle into a set of direct reciprocity cycles, i.e., nodes intentionally obtain undemanded content to use it for barter purposes. Similarly to [1], the overhead becomes an issue since the decomposition requires knowledge about existing indirect reciprocity cycle and employs a broker which provides nodes with recommendations on decomposition.

10.3   Cyclic Ranking

In the previous section we observed that such graph-based ranking algorithms as PageRank and EigenTrust actually exploit the cyclic structure of the global network graph. The computed rank of a node depends on which cycles (in the Markov chain terminology, closed random walks) the node belongs to. Based on this observation, this section shows how structural ranking algorithms can be modified to achieve better performance and scalability in P2P routing and resource sharing.

10.3.1   Local Cyclic Structure

Since a node does not know the whole network, direct application of PageRank or EigenTrust for P2P ranking requires a distributed algorithm. The later immediately meets with the following two difficulties. First, each node u relies on observations from other nodes and u uses these indirect data for analyzing the unknown part of the graph. For example in distributed PageRank algorithm [39], neighbors provide data about rank values they are currently estimated. Such third-party information certainly leads to security vulnerabilities in open environments, when some node can provide false estimates. Second, analysis of the entire network is expensive in dynamic large-scale P2P environments. The rank computation and corresponding data dissemination are slower compared to the churn rate. Rank-related data from other nodes become irrelevant when they have reached a node u.

Consider a structural ranking algorithm, e.g., PageRank or EigenTrust for concreteness. To overcome the difficulties of global graph knowledge let every node u perform the given ranking algorithm only for a known part of the network. That is, untrusted third-party information is not utilized directly. Note that a local part of the network topology is constructed on past observations the node has made, thus providing certain guarantees on the correctness based on the previous success results.

Such a local part of the network topology must have a good connectivity structure. Otherwise, the graph-based ranking algorithm is inapplicable. In this case, a pure set N_u of neighbors is not adequate. A reasonable candidate is a cycle collection that node u constructs locally as cyclic structure $\mathscr{C}_u$, similarly as it happens in CR algorithm [23].

Each cycle in $\mathscr{C}_u$ starts at u, goes through a neighbor $w_1 \in N_u$, then runs over the network and returns to u using a neighbor $w_{l-1} \in N_u$:

$$u \to w_1 \to w_2 \to \ldots \to w_i \to \ldots \to w_{l-1} \to u, \tag{10.4}$$

Recall that in the CR algorithm, information about intermediate nodes in cycles is third-party. Nevertheless, this information is verifiable since a cycle corresponds to a successful lookup (then the cycle is tested in further lookups), see Chap. 8. Therefore, $\mathscr{C}_u$ can be used as a trustable source of link structure known at u. In P2P resource sharing systems, where the EigenTrust algorithm is applicable, cycles correspond to exchange transfers (indirect reciprocity cycles), which also can be locally collected [1, 27, 29, 32, 36].

Local cyclic structure $\mathscr{C}_u$ can be represented with a cyclic directed graph consisting of all cycles known at u. This graph is a cyclic subgraph of the global topology graph. Assume that for any neighbor $v \in N_u$ there is at least one cycle that uses v as the first hop. Otherwise we simply remove such nodes (terminal nodes or leaves) from the cyclic graph. Even if the damping factor $\alpha = 1$ in (10.2) or $\varepsilon = 0$ in (10.3), all nodes in $\mathscr{C}_u$ are reachable: from an arbitrary node v the walk goes via a cycle to the host node u, then any node w is reachable from u via some other cycle. The number of steps in $v \to^+ u \to^+ w$ is finite. Hence the corresponding Markov chain is irreducible and have only positive recurrent states. Nevertheless, factors α and ε can useful for adopting rank algorithms to specific problem solutions.

Now we can summarize the key idea of cyclic ranking. Each node u maintains locally its cyclic graph, which approximates the knowledge on the global network topology. Then any graph-based ranking algorithm is applied on this subgraph, producing local node ranks. Note that different nodes can use different protocols for collecting cycles and different ranking algorithms for analysis of the cyclic graph. Below we consider applications of cyclic ranking to selection problems that appear in routing and resource sharing.

Fig. 10.1 C-PageRank
values for unweighted
forward ranking

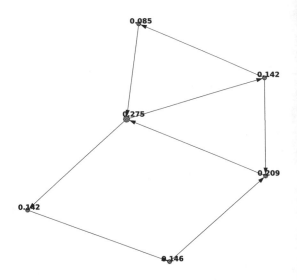

10.3.2   Relation of Routing and Ranking

The P2P routing efficiency is essentially depends on the topology structure. On the other hand, such structure is captured by the PageRank algorithm. Consider a modification of this algorithm to operate with the cyclic structure collected by CR algorithm. We call this modification the Cyclic PageRank algorithm (C-PageRank). Note that in general the cyclic structure can be constructed by another algorithm, different from the CR algorithm.

A straightforward modification employs unweighted PageRank. Let a node u run the PageRank algorithm on its local cyclic graph $\mathscr{C}_u$, see (10.1). Assume that nodes use an appropriate method for neighbor selection, thus N_u consists of good nodes only. A high rank value is interpreted as indication of node's high quality for the same reason as with web pages; a link $v \rightarrow w$ means that v perceives w to be efficient in forwarding. Those nodes are prioritized at u that participate in forwarding u's requests. Clearly, such nodes appear in cycles $u \rightarrow^+ u$.

This is the case of *forward ranking* when links in the graph and routing in the network have the same direction. A simple example is shown in Fig. 10.1. The more ingoing links from high-ranked nodes a node has the higher its rank is, i.e., many nodes connect a high-ranked node. With the assumption that all nodes are good it means that high-ranked nodes actively participate in routing. Their connectivity makes them ready to serve lookups from many nodes.

Note that the rank also positively correlates with the load the node takes. Consequently, it can be a useful metric for local decision making in load balancing. In this case, u prefers low-ranked nodes since they are less loaded. It defines a tradeoff between the routing performance and load balance.

A complementary modification is *backward ranking*. Let $\overline{\mathscr{C}}_u$ be the inversion of $\mathscr{C}_u$: every link $v \rightarrow w$ is replaced with its reversion $w \rightarrow v$. A node u runs the

Fig. 10.2 C-PageRank values for unweighted backward ranking

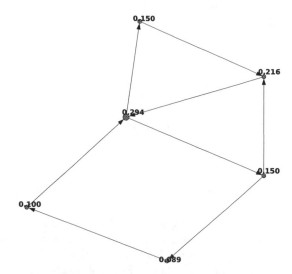

Table 10.1 C-PageRank variants in dependence on graph link direction

Direction	Purpose
Forward ranking	Perception of node quality. A high-ranked node is exploited by other high-ranked nodes: ingoing links are from many nodes and/or links are from high-ranked nodes.
Backward ranking	Node capacity for routing. A high-ranked node connects to many other nodes and/or to a few high-ranked nodes.

unweighted PageRank algorithm on $\overline{\mathcal{C}_u}$, and any link $v \to w$ of the original graph indicates v's quality. This property is useful for route and neighbor selection, since a high-ranked node connects to many other nodes. Such a node has more flexibility for forwarding requests. In other words, a node would like to contact a node that can provide as much resources for forwarding as possible. An example is shown in Fig. 10.2.

The same idea of backward ranking was used for web pages in BadRank [22,37] and in inverse PageRank [19]. BadRank aims at detecting spam web sites, based on the premise that a page is spam if it points to another spam page. Inverse PageRank gives preference to pages from which many other pages can be reached.

Table 10.1 summarizes the purpose of forward and backward ranking algorithms for decision making in P2P routing . Note that forward and backward ranking has certain similarities with HITS authority and hub score, respectively [26].

The algorithms can be augmented with weights for using the weighted PageRank (10.2). For example, weights are numbers of transported messages or amounts of transported data. It allows distinguishing between lot-used links and mostly unused links, making the former more significant in computing ranks.

10.3.3   Neighbor Selection and Malicious Nodes

Using both forward and backward ranking, let us consider how to apply cyclic ranking for the neighbor selection problem. Backward rank is used for identifying potentially contributing nodes. Forward rank is used for identifying malicious and potentially overloaded nodes.

A node v of a high backward rank is potentially interested in routing related to u, i.e., v declares to provide its resources for forwarding. However, as other nodes use the same or similar methods of neighbor selection, the best nodes can easily become overloaded. Moreover, a malicious node can also establish connections to high-ranked nodes, and then it will seldom or never forward other nodes' messages. Therefore, straightforward establishment of a connection to a high-ranked node v can be useless since high amount of messages from u will be probably dropped at v.

Forward ranking can help to identify both overloaded and malicious nodes before connecting to them. The idea is to compare the forward rank to backward rank. A normal node v will have mostly balanced ranks: v establishes connections to other nodes in accordance to resources it has and other nodes establish connections to v keeping them if they are useful. In contrast, a node v of essentially higher forward rank than backward rank has already an imbalanced share of incoming connections and thus v is unlikely to be able to process many more requests. On the other hand, a node v with lower forward rank, though being connected to several good nodes, has made quite few other nodes to consider it worthwhile to maintain outgoing connections to v.

With these notions, the neighbor selection rule can be formulated as follows. In the event of adding a node, u picks one that has high backward rank and the ratio of backward and forward ranks is near one. In addition, u employs a message drop detector for fast disconnecting from neighbors that have high rate of dropped messages or incorrect responses.

The uniform PageRank suffers when malicious nodes strategically establish connections to nodes that already have good connections with each other. As a result, a malicious node can inflate its rank. To mitigate this problem, weighted and personalized PageRank can be applied, where nodes contribution is taken into account in addition to the topology structure.

Since bad nodes (e.g., free-riders) tend to provide as little resources as possible, their ranks can be reduced by weighting links with amount of provided resources. Links to bad nodes receive low weights and a random walk is unlikely to select such link instead of a high-weight link. Consequently, PageRank value of a bad node is lowering. Note that the same idea is implemented in EigenTrust, where weights are current trust estimates.

For personalized PageRank its personalization vector π is defined such that all random jumps are to the origin node u, i.e., $\pi_u = 1$ and $\pi_v = 0 \forall v \neq u$. This property prevents a random walk to be long time in a possible cluster of malicious nodes. Such a cluster is formed when a coalition of malicious nodes cooperates and

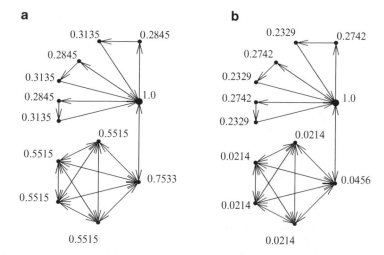

Fig. 10.3 Good nodes (on the *top*) serve ten times more messages than malicious nodes (on the *bottom*). The malicious coalition forms a tightly connected cluster. (**a**) The uniform C-PageRank assigns high rank values for the malicious nodes. (**b**) Weighed personalized C-PageRank assigns ten times lower rank values for the malicious nodes

establishes a lot of interconnections within the coalition, e.g., trying to increase each others reputation for non-coalition nodes. Note that this is an instance of the well-known Sybil attack [12].

In this case, we can consider the damping factor α as the probability of a message being dropped. For any dropped message the process of routing that message should start again at the origin. This model is very rough approximation of what is happening in P2P routing. In particular, messages in P2P networks do not travel by random paths. Nevertheless, the model provides the basic intuition. Better results are received by using these two methods together. An example is shown in Fig. 10.3, where a malicious coalition is easily differentiated by weighed personalized C-PageRank.

10.3.4   Resource Exchange

In BitTorrent-like systems [8] the exchange is based on forming *a swarm* of N nodes, which are interested in downloading the same file (file sharing), see also Chap. 6. At any time instance, each node u knows a subset of file pieces to exchange. For locally unavailable pieces u requests other nodes in the swarm. In turn, other nodes request u for pieces that u already has. Eventually, every node in the swarm is fed by all pieces.

The swarm notion implements bilateral incentives aiming in fair bandwidth allocation. The basic rule is that u uploads to v similarly to u downloaded from v.

Therefore, file download becomes slow or even impossible for stingy nodes since they refuse requests from other nodes.

Each node $u \in N$ forms a set of the n_1 best neighbors v with the highest download rates a_v (v's provision history). Besides, u selects random n_2 "optimistic" neighbors (unchoking); the aim is at discovering new neighbors in the swarm that would offer good download bandwidth. Let us call these $n_1 + n_2$ neighbors *active* since u uploads only to them (in a round-robin fashion) during the recent time slot. For downloading, u exploits all nodes in N consuming from them as much as possible.

BitTorrent incentives are simple and easily can be described in terms of ranks, as we discussed in Chap. 6: a node u arranges neighbors according to their direct provision to u. BitTorrent incentives are bilateral, which is known to be less efficient compared with multilateral case [4].

Typically, the following three types of nodes in a BitTorrent swarm are differentiated, based on their contribution patterns.

- A *good node* restricts its download rate to its upload rate; N_{good}.
- A *bad node* has prevalent download rate over its upload rate; N_{bad}.
- An *altruistic node* always keeps high upload rate; N_{alt}.

Note that in the BitTorrent terminology, *a seed* is a node who only provides file pieces to others; it corresponds to an altruistic node. *Leachers* are interested in file download; they correspond to good and bad nodes.

The basic performance metric is average file download completion time for good nodes, T_{good}. The basic fairness criteria are the following.

1. There are clear incentives to be a good node: $T_{bad} \gg T_{good}$.
2. Bad nodes make their gain mostly due to presence of altruistic nodes: T_{bad} is low if N_{alt} is small (or proportional to $|N_{alt}|$).
3. The more good nodes the more their gain from the rational cooperation: T_{good} is proportional to $|N_{good}| + |N_{alt}|$.

Consider a mechanism for a node to extend the local knowledge of BitTorrent direct reciprocity with shared history. The extension aims at preventing cheating in indirect observations. The mechanism is based on the existence of exchange cycles, a characteristic property of multilateral resource exchange. A simple example is the following provision cycle

$$u \to v \to w \to u,$$

where each link means "high" provision. If v provides low bandwidth to u, then BitTorrent exchange leads to low provision from u to v. However, if v provides generously to w and w in turn is a generous provider for u, then u prefers to provide high bandwidth to v, indirectly preserving w's generosity.

Collecting a set of provision cycles

$$u \to w_1 \to w_2 \to \ldots \to w_i \to \ldots \to w_{l-1} \to u,$$

similarly to (10.4), the node u has an additional knowledge source about participants and their provision. The more cycles where w_i participates the more gain for u to keep appropriate provision to w_i. This kind of appropriateness can be quantified based on the structural ranking algorithm applied to the cyclic graph.

An interesting problem is how to find provision cycles. It is one of most complex problems in designs that deal with exchange cycles [1, 27, 29, 32, 36]. Typically, a BitTorrent swarm has an almost fully-connected topology, i.e., each node knows most of other nodes. Nevertheless, links are not equal. Each node selects $n_1 > 0$ neighbors to upload. Thus there are active links, and a subgraph with active links only can be constructed. This case is very similar to the CR algorithm.

Based on this observation, let a node construct a reasonable coverage of this subgraph with a smaller subgraph—a cyclic graph consisting of provision cycles. Each cycle is a file exchange chain in the swarm. Each link $u \to v$ in the cycle is "good" in terms of bilateral exchange (direct reciprocity), i.e., it is v is in the top of n_1 best neighbors of u and stable in time. Such cycles exist in such P2P economies; they are short and have reasonable interpretation, see [1, 32].

If node u maintains a local collection of cycles $\mathscr{C}_u$ then u can use it to extend simple direct reciprocity of BitTorrent and other TFT-like schemes to more powerful indirect reciprocity (multilateral exchange). Computing PageRank (or EigenTrust) on this cyclic graph leads to better ranks than simple download/upload ratio. Then u applies rank-based upload bandwidth allocation. In the general case, it can be used for single-resource exchange using the ranks in the weighted round-robin strategy, as we discussed in Chap. 6.

The above ranking mechanism is gradual, similar to [27]. In the latter, u keeps rank M_{uw}^n that characterizes all n-hop paths $u \to^+ w$ known to u. For that any u's neighbor v periodically provides ranks M_{vw}^{n-1} for all $(n-1)$-paths $v \to^+ w$. A similar idea can be found in [39].

However, cycles are better structure for aggregated topological information [23] than a set of paths $u \to^+ w$. Each neighbor v provides u some of best neighbors (u's neighbors of neighbors) or best cycles $v \to^+ v$ that v knows. Based on this knowledge u maintains locally own cyclic graph and provides the knowledge periodically to its neighbors. Based on the cyclic graph, u computes ranks and uses them in its upload process.

The mechanism is illustrated in Fig. 10.4. A node u periodically receives from its neighbors what cycles they know. If v is a good provider for u then u includes them into its local cyclic graph. A good provider is a neighbor with provision/consumption ratio $\gamma_v(u) \geq 1 - \varepsilon$ for small enough ε.

Let u receive from v a $(n-1)$-cycle

$$v \to w_1 \to \cdots \to w_{n-1} \to v = w_n,$$

where $w_i \neq u \forall i$ and v expects that w_1 is its good provider, see Fig. 10.4b. The simplest case is $n = 2$ when v knows only its good neighbors, see Fig. 10.4a. Then u constructs the following n-cycle

$$u \to w_1 \to \cdots \to w_{n-1} \to v = w_n \to u.$$

Fig. 10.4 Cycle construction
at u based on
recommendations from a
good provider v.
(**a**) v recommends u to use
its good provider w as a
consumer. (**b**) v recommends
u to use $(n-1)$-cycle

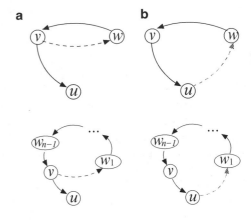

The weight of the cycle is determined by v's provision/consumption ratio $\gamma_v(u)$ relatively to u. Each link $w_i \rightarrow w_{i+1}$ increments its weight $a_{w_i w_{i+1}} := a_{w_i w_{i+1}} + \gamma_v(u)/m_v$, where m_v is the number of cycles that u takes into account based recommendations from v, including the bilateral exchange $v \rightarrow u \rightarrow v$.

The previous cycles from v are either removed or their weight is exponentially reduced (a decay function). The local cyclic graph consists of cycles collected from all good neighbors. Also, the collection includes 2-cycles $u \rightarrow v \rightarrow u$. When a link appears in several cycles its weight is a sum (or average, or maximum) over all cycles it belongs to.

Each node has an upper bound on the length of cycles it collects, $n_{\max}$. It is a tradeoff parameter between node churn and cycle dissemination. The same node w can participate in many cycles constructed at u. Recall that structural ranking algorithms, such as PageRank or EigenTrust, reward nodes that participate in many cycles.

Notably that for v there is no reason to cheat u. It is based on the cyclic property. Let v recommend a bad w_i to u. Then w_i becomes a bad node in provision cycles $u \rightarrow^+ w_i \rightarrow^+ v \rightarrow^+ u$, and consequently u will reduce v's rank.

10.4   Summary

This chapter has emphasized that network topology structure reflects the participation behavior of nodes. We focused on two cases: routing and resource exchange. When global topology graph is known then structural ranking algorithms like PageRank or EigenTrust can be used to compute global node ranks. Each node then uses these rank values to arrange nodes and solve selection problems: route selection for a given lookup, neighbor selection to fill routing table, service rates for resource requests from other node, and some others.

The assumption of global knowledge does not allow applying effectively the above structural ranking approach in open large-scale high-dynamic P2P systems. In this case, even distributed rank computation over the whole network seems unrealistic due to dynamics and security issues. We considered another approach when a node keeps locally only a part of the network topology. This part can be dynamically represented by a collection of cycles that the node has observed during its actual participation in the system. Cyclic structure provides a cyclic subgraph which allows applying known structural algorithms. This scale level reduction preserves important rank properties and allows realistic computation with tradeoff parameters between accuracy and efficiency.

References

1. Anagnostakis, K.G., Greenwald, M.B.: Exchange-based incentive mechanisms for peer-to-peer file sharing. In: Proceedings of 24th International Conference on Distributed Computing Systems (ICDCS'04), ICDCS '04, pp. 524–533. IEEE Computer Society, USA (2004)
2. Antoniadis, P., Courcoubetis, C., Mason, R.: Comparing economic incentives in peer-to-peer networks. COMNET **46**(1), 133–146 (2004). doi: http://dx.doi.org/10.1016/j.comnet.2004.03.021
3. Aperjis, C., Johari, R.: Designing aggregation mechanisms for reputation systems in online marketplaces. SIGecom Exch. **9**, 3:1–3:4 (2010). doi: http://doi.acm.org/10.1145/1980534.1980537
4. Aperjis, C., Freedman, M.J., Johari, R.: Bilateral and multilateral exchanges for peer-assisted content distribution. IEEE/ACM Trans. Netw. **19**(5), 1290–1303 (2011). doi: http://dx.doi.org/10.1109/TNET.2011.2114898
5. Balakrishnan, H., Kaashoek, M.F., Karger, D., Morris, R., Stoica, I.: Looking up data in P2P systems. Commun. ACM **46**(2), 43–48 (2003). doi: http://doi.acm.org/10.1145/606272.606299
6. Brin, S., Page, L.: The anatomy of a large-scale hypertextual web search engine. Comput. Netw. ISDN Syst. **30**(1–7), 107–117 (1998)
7. Buchegger, S., Mundinger, J., Boudec, J.Y.L.: Reputation systems for self-organized networks. IEEE Technol. Soc. Mag. **27**, 41–47 (2008)
8. Cohen, B.: Incentives build robustness in BitTorrent. In: Proceedings 1st Workshop on Economics of Peer-to-Peer Systems (2003)
9. DeFigueiredo, D., Venkatachalam, B., Wu, S.F.: Bounds on the performance of P2P networks using Tit-for-Tat strategies. In: P2P'07: Proceedings 7th IEEE International Conference on Peer-to-Peer Computing, pp. 11–18. IEEE Computer Society, USA (2007)
10. Despotovic, Z., Aberer, K.: P2P reputation management: probabilistic estimation vs. social networks. Comput. Netw. **50**(4), 485–500 (2006). doi: http://dx.doi.org/10.1016/j.comnet.2005.07.003
11. Donato, D., Paniccia, M., Selis, M., Castillo, C., Cortese, G., Leonardi, S.: New metrics for reputation management in P2P networks. In: AIRWeb '07: Proceedings of 3rd International workshop on Adversarial information retrieval on the web, pp. 65–72. ACM, New York (2007). doi: http://doi.acm.org/10.1145/1244408.1244421
12. Douceur, J.R.: The Sybil attack. In: Revised Papers from 1st International Workshop on Peer-to-Peer Systems (IPTPS '01), pp. 251–260. Springer, Berlin (2002)
13. Farahat, A., Lofaro, T., Miller, J.C., Rae, G., Ward, L.A.: Authority rankings from HITS, PageRank, and SALSA: existence, uniqueness, and effect of initialization. SIAM J. Sci. Comput. **27**(4), 1181–1201 (2006)

14. Fedotova, N., Veltri, L.: Reputation management algorithms for DHT-based peer-to-peer environment. COMCOM **32**(12), 1400–1409 (2009). doi: http://dx.doi.org/10.1016/j.comcom. 2009.03.002

15. Feldman, M., Chuang, J.: Overcoming free-riding behavior in peer-to-peer systems. ACM SIGecom Exch. **5**(4), 41–50 (2005). doi: http://doi.acm.org/10.1145/1120717.1120723

16. Feldman, M., Lai, K., Stoica, I., Chuang, J.: Robust incentive techniques for peer-to-peer networks. In: Proceedings 5th ACM Conference Electronic commerce (EC'04), pp. 102–111. ACM, New York (2004). doi: http://doi.acm.org/10.1145/988772.988788

17. Franceschet, M.: PageRank: standing on the shoulders of giants. Commun. ACM **54**(6), 92–101 (2011). doi: http://doi.acm.org/10.1145/1953122.1953146

18. Gupta, R., Somani, A.K.: Reputation management framework and its use as currency in large-scale peer-to-peer networks. In: Proceedings 4th International Conference Peer-to-Peer Computing (P2P '04), pp. 124–132. IEEE Computer Society, USA (2004). doi: http://dx.doi. org/10.1109/P2P.2004.44

19. Gyöngyi, Z., Garcia-Molina, H., Pedersen, J.: Combating web spam with TrustRank. In: Proceedings 30th International Conference Very large data bases (VLDB '04), pp. 576–587. VLDB Endowment, USA (2004). doi: http://dl.acm.org/citation.cfm?id=1316689.1316740

20. Jin, X., Chan, S.H.G.: Reputation estimation and query in peer-to-peer networks. IEEE Commun. Mag. **48**, 122–127 (2010)

21. Kamvar, S.D., Schlosser, M.T., Garcia-Molina, H.: The Eigentrust algorithm for reputation management in p2p networks. In: Proceedings 12th International Conference World Wide Web (WWW '03), pp. 640–651. ACM, New York (2003). doi: http://doi.acm.org/10.1145/775152. 775242

22. Kolda, T.G., Procopio, M.J.: Generalized BadRank with graduated trust. Technical Report SAND2009-6670, Sandia National Laboratories, Albuquerque, NM and Livermore, CA (2009)

23. Korzun, D., Nechaev, B., Gurtov, A.: Cyclic routing: Generalizing lookahead in peer-to-peer networks. In: AICCSA2009: Proceedings 7th IEEE/ACS International Conference Computer Systems and Applications, pp. 697–704. IEEE Computer Society, USA (2009). doi: http://doi. ieeecomputersociety.org/10.1109/AICCSA.2009.5069403

24. Landa, R., Griffin, D., Clegg, R.G., Mykoniati, E., Rio, M.: A sybilproof indirect reciprocity mechanism for peer-to-peer networks. In: Proceedings of IEEE INFOCOM'09, pp. 343–351. IEEE, USA (2009). doi: http://dx.doi.org/10.1109/INFCOM.2009.5061938

25. Langville, A.N., Meyer, C.D.: Deeper inside pagerank. Internet Math. **1**(3), 335–380 (2004). URL: http://akpeters.metapress.com/content/bn22r01j43g6q8g6/

26. Langville, A.N., Meyer, C.D.: A survey of eigenvector methods for web information retrieval. SIAM Rev. **47**(1), 135–161 (2005). doi: http://dx.doi.org/10.1137/S0036144503424786

27. Lian, Q., Peng, Y., Yang, M., Zhang, Z., Dai, Y., Li, X.: Robust incentives via multi-level Tit-for-Tat. Concurr. Comput. Pract. Exper. **20**, 167–178 (2008). DOI 10.1002/cpe.v20:2

28. Liang, Z., Shi, W.: Analysis of ratings on trust inference in open environments. Perform. Eval. **65**(2), 99–128 (2008). doi: http://dx.doi.org/10.1016/j.peva.2007.04.001

29. Liu, Z., Hu, H., Liu, Y., Ross, K.W., Wang, Y., Mobius, M.: P2P trading in social networks: the value of staying connected. In: Proceedings of IEEE INFOCOM'10, pp. 2489–2497. IEEE, USA (2010)

30. Manku, G.S., Naor, M., Wieder, U.: Know thy neighbor's neighbor: the power of lookahead in randomized P2P networks. In: STOC '04: Proceedings 36th Annual ACM Symposium on Theory of computing, pp. 54–63. ACM, New York (2004). doi: http://doi.acm.org/10.1145/ 1007352.1007368

31. Marti, S., Garcia-Molina, H.: Taxonomy of trust: categorizing P2P reputation systems. Comput. Netw. **50**(4), 472–484 (2006). doi: http://dx.doi.org/10.1016/j.comnet.2005.07.011

32. Menasché, D.S., Massoulié, L., Towsley, D.: Reciprocity and barter in peer-to-peer systems. In: Proceedings of IEEE INFOCOM'10, pp. 1505–1513. IEEE, USA (2010)

33. Naor, M., Wieder, U.: Know thy neighbor's neighbor: Better routing for skip-graphs and small worlds. In: IPTPS '04: Proceedings 3rd International Workshop on Peer-to-Peer Systems. Lecture Notes in Computer Science, vol. 3279. Springer, Berlin (2004)

34. Nechaev, B., Korzun, D., Gurtov, A.: CR-Chord: Improving lookup availability in the presence of malicious DHT nodes. Comput. Netw. **55**, 2914–2928 (2011)

35. Page, L., Brin, S., Motwani, R., Winograd, T.: The PageRank citation ranking: Bringing order to the Web. Technical Report 1999-66, Stanford InfoLab (1999). URL http://ilpubs.stanford.edu:8090/422/. Previous number = SIDL-WP-1999-0120

36. Piatek, M., Isdal, T., Krishnamurthy, A., Anderson, T.E.: One hop reputations for peer to peer file sharing workloads. In: Proceedings 5th USENIX Symposium on Networked Systems Design & Implementation (NSDI 2008), pp. 1–14. USENIX Association, USA (2008)

37. Sobek, M.: PR0 — Google's PageRank 0 penalty (2003). URL: http://pr.efactory.de/e-pr0.shtml. Accessed 26 June 2012

38. Stroock, D.W.: An Introduction to Markov Processes. Springer, Berlin (2005)

39. Yamamoto, A., Asahara, D., Itao, T., Tanaka, S., Suda, T.: Distributed pagerank: A distributed reputation model for open peer-to-peer networks. In: Proceedings 2004 International Symposium on Applications and the Internet Workshops (SAINTW'04), pp. 389–394. IEEE Computer Society, USA (2004). doi: 10.1109/SAINTW.2004.1268664

Summary of Part III

This part has discussed some models and methods that allow a node to collect additional knowledge about the global network. The local knowledge scope can be too limited for some problems, especially in a large-scale system. On the other hand, the global knowledge approach is inappropriate since such information typically requires $\Theta(N^2)$ storage (representation of the topology graph) and subject to frequent changes due to churn and other dynamic events.

The models and methods we considered provide tradeoffs between local and global knowledge. The problem is often called *the partial knowledge problem*. Typically, an effective solution to this problem requires that the local knowledge extension is incremental and started from a base P2P protocol. The latter provides default amount of local knowledge per node. If a node is interested it can append its local knowledge with additional information. This extension, as previously, follows some arrangement model to embed effective composition structure into the overall system of many nodes.

As particular cases we considered the following extensions. HDHT architectures is the evolution result of hierarchical routing schemes. Extended knowledge in HDHT is structured with a layered topology structure, an excellent evidence of the generic decomposition principle to prioritize potentially available knowledge. The CR method supports a wide family of DHT routing protocols when nodes can benefit from additional look-ahead information. Local cyclic structure is a snapshot of what is happening beyond the neighbors. Diophantine routing provides the theoretical framework for P2P routing in general. The framework allows formulating particular models that compactly describe certain routes in the P2P overlay, including routes that consists of multiple paths because of multipath routing. Structural ranking is a generic method that a node can locally use for ranking other nodes based on additional information about the network topology. We considered cyclic ranking when this information is represented as a cyclic structure, similarly to the CR method.

Part IV
Applications

Overview of Part IV

P2P algorithms are not just an academic exercise but found their use in many Internet applications and enterprise-grade systems. In this book part, we describe CR-Chord P2P protocol to address growing security concerns in the real-world. We also introduce the Host Identity Indirection Infrastructure, a potential candidate architecture for Future Internet that supports host mobility, multihoming, protection against Denial-of-Service attacks and rendezvous service. Finally, we conclude this part and the book with a chapter on P2P use in several real-world applications.

Chapter 11 presents extensions to Chord utilizing the Cyclic Routing approach that enhances lookup routing in presence of malicious DHT nodes that drop packets. Thanks to additional information collected from intermediate nodes, the sender node is able to choose lookup paths that travel around suspicious DHT nodes.

Chapter 12 introduces an extension of Internet Indirection Infrastructure (i^3) that supports privacy-preserving host location and security association establishment. i^3 is a distributed system that enables request routing using triggers and runs Chord DHT internally. Hi^3 adds a possibility for delegation of some processing from end hosts to i^3 servers that improves its scalability and attack resistance.

Chapter 13 describes several uses of P2P algorithms in popular Internet applications, such as BitTorrent and their proposed use in Future Internet architectures such as host identifier and locator resolution with OpenDHT. We also introduce several distributed systems that use P2P algorithms although are targeted for large-scale commercial systems with higher reliability requirements and less dynamic membership. These include Amazon's Dynamo, Facebook's Cassandra, Linked-in's Voldemort.

Chapter 11
CR-Chord

Abstract Without additional mechanisms conventional DHTs are vulnerable to attacks. In particular, previous research showed that Chord is not well resistant to malicious nodes that joined the DHT. This chapter describes the CR-Chord protocol, our implementation of the cyclic routing algorithm. Using simulations we compare the lookup availability of basic Chord and CR-Chord. The results suggest that CR-Chord improves the lookup availability on the average by 1.4 times. When the number of malicious nodes is small, such as 5%, CR-Chord has almost twice lower lookup failure rate.

11.1 Introduction

The field[1] of structured P2P systems has seen a fast growth and evolutions upon introduction of first Distributed Hash Tables (DHTs) in the early 2000s. The first proposals including Chord, Pastry, Tapestry were gradually improved to cope with scalability, locality and security issues. Deployable as an overlay on the application layer without the need to change the network infrastructure, P2P approach has opened great opportunities for innovation for developers. By utilizing the processing and bandwidth resources of end users, P2P approach enables high performance of data distribution which is hard to achieve with traditional client-server architectures. That enables commercial use of P2P systems such as distributing updates to the World-of-Warcraft virtual world, where patches over 100 MB are applied simultaneously to all users using P2P technology. Many popular social networks, such as Facebook, utilize the DHT internally to store tremendous numbers of key-value pairs.

---

[1]Reprinted from Computer Networks, 55(13), B. Nechaev, D. Korzun, A. Gurtov, CR-Chord: Improving Lookup Availability in the Presence of Malicious DHT Nodes, 2914–2928, Copyright (2011), with permission from Elsevier. Some material is also applied in Chap. 8.

D. Korzun and A. Gurtov, *Structured Peer-to-Peer Systems: Fundamentals of Hierarchical Organization, Routing, Scaling, and Security*, DOI 10.1007/978-1-4614-5483-0_11,   293
© Springer Science+Business Media New York 2013

Now P2P computing is a fast research field with multiple conferences and research groups in the area. The P2P computing is being actively utilized in the Internet for software updates, P2PSIP VoIP, video-on-demand, and distributed backups. Recent introduction of identifier-locator split proposal for Future Internet architectures poses another important application for DHTs, namely mapping between host permanent identity and changing IP address. The growing complexity and scale of modern P2P systems requires introduction of hierarchy and intelligence in routing of requests. Additionally, researchers proposed several anti-cheating mechanisms to ensure fair resource distribution and avoid the "tragedy of commons." Popular P2P systems have been a subject of various attacks thus bringing security and resilience issues to the front.

Nowadays Distributed Hash Tables (DHT) are a part of many peer-to-peer (P2P) applications in the Internet. To mention a few examples, DHTs are used to track the upload/download ratings in BitTorrent and resolve host identifiers to IP addresses for Host Identity Protocol (HIP) [24]. Each DHT node maintains a routing table of its neighbors containing node identifiers (IDs) and IP addresses. The main service provided by DHTs is routing a lookup query for a certain key to a DHT node that stores the value for that key.

As Internet applications increasingly depend on DHTs to operate, DHT should be resilient to all kinds of attacks. One of the most dangerous scenarios is when adversaries are able to become a part of DHT by joining as regular nodes. Then attackers can corrupt, drop, or misroute lookup messages. In this study we limit ourselves to dropped lookups only.

Let *lookup failure rate* be the probability that an arbitrary lookup is dropped. Complementary, *lookup availability* is the probability that a lookup arrives at the destination. Several techniques for improving lookup availability in the presence of malicious nodes were proposed, including iterative routing and lookup progress monitoring [28, 29], self-certifying data [3, 6, 26], routing failure tests and root verification [3, 33]. In this chapter we apply cyclic routing (CR) [16] as a way to enhance robustness of existing DHTs in the presence of malicious nodes. Our focus is on integration of cyclic routing with Chord [30], one of the first and still most popular DHTs. The resulting system is called CR-Chord. Note that [16] focuses on generic CR properties, while this chapter advances [16] further by introducing and evaluating a concrete design and its implementation.

In CR-Chord, a node maintains a collection of cycles additionally to its routing (finger) table. A cycle is a path that starts from the node to its finger, then runs through the network and returns to the node. Only IDs of cycle nodes, but not IP addresses, are stored. Cycles present global knowledge about good paths in a DHT. If there is a cycle containing a node close to the destination, the message is sent along this cycle. Otherwise, Chord is used to select the next hop. We implemented CR-Chord as an extension of the MIT Chord simulator. With extensive simulations, we analyzed and compared the Chord and CR-Chord lookup availability. The results suggest that CR-Chord improves the lookup availability by 1.4 times on the average.

The rest of the chapter is organized as follows. Section 11.2 provides background on DHT security and defines the problem of dropped lookups. Section 11.3

describes the CR-Chord protocol and estimates the load increase that CR-Chord introduces. In Sect. 11.4, we define our simulation methodology. Section 11.5 explains the simulation results of CR-Chord vs. Chord. Section 11.6 compares CR-Chord to existing work and discusses the most important findings.

11.2 Background and Motivation

Many DHTs and in particular the Chord DHT are not well resistant to dropped lookups in the presence of malicious and faulty nodes. In this section, we consider this problem and introduce the basics of the cyclic routing concept as a promising remedy.

11.2.1 DHT Routing

Consider a DHT consisting of N nodes. Node IDs are assigned from an identifier space with a distance metric. Each node s maintains a routing table T_s of entries (u, IP_u), where u is a neighbor and IP_u is its IP address. Hence, s forwards messages to u via the underlying IP network. A message to a destination node d goes sequentially to nodes whose IDs are progressively closer to d according to the distance metric. If $v \in T_s$ then s can forward a message to v forming the one-hop path $s \to v$. Routing from s to d takes several hops forming a multi-hop path $s \to^+ d$.

DHT routing is divided onto *global* and *local* parts. In global routing, a message is delivered close to the destination. In local routing, the destination is at a nearby node. The reasons for division are as follows. (a) Since a node is responsible for the keys closest to its ID, let a lookup message arrive to a node close to the key. (b) Various replication techniques support routing into an area of neighboring nodes. (c) A DHT node knows its neighborhood well, keeping close nodes to its routing table when possible. Obviously, global routing is more vulnerable to attacks.

DHT routing is either *iterative* or *recursive*. With iterative routing each node on the lookup path returns the next-hop node v to the querying node. The latter then contacts v to get iteratively closer to the destination. With recursive routing, each node forwards lookups directly to the next hop nodes, and the querying node receives a response from the destination. Iterative routing is more secure since a querying node can control the routing progress. Nevertheless, more network resources are consumed, and iterative routing is not possible when a querying node cannot directly contact some nodes on the path, e.g., due to NATs. In this study, we consider recursive routing only.

The lookup availability depends on the number of alternative paths between a source and destination [21, 27, 29]. The more paths, the more chances to go around malicious nodes. However, a mechanism for finding paths consisting of good nodes is needed. The straightforward approach exploits multi-path routing when nodes

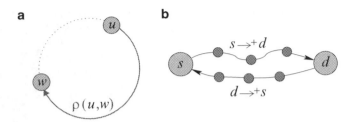

Fig. 11.1 (**a**) Asymmetrical distance in Chord. (**b**) Cycle in bidirectional communication

multicast messages to several neighbors [3, 14, 19, 37]. It has two disadvantages. First, a lot of duplicate messages are generated; many of them are redundant due to the local route selection (non-optimal). Second, paths converge in some DHTs [13, 27, 37], and bypassing malicious nodes becomes impossible.

11.2.2   Chord

Chord [30] uses an identifier circular space of n-bit integers, and participating nodes form a ring taking IDs from $[0, 2^n - 1]$. A node forwards messages in clockwise direction. The distance $\rho(u, w)$ is length of the clockwise ring arc between u and w, see Fig. 11.1a. Key k is assigned to the first node whose ID is equal or follows k clockwise.

A Chord routing table for a node s consists of three types of neighbors: (1) successors, (2) fingers, and (3) predecessors. Successors and predecessors are several closest nodes to s, clockwise and counterclockwise, respectively. Successors are node's short-distance contacts aiming at local routing. The ith finger of s is the node that succeeds s by at least 2^{i-1} on the ring, where $i = 1, 2, \ldots, n$. For $i = 1$, the finger is the immediate successor. In a sparse network, several first and last fingers coincide typically with some successors and predecessors, respectively. Fingers are node's long-distance contacts aiming at global routing. Predecessors are mainly for routing table maintenance.

In a lookup for a key k, each node finds the closest preceding neighbor v. When v is a finger, each hop at least halves the distance. Eventually, a lookup arrives at the node whose immediate successor is responsible for k.

Basic Chord restricts a finger table to n entries. They are updated systematically. More efficient routing is achieved when a node can keep *additional fingers* having more knowledge about the network [20]. Let n_{af} denote the number of additional fingers. When Chord occasionally discovers a node, it is inserted as an additional finger. Periodically, Chord removes fingers that have not been used.

11.2.3 Attacks by Dropping Lookups

In attacks on routing a malicious node drops messages, modifies them, or forwards incorrectly. We consider only the case of dropped lookups. Such a lookup fails to reach the destination, and no response is sent.

Dropped lookups concern routing security as well as fault-tolerance. In recursive routing detection of lookup failures is more difficult than in iterative case. Dropped lookups are harder to reveal compared to incorrect lookups; a querying node has no response, and the lookup validity cannot be checked. In fact, only straightforward timeout techniques detect such failures.

Multicast mitigates dropped lookups. Instead of the only next-hop, a node uses several neighbors to forward a message (several alternate paths). The efficiency, however, depends on the number of disjointed paths m [29]:

$$f^m \leq \Pr(\text{lookup failure}) \leq \left(1 - (1-f)^l\right)^m, \tag{11.1}$$

where f is the fraction of malicious nodes (uniform distribution), l is the number of hops in a lookup. Equation (11.1) is valid for any DHT, but Chord is more sensitive than many others. Its finger selection is restrictive, and the ith finger must immediately succeed $s + 2^{i-1}$. This is avoided in randomized-Chord [13, 36], where the ith finger is taken randomly from $[s + 2^{i-1}, s + 2^i)$. Moreover, introducing additional fingers allows s to know several neighbors in $[s + 2^{i-1}, s + 2^i)$.

In Chord, any path $s \rightarrow^+ d$ goes through p, the d's immediate predecessor [27, 29], defining the so called *shield problem*. Although Chord has alternative paths $s \rightarrow^+ p \rightarrow d$, they are not disjointed, and p becomes a single point of failure for all lookups to d. Therefore, Chord satisfies (11.1) only for $m = 1$, and the lower bound can be refined (valid for $N \geq 16$) [27]:

$$1 - (1-f)^2 \leq \Pr(\text{lookup failure}) \leq 1 - (1-f)^l. \tag{11.2}$$

Multicast helps even when there are no disjointed paths. However, the problem is in finding a good path among available. DHT routing works locally, and a node just selects several next hops without much knowledge about the remaining paths. In this chapter, we offer a more systematic way that uses the results of previous lookups. Having a path that a successful lookup followed lately, a node uses it for subsequent lookups.

With global routing each hop typically halves the distance, $\rho(v_{j-1}, d) \approx 2\rho(v_j, d)$ in $s \rightarrow v_1 \rightarrow \cdots \rightarrow v_{j-1} \rightarrow v_j \rightarrow \cdots \rightarrow v_{l-1} \rightarrow d$. That is, the distance decreases geometrically ($\rho(s, d) = \delta, \delta/2, \ldots, \delta/2^j, \ldots$) while the number of remaining hops decreases linearly ($l, l-1, \ldots, l-j, \ldots$). With the uniform distribution of nodes, there are twice as little nodes in an interval (v_j, d) than in (v_{j-1}, d), and good and malicious nodes decrease equally. Nevertheless, attacks on ID assignment and routing tables can make the malicious node population slower decreasing from (v_{j-1}, d) to (v_j, d) than of good nodes. Then at each hop the probability of failure becomes higher, and the lookups failure rate is worse than dictated by the above bounds.

Algorithm 2 Cyclic routing a message to a node d.

Require: Message p arrives at $u \neq d$.
  The node u maintains T_u and $\mathscr{C}_u$.

  Find $c \in \mathscr{C}_u$ such that
      $c = \left(u \to v_1 \to^+ \tilde{d} \to^+ u\right)$ where $\tilde{d}$ is close to d;
  if c is found **then**
      Let the next-hop node v be v_1;
  else
      Find the next-hop node $v \in T_u$ according to the underlying DHT;
  end if
  Forward p to v;

These problems are likely to happen in a real environment, and they motivate our study on improving Chord security. Our solution exploits the concept of cyclic routing.

11.2.4   Cyclic Routing (CR)

Formal background of CR is introduced in [16] and Chap. 8. Let a node s send a message to a node d ($s \to^+ d$). Then d replies to s ($d \to^+ s$), see Fig. 11.1b. The paths form the cycle $s \to^+ d \to^+ s$. Taking intermediate nodes, $c = \left(s \to v_1 \to^+ v_2 \to^+ \cdots \to^+ v_{l-1} \to^+ s\right)$, where $v_1 \in T_s$. Nodes $v_1, \ldots, v_{l-1}$ may represent some but not all the nodes visited.

Let each node s maintain a collection of cycles additionally to its routing table T_s, $\mathscr{C}_s = \{c_1, \ldots, c_q\}$, where $c_j = (s \to v_{j,1} \to^+ \cdots \to^+ v_{j,l-1} \to s)$ for $j = 1, \ldots, q$. The collection $\mathscr{C}_s$ is a network *cyclic structure* known to s. In Algorithm 2, each node u of $s \to^+ d$ can construct the remaining route using $\mathscr{C}_u$ when appropriate. Otherwise, the underlying DHT is called to find the next-hop node.

A simple reactive strategy is used for constructing cycles. When receiving a successful lookup response, s keeps the cycle if it is efficient (e.g., a low number of hops). The underlying DHT facilities are used more optimally; only good and efficient paths are collected among available ones. Note, however, that the strategy should be considered as a catalyst; it fails when the underlying DHT cannot produce good paths, and another way for constructing cycles is needed.

11.3   Integration of Cyclic Routing with Chord

Algorithm 2 works on top of any DHT. We adapt cyclic routing to the Chord DHT. The resulting system is called CR-Chord and follows Algorithm 3. In general, CR aims at routing a packet p from a source node s to a destination node d. In Chord, u usually does not know d's ID since most lookups are done to documents, and the goal is to route p to such d that stores the requested document knowing only the document key.

Algorithm 3 The CR-Chord pseudocode.

Require: A node u receives a lookup packet p for a key k.
   Let $(v_1, \ldots v_{m_d})$ be the closest to k fingers in T_u

---

if $u = s$ **then**
      Send secondary lookups to $(v_1, \ldots v_{m_d})$ {Multicast}
end if
Let c_p be a cycle piggybacked in p (if any)
$c_p = BestCycle(\mathscr{C}_u \cup \{c_p\}, k)$
$v_{\text{cycle}} = NextCycleHop(c_p, k)$
if $v_{\text{cycle}} \neq \varnothing$ **then**
      Piggyback c_p into p
      Send p to v_{cycle} {Forward along the cycle}
else
      Send p to v_1 {Forward via the underlying Chord}
end if

$BestCycle(\mathscr{C}, k)$
   Find c in $\mathscr{C}$ such that
      $CycleDist(c, k)$ is minimal and $CycleDist(c, k) < \infty$
   return c if found, and $\varnothing$ otherwise

$NextCycleHop(c, k)$
   Find in c the closest node v to k such that
      $v \in T_u$ and $\rho(v, k) < \rho(u, k)$
   return v if found, and $\varnothing$ otherwise

$CycleDist(c, k)$
   Find in c the closest node $\widetilde{d}$ to k such that $\rho(\widetilde{d}, k) < \rho(u, k)$
   return $\rho(\widetilde{d}, k)$ if found, and ∞ otherwise

Nodes maintain cyclic structures exploiting regular Chord lookups. Let s initiate a lookup for a key k. The lookup arrives to the destination d according to Chord routing. Then d acknowledges to s. Nodes of $s \rightarrow^+ d \rightarrow^+ s$ form a candidate cycle c for $\mathscr{C}_s$.

There are two types of lookup messages in CR-Chord, *primary* and *secondary*. Primary lookups may exploit cycles. Secondary lookups are used for constructing cycles and do not use cycles. Both lookup types allow finding requested documents. A source sends a primary lookup and multicasts secondary lookups with the *multicast degree* m_d. Thus, a primary lookup is duplicated with m_d Chord-based ones. Since at every node cyclic routing is not applied to secondary lookups, cycles reflect good paths available in a regular Chord network.

Although the lookup success is our primary goal, CR-Chord remains efficient by storing only those cycles that satisfy *the performance criterion*. Let l be the number of hops in c and f_s be the number of fingers at s. Then c is stored when

$$l \leq k_h \min(f_s, n), \tag{11.3}$$

where $\min(f, n)$ approximates $\log_2 N$ (the mean number of hops in a Chord cyclic path $s \rightarrow^+ d \rightarrow^+ s$), k_h is a tradeoff parameter between performance and security. In our simulations, we allow $k_h = 2$ for bypassing malicious nodes.

Let f_c be the number of fingers in c. Then c is stored when

$$f_c \leq k_c l, \tag{11.4}$$

where k_c defines an upper bound to the number of "local" nodes that a cycle includes. The fewer such nodes in a cycle the more global network info the cycle contains—*the network coverage criterion*. In our simulations we use $k_c = 1/2$.

At each routing step, a node u selects a next-hop v to forward p. There are three options. (1) A cycle c_p piggybacked at p is used. (2) A better cycle is available in $\mathscr{C}_u$. (3) None of cycles in $\mathscr{C}_u \cup \{c_p\}$ is appropriate, and the Chord-provided choice is used.

In Algorithm 3, u calls *BestCycle* to decide either to use the previously fixed cycle c_p or there is a better cycle in $\mathscr{C}_u$. The criterion is the closeness according to the Chord distance ρ (see *CycleDist*). If a cycle contains a node $\tilde{d}$ that is closer to k than u, then the cycle is appropriate. Also a cycle must have at least one node $\tilde{v}$ such that $\tilde{v} \in T_u$. Therefore, u searches for the best cycle among appropriate ones. When no appropriate cycle exists, u performs regular Chord forwarding. Otherwise, u finds in c_p the finger v_{cycle} that is closest to k, piggybacks c_p into p, and forwards p.

A cycle piggybacked into a packet does not consume much space. Chord node ID is 160-bit long (20 bytes). In a Chord network of $N = 10^6$ nodes, the mean cycle length l does not exceed 20 hops, which results in 400 bytes-long cycles ($n = 20$ bits in ID). IP packets with a typical Maximum Transmission Unit of 1,500 bytes can easily include such a cycle. The number of cycle nodes piggybacked into a packet can be reduced by storing not a full cycle, but a contiguous subset of its nodes. After a lookup traverses all these nodes, according to Algorithm 3 it will follow the regular Chord algorithm until it finds a new cycle or reaches the destination.

The storage space consumed by cycles at a node can be approximated by $l_b|\mathscr{C}|$ where l_b is the cycle length in bytes and $|\mathscr{C}|$ is the number of cycles per node. As discussed above, $l_b \approx 400$ bytes is enough in a network of 10^6 nodes. In our experiments (see Sect. 11.5.3 and Figs. 11.11–11.13), we have $|\mathscr{C}| \approx 60$ in the worst case of various scenarios. Thus $400 \times 60 = 24{,}000$ bytes or about 24 KB are needed to store cycles. We consider this to be negligible for modern computers.

The same node may appear in different cycles, and efficient data structures can be applied to store the cyclic structure, e.g., a digraph (a local representation of known network topology). Hence $O(N_u)$ space instead of $O(l|\mathscr{C}_u|)$ is needed at u, where N_u is the number of nodes whose IDs u knows. Note that u controls N_u independently on other nodes and may vary from $N_u = 0$ (no CR) to $N_u = N$ (total knowledge about the network) according to locally available capacity. Moreover in practice, the maintenance cost for node IDs is less than for fingers since the former (1) does not involve expensive IP address maintenance and (2) uses piggybacking to regular lookups.

The processing time at u is mostly determined by $BestCycle(\mathscr{C}_u, k)$ depending on the data structure that implements $\mathscr{C}_u$. First, finding the closest node to k takes time $O(l|\mathscr{C}_u|)$ or $O(N_u)$. The efficiency can be improved by standard searching optimization technique like indexing. Second, finding the cycle c and the next hop v takes time $O(l)$.

11.4 Simulation Methodology

The only security parameter, the fraction of malicious nodes $0 \leq f < 1$, is used. For readability we express f also as percentage. Although the model is simple it allows comparing key properties of the lookup availability in Chord and CR-Chord.

11.4.1 Attack Model

Let $M = fN$ be the number of malicious nodes. Then $G = N - M$ is the number of good nodes. Malicious and good nodes are distributed uniformly in the Chord ring.

Malicious nodes attack the network by dropping lookups. Since such a node pretends to be good as much as possible, we assume that it processes correctly all routing maintenance traffic. Otherwise, its malicious activity can be detected easier, and we do not consider this case here. To some degree this behavior is similar to a faulty or overloaded node; its malicious but imperceptible activity includes the following.

1. Dropping lookup and acknowledgment messages (no forwarding).
2. Processing data placement requests as a good node but no real data are stored.
3. Ignoring lookup requests for data items for which the node is responsible.

Note that a malicious node can perform this activity only with a certain probability p. However, such a case is similar to the previous one with the fraction $f' = fp$. Without loss of generality we also assume that malicious nodes do not send lookup requests.

We assume that the network has reached a stable state where data items are distributed uniformly among nodes. If any data happen to be at a malicious node then the data are lost. There are no more data insertion and deletion. Routing maintenance goes as if all nodes are good. In this stable state the network provides the lookup service for data by keys. A random good node u queries a lookup for a given key k. Assuming that u selects k uniformly, the lookup failure and success rates characterize the availability of the lookup service.

11.4.2 Simulation Scheme

Our simulation of a Chord network with malicious nodes follows Algorithm 4. We used the MIT original Chord simulator (http://cvs.pdos.csail.mit.edu/cvs/sfsnet/simulator/) for networks with $N = 1,000, 2,000, 3,000$ and IDs with $n = 24$. The simulator implements the basic Chord [30]; the ith finger of s is the first node in $[s + 2^{i-1}, s + 2^i)$ for $i = 1, 2, \ldots, n$. The randomized-Chord [13, 36] would lead to better results for CR-Chord since nodes have more freedom in selecting fingers, and more good paths are available.

Algorithm 4 Simulation steps

---

1. *Initially a good network.* A network of G good nodes is constructed. Nodes are joining randomly.
2. *Data placement.* D data items are distributed in a good network. In all simulations $D = 100N$ assuming that the mean number of data items per node does not affect much the lookup availability.
3. *Introducing malicious nodes.* M malicious nodes join the network at the same time. Some data items are lost since each malicious node becomes responsible for a certain part (arc) of the Chord ring.
4. *Serving lookups.* L requests are performed. For a request a pair (u, k) is selected randomly, where u is a good node and k is the key for a data item stored at step 2. There are two phases, *Stabilization* and *Analysis*, $L = L_{\text{stab}} + L_{\text{rate}}$ requests.
   Stabilization (for CR-Chord only). Good nodes initiate L_{stab} requests without multicast. The requests are used for constructing cycles but they are not counted in the lookup availability estimation.
   Analysis. Good nodes initiate $L_{\text{rate}} = 0.01N^2$ requests. They are analyzed for the success or failure. In CR-Chord, a request induces one primary lookup and m_{d} secondary lookups. In Scenario 7, churn of both good and malicious nodes is performed during this phase.

We enhanced the MIT simulator with CR-Chord implementation (see Algorithm 3 above). In current version, the size $|\mathscr{C}_u|$ is unbounded and no maintenance for cycles is implemented. It does not, however, affect the results much. First, the attack model assumes a stable network. Second, the number of cycles constructed per node is moderate or even small for large M (see Figs. 11.11–11.13).

Variation of N provides a more adequate analysis of the lookup availability [27]. For every value of N, the fraction of malicious nodes f varies from 5% or 10% up to 50%; the incremental step is either 5 or 10. When varying N, parameters D (Step 2) and L_{rate} (Step 4) are proportional to N and N^2, respectively. Table 11.1 summarizes our simulation scenarios. Each scenario consists of two parts, for Chord and CR-Chord, respectively. Each part takes ten executions of Algorithm 4 for statistical averaging.

11.4.3   Churn Model

To assess the CR-Chord behavior in a dynamic environment we introduce churn. In simulations, nodes join and leave the network at a constant rate R. At the same time good nodes generate lookup requests. The ratio of good and malicious nodes joining network (parameter f) is preserved constant; each join event is accompanied by a leave event. Thus during churn total number of nodes in the network and fraction of malicious nodes on average remain constant for the whole simulation.

This churn model comprises only dynamics of nodes, not documents. The latter are inserted before churn starts. During churn, documents are moved to appropriate nodes. When a node joins the network it acquires from its successor all the documents it becomes responsible for. When a node, either good or malicious,

Table 11.1 Summary of simulation scenarios

Scenario	N	f	L_{stab}	n_{af}	m_d	R
Typical simulation configuration						
1	1,000	5%, 10%, ..., 50%	0	12	3	0
Requesting data when a responsible node and its immediate predecessor are good						
2	1,000	10%, 20%, ..., 50%	0	12	3	0
Variation of the network size, N						
3	1,000, 2,000, 3,000	10%, 20%, ..., 50%	0	12	3	0
Variation of the number of additional fingers, n_{af}						
4	1,000	10%, 20%, ..., 50%	0	0, 12, 24, 48	3	0
Variation of the multicast degree, m_d						
5	1,000	10%, 20%, ..., 50%	0	12	1, 2, ..., 10	0
Variation of the stabilization period, L_{stab}						
6	1,000	10%, 20%, ..., 50%	0, 1,000, 5,000, 10,000	12	3	0
Variation of churn rate, R						
7	1,000	10%, 20%, ..., 50%	0	12	3	0.01, 0.02, 0.05, 0.1, 0.2

leaves the network it hands over all its documents to the successor. For simplicity we omit a more realistic model where documents stored at malicious nodes are lost completely. To masquerade their activity malicious nodes also perform these operations.

According to [30] the churn rate R is the number of joins/leaves per second. We vary churn rate from 0.01 to 0.2. Note that in our simulations lookups are generated with interval of 1 s and stabilization of nodes is done every 30 s on average. Thus, for boundary values of R each churn event is performed after every 100 and 5 lookups respectively. Such rates are consistent with the previous studies. In [25] a node generates a lookup every 0.1 s, while churn events occur at the rate of 0.067 to 8 per second, which corresponds to a churn event happening every 150 to 1.25 lookups. In [30] for the same lookup generation rate as ours, churn rate varies from 0.05 to 0.4, which is equivalent to 20 and 2.5 lookups between two churn events.

11.4.4   Success and Failure Types

A request for a key includes a primary lookup and m_d secondary lookups. Either of the lookups or both of them can succeed by reaching the responsible node. To find out the share of each we compute detailed successes and failures metrics, see Table 11.2 for Scenario 1, which is most typical in our simulations.

We distinguish three success cases:

1. Only primary lookup succeeds, all secondary lookups fail (*Primary lookups success*).
2. At least one of secondary lookups succeeds, primary lookup fails (*Multicast lookups success*).
3. Both primary and at least one of secondary lookups succeed (*Joint success*).

The total success rate (*Successful requests*) is the sum of the above values. Note that only one (first reached) successful secondary lookup contributes to the success rate regardless of the other $m_d - 1$ secondary lookups.

The metrics below show how actively successful primary lookups (*Primary lookups success*) use cycles:

1. A successful primary lookup cycle was first inserted at the initiator node (*Initiator cycle success*).
2. A successful primary lookup's cycle was first inserted at an intermediate node (*Intermediate cycle success*).
3. A successful primary lookup does not contain a cycle (*No cycle success*).

Besides success metrics we estimate primary lookup failure rates (*Primary lookups failure*):

Table 11.2 Locations of lookup failures in Chord and CR-Chord (Scenario 1)

Percent of malicious nodes	10%		20%		30%		40%		50%	
	Chord	CR-Chord	Chord	CR-Chord	Chord	CR-Chord	Chord	CR-Chord	Chord	CR-Chord
Successful requests	6,724.9	7,958.4	4,353.8	5,985.4	2,673.9	4,197.9	1,501.1	2,617.3	808.2	1,518.5
Primary lookups success	0.0	55.3	0.0	88.2	0.0	89.2	0.0	64.8	0.0	39.9
Multicast lookups success	0.0	859.0	0.0	1,161.9	0.0	1,121.5	0.0	847.8	0.0	534.7
Joint success	0.0	7,044.1	0.0	4,735.3	0.0	2,987.2	0.0	1,704.7	0.0	943.9
Primary lookups success	6,724.9	7,099.4	4,353.8	4,823.5	2,673.9	3,076.4	1,501.1	1,769.5	808.2	983.8
Initiator cycle success	0.0	3,287.2	0.0	1,912.5	0.0	963.8	0.0	373.3	0.0	137.1
Intermediate cycle success	0.0	1,105.8	0.0	626.6	0.0	316.4	0.0	127.9	0.0	45.1
No cycle success	0.0	2,706.4	0.0	2,284.4	0.0	1,796.2	0.0	1,268.3	0.0	801.6
Primary lookups failure	3,275.1	2,900.6	5,646.2	5,176.5	7,326.1	6,923.6	8,498.9	8,230.5	9,191.8	9,016.2
– At responsible node	743.2	789.4	1,037.8	1,144.2	1,130.6	1,300.9	1,017.4	1,219.0	834.8	1,010.0
– At predecessor	814.1	862.5	1,375.6	1,521.2	1,587.0	1,817.7	1,670.2	1,956.0	1,601.6	1,916.0
– In the middle	1,717.8	1,248.7	3,232.8	2,511.1	4,608.5	3,805.0	5,811.3	5,055.5	6,755.4	6,090.2

1. The responsible node d is malicious (*Primary lookups failure at responsible node*).
2. The predecessor node of a key k is malicious (*Primary lookups failure at predecessor*).
3. A primary lookup failed at an intermediate node (*Primary lookups failure in the middle*).

There are also two primary lookup failure types that are solely due to churn and are relevant only for Scenario 7:

1. A document cannot be found at the responsible node due to churn (*Primary lookup document churn failures*).
2. A primary lookup failed along the route due to churn (*Primary lookup routing churn failures*).

A primary lookup document churn failure happens when the lookup arrives at the predecessor after the successor leaves the network, but before predecessor is notified about that. Similarly, when a new node joined between the predecessor and former successor, but the Chord stabilization is not completed.

There is no timeout mechanism to retransmit lost packets. It leads to the primary lookup routing churn failures. At each moment a node has a number of requests to forward. If the node leaves the network, all unsent requests are lost. Sender node does not know that its request was lost and since there is no retransmission, we can count it a lookup failure. Similarly, a primary lookup can fail along the route due to churn since nodes do not ping fingers before forwarding. In this case a lookup can be forwarded to a node that is already not a part of the network. Eventually such a finger will be removed from the finger table by the stabilization procedure, but before this sending a packet to it will result in a routing churn failure.

11.5   Analysis of Lookup Availability

Table 11.2 presents its key metrics for Scenario 1. Other scenarios are focused on particular aspects of lookup availability. Instead of lookup availability we sometimes equivalently prefer an opposite metric, the lookup failure rate. The lower bound of Chord lookup failure rate is defined in (11.2). Standard statistical averaging is used for computing typical values.

11.5.1   Basic Facts

Figure 11.2 confirms [27, 29] that Chord is not well resistant to the presence of malicious nodes. The lookup failure rate exceeds the lower bound significantly. The peak is for moderate malicious cases ($20\% \leq f \leq 35\%$). For $f = 10\%, 20\%, \ldots, 50\%$ the lookup availability is 32% on average.

Fig. 11.2 Lookup failure rates (Scenario 1)

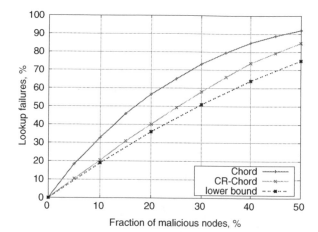

Let $S_{chord}(f,R)$ and $F_{chord}(f,R) = L - S_{chord}(f,R)$ be counters of successful and failed lookups for Chord. Similarly, $S_{cr}(f,R)$ and $F_{cr}(f,R) = L - S_{cr}(f,R)$ are for CR-Chord.

Table 11.3 partially overlaps with Table 11.2 in showing absolute success and failure values, but also includes relative lookup success improvement and relative lookup failure reduction offered by CR-Chord as well as performance under churn.

The relative lookup success improvement and relative lookup failure reduction,

$$\Delta(S) = \frac{S_{cr} - S_{chord}}{S_{chord}} \quad \text{and} \quad \Delta(F) = \frac{F_{chord} - F_{cr}}{F_{chord}},$$

are also given. In these relative estimates, Chord is the base.

Two cases—with churn ($R = 0.2$) and without churn ($R = 0$)—are considered. The effect of churn is estimated with the relative variation

$$\Delta_c(x) = \frac{x(f,0) - x(f,0.2)}{x(f,0)} \quad \text{for } x = S, F,$$

where $R = 0$ is the base.

Both in Chord and CR-Chord, the lookup success (failure) rate decreases (increases) with more malicious nodes and churn. CR-Chord always has better values for these rates. Interestingly that CR-Chord under churn ($R = 0.2$) outperforms Chord in stable networks ($R = 0$).

When the fraction of malicious nodes is small (5–10%) the CR-Chord lookup failure rate is almost twice lower. When the fraction of malicious nodes is large (40–50%) the CR-Chord lookup success rate is about twice higher.

Churn affects the CR-Chord lookup availability less than the one of Chord. In particular, $\Delta_c(S_{chord}) > \Delta_c(S_{cr})$ for all fractions of malicious nodes, and $\Delta_c(S_{chord})/\Delta_c(S_{cr}) \approx 2$ on the average. In terms of the lookup failure rate, $\Delta_c(F_{chord})$ becomes slightly lower than $\Delta_c(F_{cr})$ only for large f. On average, $\Delta_c(F_{chord})/\Delta_c(F_{cr}) \approx 1.5$.

Table 11.3 Lookup success improvement and lookup failure reduction (Scenario 7)

Malicious nodes $M = f/N$, %	10%		20%		30%		40%		50%		Average	
R, s^{-1}	0	0.2	0	0.2	0	0.2	0	0.2	0	0.2	0	0.2
S_{chord}	6,724.9	5,141.0	4,353.8	3,110.1	2,673.9	1,900.4	1,501.1	994.7	808.2	534.9	3,212.4	2,336.2
$\Delta_c(S_{chord})$, %	23.6		28.6		28.9		33.7		33.8		29.7	
S_{cr}	7,958.4	7,557.9	5,985.4	5,262.2	4,197.9	3,526.2	2,617.3	2,042.3	1,518.5	1,154.9	4,455.5	3,908.7
$\Delta_c(S_{cr})$, %	5.0		12.1		16.0		22.0		23.9		15.8	
$\dfrac{S_{cr} - S_{chord}}{S_{chord}}$, %	18.3	47.0	37.5	69.2	57.0	85.6	74.4	105.3	87.9	115.9	55.0	84.6
F_{chord}	3,275.1	4,859.0	5,646.2	6,889.9	7,326.1	8,099.6	8,498.9	9,005.3	9,191.8	9,465.1	6,787.6	7,663.8
$\Delta_c(F_{chord})$, %	48.4		22.0		10.6		6.0		3.0		18.0	
F_{cr}	2,041.6	2,442.1	4,014.6	4,737.8	5,802.1	6,473.8	7,382.7	7,957.7	8,481.5	8,845.1	5,544.5	6,091.3
$\Delta_c(F_{cr})$, %	19.6		18.0		11.6		7.8		4.3		12.3	
$\dfrac{F_{chord} - F_{cr}}{F_{chord}}$, %	37.7	49.7	28.9	37.0	20.8	20.1	13.1	11.6	7.7	6.6	21.6	23.8

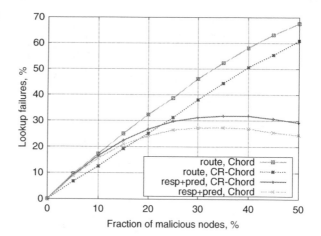

Fig. 11.3 Lookup failure rate at responsible+predecessor nodes and en route (Scenario 1)

The CR-Chord lookup failure rate is lower. For $f = 5\%$ it is 1.77 times less than for Chord (10.4% and 18.4% respectively). In contrast to Chord, the difference with the lower bound monotonically grows when increasing f. The lookup availability of CR-Chord is 45% on average. Comparing with basic Chord, CR-Chord has 7–16% better absolute lookup availability; the best result is for moderate malicious cases. On average CR-Chord performs 1.4 times better; the absolute effect is 13%. For 50% of malicious nodes lookup availability of Chord and CR-Chord is 8.1% and 15.2% respectively; thus for big values of f lookup availability is almost doubled.

A lookup fails either at a responsible node, at its immediate predecessor, or in the middle of the route. Figure 11.3 demonstrates share of lookups that failed at these locations among all lookups. While CR-Chord has more lookup failures en route than Chord for all f, the picture is opposite for failures at shield and destination nodes. This suggests that CR-Chord makes it easier for lookups to bypass malicious nodes along the route. Another interesting phenomenon is that share of responsible+predecessor nodes falls for high f. This is explained by more failures at middle nodes—with growth of f it gets harder for packets to reach last two nodes and fail there.

According to the widespread opinion the shield problem is most essential for Chord lookup availability. Figure 11.3 and the last rows in Table 11.2 show, however, that failures in the middle constitute a comparable part or even bigger part. It means that a solution to the shield problem does not necessarily lead to high lookup availability.

In fact, the shield problem relates to local routing but failures in the middle affect global routing. CR-Chord has more failures at responsible nodes and immediate predecessors than Chord (average value of absolute difference is 3%) because of reducing failures in the middle (average absolute difference is 7%). Consequently, CR-Chord optimizes global routing.

Scenario 2 provides further clarification. Any request satisfies two extra requirements: (1) it is for data stored at a good node and (2) its immediate predecessor

Fig. 11.4 Lookup failure rate
without failures at responsible
node and predecessor
(Scenario 2)

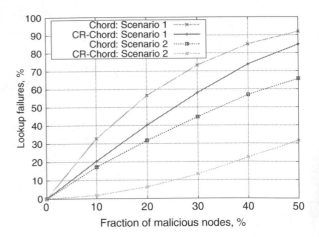

is also good. The scenario focuses on global routing since all lookups can fail only in the middle. Figure 11.4 shows the results; Scenario 1 is for comparison. Chord suffers much from failures in the middle. In contrast, CR-Chord reduces such failures up to three to four times.

Most failures in Chord happen at the few first hops, see Fig. 11.5a–c (all failed lookups are 100%). CR-Chord routes better around malicious nodes, and less failures happen at the beginning. The trend is seen for all f even despite the fact that for high fractions of malicious nodes it gets easier for lookups to hit a malicious node, which explains the shift of failures to smaller number of hops as f increases. This is yet another confirmation of optimized global routing in CR-Chord.

11.5.2   Variations

Previous research showed that lookup availability varies depending on the network size [27]. To validate how this affects CR-Chord we ran scenario 3 simulations that estimate lookup availability for different values of N (Fig. 11.6). Clearly, growing N increases the number of hops in lookups, and the lookup failure rate becomes higher (see also (11.1)). However since CR-Chord improves global routing, the effect of higher number of hops is mitigated by the use of cycles. This is proved by lookup availability ratio. For $N = 3,000$ the lookup availability of Chord and CR-Chord is 25% and 39% respectively. Thus CR-Chord performs 1.56 times better than Chord, while for $N = 1,000$ if does only 1.4 times better. The failure rate seems to grow sublinearly, and the reason is $O(\log N)$-paths.

Scenario 4 focuses on the effect of additional fingers (Fig. 11.7). Increasing n_{af} converges the failure rate to the lower bound. However, introducing too many fingers does not help much; doubling n_{af} improves the failure rate only by a constant.

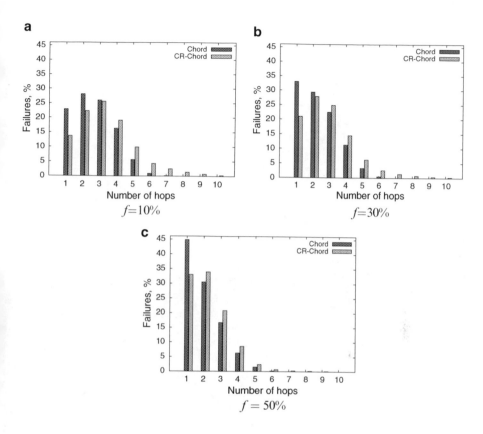

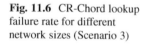

Fig. 11.5 Distribution of lookup failures according to hop length (Scenario 1). (**a**) $f = 10\%$. (**b**) $f = 30\%$. (**c**) $f = 50\%$

Fig. 11.6 CR-Chord lookup failure rate for different network sizes (Scenario 3)

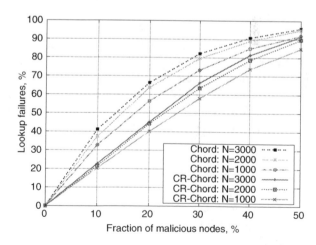

Fig. 11.7 Lookup failure rate
for different numbers of
additional fingers
(Scenario 4)

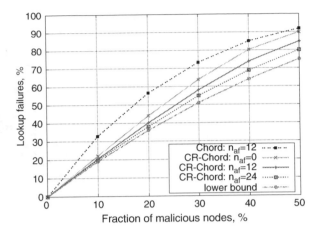

Fig. 11.8 Lookup failure rate
for different multicast degrees
(Scenario 5)

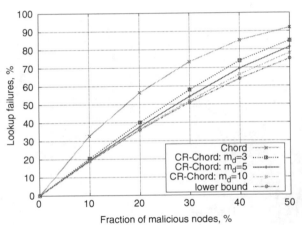

Scenario 5 aims at improving the lookup availability by multicast (Fig. 11.8).
Basic Chord does not support multicast ($m_d = 0$). In CR-Chord, increasing m_d
converges the failure rate to the lower bound (as with additional fingers). The small
multicast degree ($m_d = 3$) reduces the failure rate significantly. Further increasing
does not affect much.

Scenario 6 considers the effect of stabilization period (for CR-Chord only).
Intuitively, with larger L_{stab} more cycles are collected before the actual measurement, and better values are estimated for the lookup availability. Actually, this
dependence is negligible since the average difference in lookup availability is about
0.1% of L_{rate}. Hence $L_{stab} = 0$ is used in other simulation scenarios. Moreover, it is
an indicator of good dynamic properties of CR-Chord since only a short stabilization
period is required to adopt cycles to the current network state.

Scenario 7 analyzes dynamic properties of CR-Chord under churn. Figure 11.9
shows that Chord is weakly resistant to the high churn rate $R = 0.2$. In Sect. 11.4.4
we explained two churn-related failure types. Churn-related losses along the route

Fig. 11.9 Lookup failure rate for different churn rates (Scenario 7)

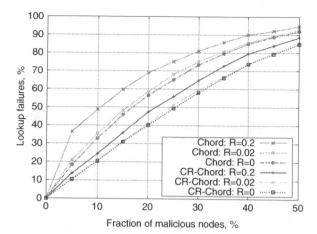

are much more prevalent than inability to find target document. For example, for $R = 0.2$ and $f = 10\%$ Chord has only 35 document failures but 2,077 along-the-route loss failures for the total of 10,000 requests.

CR-Chord helps to mitigate the negative effect of churn. The mean difference in number of lookup failures among $R = 0$ and $R = 0.2$ is lower for CR-Chord than for Chord. CR-Chord has less loss along the route. This is yet another confirmation that CR-Chord improves global routing. For moderate churn rates CR-Chord has lookup availability very close to that shown by simulations without churn, which proves its good dynamic properties.

11.5.3   Analysis of Cycles

Cyclic routing allows transitions when a node changes the cycle obtained previously. Figure 11.10 shows the cycle usage in lookups. Many successful lookups follow one or two cycles only; the number of lookups with more cycles is essentially less. It validates that CR is stable to changing a cycle, and previous cycle is likely to be preserved.

There are lookups without cycles (Figs. 11.11–11.13) when nodes cannot find appropriate cycles because of the small number of cycles at a node for large f. Additional fingers (Fig. 11.11) and multicast (Fig. 11.12) do not help much in constructing cycles. Additional fingers increase the number of cycles almost uniformly for all f while the multicast efficiency decreases for large f. Churn has negative effect since some cycles become incorrect (Fig. 11.13). A more sophisticated cycle construction is needed in this case.

Figure 11.14 is similar to Fig. 11.10, but also shows the share of lookup failures due to churn. The share of churn failures is mostly independent of f. Though as f grows, malicious failures start vastly prevailing over churn failures, which suggests that it is more important to mitigate malicious presence in the network than churn.

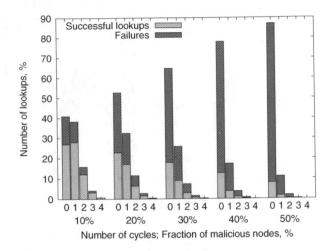

Fig. 11.10 Distribution of lookups among the number of cycles used (Scenario 1)

Fig. 11.11 Number of cycles per node for different numbers of additional fingers (Scenario 4)

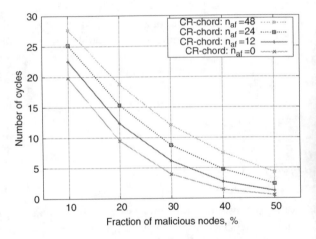

CR-Chord lookup hops are either along cycles or use regular Chord. Figure 11.15 shows the cycle hops share in successful lookups. For instance, for $f = 10\%$ about 16% of lookups use cycles for $21\% \ldots 30\%$ of hops. There are no successful lookups with $1\% \ldots 10\%$ cycle share of hops.

Many successful lookups do not use cycles. In particular, their share is 45%, 58% and 70% for $f = 10\%, 20\%, 30\%$ respectively (not shown in Fig. 11.15). Nevertheless, when there are enough cycles available the essential part of a successful lookup is due to cycles. Thus the problem is in cycle construction, not in CR.

In Scenario 3 we studied how varying number of nodes in the network affects average cycle length in number of hops (Fig. 11.16). For small f cycle length varies as expected—logarithmically in respect to N. This also implies that embedding

Fig. 11.12 Number of cycles per node for varying multicast degree (Scenario 5)

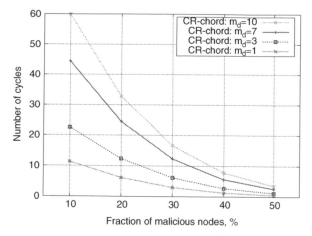

Fig. 11.13 Number of cycles per node for varying churn rate (Scenario 7)

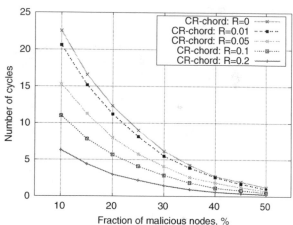

Fig. 11.14 Distribution of lookups among the number of cycles used, $R = 0.1$ (Scenario 7)

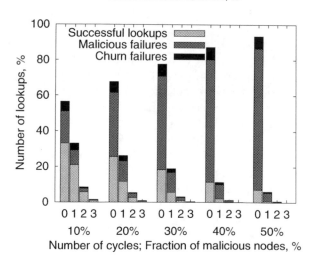

Fig. 11.15 Distribution of
successful lookups along the
cycle share in hops
(Scenario 1)

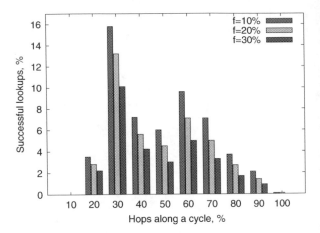

Fig. 11.16 Average cycle
length (Scenario 3)

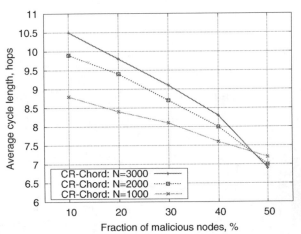

cycles into packets routed in the network is feasible even for high N, since cycle
length doesn't grow linearly in respect to number of nodes. Average cycle length
falls as we increase f since it gets increasingly harder to construct long cycles.

Besides the hop count, the lookup length can be measured with Chord distance.
The simulation (Scenario 1) identifies two extreme cases. Either a successful lookup
does not use cycles, or 91%...100% of its distance is along cycles (41%, 30% and
19% for $f = 10\%, 20\%, 30\%$ respectively).

Our experiments also assess the number of cycles per node for various scenarios.
In particular, Fig. 11.12 shows that the highest number of cycles is 60 for 10% of
malicious nodes and $m_d = 10$. Note that such a value is enough for the reasonable
improvement of lookup availability while it keeps the increase in storage and
processing time moderate at node (see Sect. 11.3).

11.6 Related Work, Comparison, and Discussion

There has been a substantial amount of work on various approaches to enhance the basic Chord routing mechanism. In this section, we discuss these approaches in comparison with CR-Chord and show that most of them can complement our design. Table 11.4 summarizes some relevant solutions proposed in the literature. Inspection of the whole lookup path (cycle) is one of the main CR-Chord features; it is present in neither of the listed approaches, and thus we skip it in the key differences column.

Leong et al. [19] parallelize Chord lookups with reactive routing state management (EpiChord). CR-Chord similarly exploits multi-path routing; cycles maintenance is reactive, amortizing the cost into regular lookups. EpiChord, however, uses iterative routing that suffers when connectivity is non-transitive [10] or other IP-level reachability restrictions happen [16]. In contrast to CR-Chord, the EpiChord next hop selection and similar multicast schemes do not analyze the rest of paths in route selection.

EpiChord exploits the approach of large routing tables when a node maintains a lot of neighbors and lookups consist of few hops. Tang et al. [31] compare DHTs with $O(1)$ to $O(N)$ routing tables. Extreme cases are [31] 1h-Calot (one hop routing) and 2h-Calot (two-hop routing) with $O(N)$ and $O(\sqrt{N})$ routing tables, correspondingly.

IP connectivity problems can prevent keeping some neighbors in a routing table (as it also happens in iterative lookups). Moreover, nodes of low capacity cannot store all or even $O(\sqrt{N})$ nodes. EpiChord [19] and Accordion [20] utilize a flexible approach when a node independently varies the state between $O(\log N)$ and $O(N)$. CR-Chord also allows varying the number of additional fingers. In contrast, even if a node u fails to insert a finger v (no direct IP contact, e.g., v does not trust u), u can still progress further and make more knowledge via cycles to cache indirect contacts.

In a dynamic environment a large routing table needs much traffic for keeping neighbors up-to-date. IP addresses change more frequently than node IDs, e.g., for mobile hosts. If nodes probe their neighbors periodically (proactive strategy), then high IP traffic is generated (many ping packets). Even if nodes use overlay multicast of regular lookups (reactive strategy) instead of IP probing, the IP change rate can be too high, and routing tables contain many neighbors with incorrect IP addresses.

CR-Chord does not focus on increasing the number of neighbors; a node makes more knowledge keeping non-neighboring nodes in cycles. No IP address is stored for such nodes, hence no IP probing is needed. Although availability checks use existing lookups with multi-path routing, CR is less sensitive to the dynamic level: even if one or more nodes in a cycle become unavailable, other nodes are still servable.

DHT nodes can optimize lookups and its availability by careful selection of neighbors. Chiola et al. [4] reduce the Chord number of hops and lookup availability by optimizing fingers based on Fibonacci distances (F-Chord). Kunzmann and Schollmeier [18] propose a scheme that extends a routing table with additional

Table 11.4 Comparison of CR-Chord with existing extensions to basic Chord

Chord-based design	Parameters being improved	Means of improvement	Key differences with CR-Chord
EpiChord [19]	Lookup latency, number of hops, availability under churn.	Increased finger table size, parallel lookups.	Maintenance of many neighbors. Iterative lookups.
F-Chord [4]	Number of hops, availability.	Modified finger selection procedure.	No independent variation of routing state. No multi-path routing.
Freebie fingers [18]	Number of hops.	Additional $O(\log N)$ fingers without causing additional traffic.	No independent variation of routing state. No multi-path routing.
Bidirectional latency-sensitive Chord [32], LLChord [15], topology matching [8]	Lookup latency. ([32] also reduces the number of hops via bidirectional routing similarly to [23].)	Proximity neighbor selection (PNS).	No independent variation of routing table sizes. No multi-path routing.
Fiat's et al. S-Chord [9]	Resilience to Byzantine join attack.	Increased finger table size, more messages sent during lookup and join procedures.	No multi-path routing.
Zig-zag routing [7, 27]	Resilience to sybil attack.	Trust profile for individual nodes, altered routing algorithm.	Iterative lookups. No independent variation of routing table sizes. No multi-path routing.

Cyclone [1]	Resilience to data-forwarding attack.	Increasing the number of disjoint paths. Multi-path routing.	No independent variation of routing table sizes.
Mesaros's et al. S-Chord [23]	Lookup performance in number of hops.	Bidirectional routing.	No independent variation of routing table sizes. No multi-path routing.
RChord [35]	Lookup performance in number of hops. Resilience to routing attacks.	Bidirectional routing.	No independent variation of routing table sizes. No multi-path routing.
P-Chord, PL-Chord [17]	Lookup availability depending on node lifetime, number of hops.	Bidirectional routing, supernodes.	No independent variation of routing table sizes. Doesn't consider malicious nodes.
H-Chord [5]	Number of hops.	Neighbor-of-neighbor routing.	No independent variation of routing table sizes. No resistance to malicious nodes and churn. No multi-path routing.
Chord enhancement by [2]	Number of hops. Lookup latency.	Neighbor-of-neighbor routing. Proximity route selection.	No independent variation of routing table sizes. No resistance to malicious nodes and churn. No multi-path routing.

fingers without causing additional traffic (existing maintenance requests are used). In randomized Chord [13], a node u takes its ith finger randomly in $[u + 2^{i-1}, u + 2^i)$, decreasing in result the fraction of failed lookups (by 29% in the Gnutella trace [12]). The proximity neighbor selection scheme (PNS) [13] allows reducing the lookup latency; concrete Chord extensions include bidirectional latency-sensitive Chord [32], LLChord [15],[2] and topology matching [8]. All these approaches also apply to CR-Chord since it does not fix a strategy of neighbor selection.

Experimental study of various neighbor selection strategies under churn can be found in [12]. Another analysis of the Chord performance under churn is given in [34]. Among other results authors derive a closed-form analytic expression for performance of lookup parallelism. Their approach can be used to select an efficient value of multicast degree for CR-Chord.

Fiat et al. [9] propose S-Chord,[3] aiming at robustness to Byzantine join attacks when a large number of malicious nodes can join the network. The number of malicious nodes is assumed to be no more than 1/4 of all peers, while in CR-Chord we considered the fraction up to 1/2. For robustness S-Chord relies on the concept of swarms that are sets of adjacent nodes; it is very similar to our idea of local routing. S-Chord requires each node to store $O(\log^2(N))$ fingers. The number of messages sent is also increased: during a lookup operation to $\Theta(\log^2(N))$ and during a node join to $\Theta(\log^3(N))$. Note that [9] lacks extensive experimentation, and we cannot compare its efficiency with CR-Chord numerically.

Seedorf and Muus [27] look at the analytical and simulation methodology to study the Chord lookup availability and the shield problem. They also experiment with zig-zag routing [7] for diminishing Sybil attacks. The algorithm alternates basic Chord routing and trust diversity routing (in the bootstrap graph that describes which node introduced which to the network), hence balancing distance progress and social properties. Our simulation model is close to [27] but we capture more factors, including additional fingers, multi-path routing, and churn. The absolute gain in availability for CR-Chord is 7–16% compared to at most by 6% for zig-zag routing ($10\% \le f \le 50\%$).

Artigas et al. [1] propose an elegant method—a binary-equivalence relationship makes each node available via n different predecessors. It fortifies Chord routing to deal with the shield problem and to defend against data-forwarding attacks. Bidirectional variants of Chord diminish the shield problem allowing to contact a destination both clockwise and anticlockwise, e.g. S-Chord [23],[4] RChord [35],[5] and Bidirectional Chord [32]. Theoretical analysis of Chord bidirectional schemes and their optimal values are studied in [11]. Ktari et al. [17] combine bidirectional routing with an altered finger selection procedure that allows to create supernodes

---

[2]In LLChord the abbreviation "LL" stands for "low latency."

[3]In Fiat's et al. S-Chord the abbreviation "S" stands for "secure routing."

[4]In Mesaros's S-Chord the abbreviation "S" stands for "routing symmetry."

[5]In RChord the abbreviation "R" stands for "with reverse edges."

with a large number of fingers following a power law. These techniques tend to reduce the path hop length even in presence of malicious nodes, since there are more paths for bypassing malicious nodes. In turn, the lookup availability is higher. CR-Chord can adapt these techniques since they improve local routing and provide more good paths for global routing.

Manku et al. [22] show that neighbor-of-neighbor (NoN) routing (one step lookahead to the rest of path) optimizes the path hop length in some networks. Evaluation of NoN-routing for Chord networks can be found in [2, 5]. In fact, this type of lookahead is very close to CR-Chord and may be considered as a special case of cyclic routing [16].

In Chap. 8 we introduced the general CR method and analytically studied its properties. This chapter develops the topic by (1) designing a particular protocol on top of Chord, (2) simulation modeling a Chord network with malicious nodes for the lookup availability analysis, and (3) identifying the key factors that influence the lookup availability.

The improvement that CR-Chord provides depends on the number of cycles stored by nodes. For large f and R, cycles are constructed insufficiently, since our simple construction method is based on Chord lookups. It aims merely at more efficient usage of Chord routing facilities. As a result, there are only few cycles per node and the difference between Chord and CR-Chord is smaller.

A core problem of Chord (as of other DHTs) is that lookups can fail although good paths do exist. In CR-Chord, good paths are stored for future use, but an efficient method for cycle construction is needed. We refer to [16] where an interesting algorithm is proposed that uses a kind of look-ahead. This is a topic of our further research.

We suggested dividing DHT routing to global and local parts. Weak local routing in Chord does not allow better routing than dictated by the lower bound in (11.2). Recall that CR is not limited to the Chord DHT but can be implemented on top of others. If the underlying DHT has better local routing properties then its basic routing is more efficiently utilized in CR.

CR-Chord optimizes global routing compared to Chord. Combining CR-Chord with a mechanism for secure local routing would improve the lookup availability. The better local routing properties a DHT has, the more efficiently its routing is utilized by CR. Secure local routing seems an easier problem compared to global routing. Node's neighborhood is limited and its exhaustive secure maintenance is possible.

11.7 Summary

We have compared the resilience of CR-Chord and Chord in the presence of malicious nodes. The lookup availability of CR-Chord is on the average 1.4 times higher compared to Chord, and up to two times higher for big number of malicious nodes. The absolute gain in availability of 7–16% for $10\% \leq f \leq 50\%$ is reasonable compared to existing enhancements for Chord.

Our analysis shows that the increase in packet and node storage space as well as processing time at nodes is moderate. It depends linearly on the number of nodes (IDs only) a given node would like to know. Our simulations suggest that CR-Chord can not only mitigate malicious activity, but also compensate routing churn losses caused by unconcerned nodes routing policy.

We also confirm the idea of combining additional fingers and cyclic routing [16]. Some nodes have resources to maintain more fingers than in basic Chord. When inserting a finger is not possible due to IP addressing restrictions, CR helps in keeping additional information about the network. That is, these two techniques efficiently complement each other.

References

1. Artigas, M.S., Lopez, P.G., Skarmeta, A.F.G.: A novel methodology for constructing secure multipath overlays. IEEE Internet Comput. **9**(6), 50–57 (2005). doi: http://dx.doi.org/10.1109/MIC.2005.117
2. Bin, D., Furong, W., Ma, J., Jian, L.: Enhanced chord-based routing protocol using neighbors' neighbors links. In: AINAW '08: Proceedings 22nd International Conference on Advanced Information Networking and Applications, pp. 463–466. IEEE Computer Society (2008). doi: http://dx.doi.org/10.1109/WAINA.2008.53
3. Castro, M., Drushel, P., Ganesh, A., Rowstron, A., Wallach, D.S.: Secure routing for structured peer-to-peer overlay networks. In: Proceedings 5th USENIX Symposium on Operating System Design and Implementation (OSDI 2002), pp. 299–314. ACM, Boston (2002)
4. Chiola, G., Cordasco, G., Gargano, L., Negro, A., Scarano, V.: Optimizing the finger tables in Chord-like DHTs. Concurr. Comput. Pract. Exper. **20**(6), 643–657 (2008). doi: http://dx.doi.org/10.1002/cpe.v20:6
5. Cordasco, G.: Degree-optimal deterministic routing for P2P systems. In: ISCC '05: Proceedings 10th IEEE Symposium on Computers and Communications, pp. 158–163. IEEE Computer Society (2005). doi: http://dx.doi.org/10.1109/ISCC.2005.45
6. Dabek, F., Kaashoek, M.F., Karger, D., Morris, R., Stoica, I.: Wide-area cooperative storage with CFS. In: Proceedings 18th ACM Symposium Operating systems principles (SOSP '01), pp. 202–215. ACM, Boston (2001). doi: http://doi.acm.org/10.1145/502034.502054
7. Danezis, G., Lesniewski-Laas, C., Kaashoek, M.F., Anderson, R.: Sybil-resistant DHT routing. In: Proceedings 10th European Symposium on Research in Computer Security, pp. 305–318. Springer, Berlin (2005)
8. Dang, N.B., Vu, S.T., Nguyen, H.S.: Building a low-latency, proximity-aware DHT-based P2P network. In: Proceedings International Conference on Knowledge and Systems Engineering, pp. 195–200. IEEE Computer Society (2009). doi: http://doi.ieeecomputersociety.org/10.1109/KSE.2009.49
9. Fiat, A., Saia, J., Young, M.: Making Chord robust to byzantine attacks. In: Brodal, G., Leonardi S. (eds.) Algorithms — ESA 2005. Lecture Notes in Computer Science, vol. 3669, pp. 803–814. Springer, Berlin (2005). URL http://dx.doi.org/10.1007/11561071_71
10. Freedman, M.J., Lakshminarayanan, K., Rhea, S., Stoica, I.: Non-transitive connectivity and DHTs. In: Proceedings of the 2nd USENIX Workshop on Real, Large Distributed Systems (WORLDS'05), San Francisco, pp. 55–60 (2005)
11. Ganesan, P., Manku, G.S.: Optimal routing in Chord. In: SODA '04: Proceedings 15th annual ACM-SIAM Symp. Discrete algorithms, pp. 176–185. Society for Industrial and Applied Mathematics (2004)

12. Godfrey, P.B., Shenker, S., Stoica, I.: Minimizing churn in distributed systems. SIGCOMM Comput. Commun. Rev. **36**(4), 147–158 (2006). doi: http://doi.acm.org/10.1145/1151659. 1159931
13. Gummadi, K., Gummadi, R., Gribble, S., Ratnasamy, S., Shenker, S., Stoica, I.: The impact of DHT routing geometry on resilience and proximity. In: Proceedings of ACM SIGCOMM'03, pp. 381–394. ACM, New York (2003). doi: http://doi.acm.org/10.1145/863955.863998
14. Hildrum, K., Kubiatowicz, J.: Asymptotically efficient approaches to fault-tolerance in peer-to-peer networks. In: Proceedings 17th International Symposium on Distributed Computing (DISC '03), pp. 321–336 (2003)
15. Jiang, Y., You, J.: A low latency Chord routing algorithm for DHT. In: Proceedings 1st International Symposium on Pervasive Computing and Applications, pp. 825–830 (2006). doi: http://dx.doi.org/10.1109/SPCA.2006.297539
16. Korzun, D., Nechaev, B., Gurtov, A.: Cyclic routing: Generalizing lookahead in peer-to-peer networks. In: AICCSA2009: Proceedings 7th IEEE/ACS International Conference Computer Systems and Applications, pp. 697–704. IEEE Computer Society (2009). doi: http://doi. ieeecomputersociety.org/10.1109/AICCSA.2009.5069403
17. Ktari, S., Hecker, A., Labiod, H.: Empowering Chord DHT overlays. In: HPSR'09: Proceedings 15th International Conference on High Performance Switching and Routing, pp. 88–93. IEEE Press (2009)
18. Kunzmann, G., Schollmeier, R.: Exploiting the overhead in a DHT to improve lookup latency. In: Kloos, C., Marin, A., Larrabeiti, D. (eds.) EUNICE 2005: Networks and Applications Towards a Ubiquitously Connected World, IFIP International Federation for Information Processing, vol. 196, pp. 247–254. Springer, Boston (2006). doi: http://dx.doi.org/10.1007/ 0-387-31170-X_18
19. Leong, B., Liskov, B., Demaine, E.: Epichord: parallelizing the Chord lookup algorithm with reactive routing state management. In: ICON 2004: Proceedings 12th International Conference on Networks, pp. 270–276 (2004)
20. Li, J., Stribling, J., Morris, R., Kaashoek, M.F.: Bandwidth-efficient management of DHT routing tables. In: Proceedings of the 2nd Symposium on Networked Systems Design and Implementation (NSDI '05), pp. 99–114 (2005)
21. Loguinov, D., Kumar, A., Rai, V., Ganesh, S.: Graph-theoretic analysis of structured peer-to-peer systems: routing distances and fault resilience. IEEE/ACM Trans. Networking **13**(5), 1107–1120 (2005)
22. Manku, G.S., Naor, M., Wieder, U.: Know thy neighbor's neighbor: the power of lookahead in randomized P2P networks. In: STOC '04: Proceedings 36th Annual ACM Symposium on Theory of computing, pp. 54–63. ACM, New York (2004). doi: http://doi.acm.org/10.1145/ 1007352.1007368
23. Mesaros, V.A., Carton, B., Roy, P.V.: S-Chord: Using symmetry to improve lookup efficiency in Chord. In: Proceedings International Conference Parallel and Distributed Processing Techniques and Applications (PDPTA'03) (2003)
24. Moskowitz, R., Nikander, P., Jokela, P., Henderson, T.: Experimental Host Identity Protocol (HIP). IETF RFC 5201, http://tools.ietf.org/html/rfc5201 (2008)
25. Rhea, S., Geels, D., Roscoe, T., Kubiatowicz, J.: Handling churn in a DHT. In: Proceedings of the USENIX Annual Technical Conference (2004)
26. Rowstron, A., Druschel, P.: Storage management and caching in past, a large-scale, persistent peer-to-peer storage utility. SIGOPS Oper. Syst. Rev. **35**(5), 188–201 (2001). doi: http://doi. acm.org/10.1145/502059.502053
27. Seedorf, J., Muus, C.: Availability for structured overlay networks: Considerations for simulation and a new bound on lookup success. In: Proceedings 12th Nordic Workshop on Secure IT Systems, pp. 23–34 (2007)
28. Sit, E., Morris, R.: Security considerations for peer-to-peer distributed hash tables. In: IPTPS '02: Proceedings 1st International Workshop on Peer-to-Peer Systems, Lecture Notes in Computer Science, vol. 2429, pp. 261–269. Springer, Berlin (2002)

29. Srivatsa, M., Liu, L.: Vulnerabilities and security threats in structured overlay networks: A quantitative analysis. In: ACSAC '04: Proceedings 20th Annual Computer Security Applications Conference, pp. 252–261. IEEE Computer Society (2004). doi: http://dx.doi.org/10.1109/CSAC.2004.50

30. Stoica, I., Morris, R., Liben-Nowell, D., Karger, D., Kaashoek, M.F., Dabek, F., Balakrishnan, H.: Chord: A scalable peer-to-peer lookup service for Internet applications. IEEE/ACM Trans. Networking **11**(1), 17–32 (2003)

31. Tang, C., Buco, M.J., Chang, R.N., Dwarkadas, S., Luan, L.Z., So, E., Ward, C.: Low traffic overlay networks with large routing tables. SIGMETRICS Perform. Eval. Rev. **33**(1), 14–25 (2005). doi: http://doi.acm.org/10.1145/1071690.1064216

32. Wang, J., Yang, S., Guo, L.: A bidirectional query Chord system based on latency-sensitivity. In: Proceedings 5th International Conference on Grid and Cooperative Computing, pp. 164–167 (2006)

33. Wang, P., Hopper, N., Osipkiv, I., Kim, Y.: Myrmic: Secure and robust DHT routing. DTC Research Report 2006/20, Digital Technology Center, University of Minnesota (2006)

34. Wu, D., Tian, Y., Ng, K.W.: Analytical study on improving DHT lookup performance under churn. In: P2P '06: Proceedings 6th IEEE International Conference on Peer-to-Peer Computing, pp. 249–258. IEEE Computer Society (2006). doi: http://dx.doi.org/10.1109/P2P.2006.4

35. Xuan, D., Chellappan, S., Krishnamoorthy, M.: RChord: An enhanced Chord system resilient to routing attacks. In: ICCNMC '03: Proceedings International Conference Computer Networks and Mobile Computing, pp. 253–260. IEEE Computer Society (2003)

36. Zhang, H., Goel, A., Govindan, R.: Incrementally improving lookup latency in distributed hash table systems. In: Proceedings 2003 ACM SIGMETRICS International Conference Measurement and modeling of computer systems, pp. 114–125. ACM, New York (2003). doi: http://doi.acm.org/10.1145/781027.781042

37. Zhao, B.Y., Huang, L., Stribling, J., Joseph, A.D., Kubiatowicz, J.D.: Exploiting routing redundancy via structured peer-to-peer overlays. In: ICNP '03: Proceedings 11th IEEE International Conference on Network Protocols, pp. 246–257 (2003)

Chapter 12
Indirection Infrastructures

Abstract The Host Identity Indirection Infrastructure (*Hi*3) is a general-purpose networking architecture, derived from the Internet Indirection Infrastructure (*i*3) and the Host Identity Protocol (HIP). *Hi*3 combines efficient and secure end-to-end data plane transmission of HIP with robustness and resilience of *i*3. The architecture is well-suited for mobile hosts given the support for simultaneous host mobility, rendezvous and multi-homing. Although an *Hi*3 prototype is implemented and tested on PlanetLab, scalability properties of *Hi*3 for a large number of hosts are unknown. In this chapter, we propose a simple model for bounds of size and latency of the *Hi*3 control plane for a large number of clients and in the presence of DoS attacks. The model can be used for a first approximation study of a large-scale Internet control plane before its deployment. We apply the model to quantify the performance of the *Hi*3 control plane. Our results show that the *Hi*3 control plane can support a large number of mobile hosts with acceptable latency.

12.1 Introduction

The original Internet Protocol stack was designed without explicit consideration for address agility or IP-layer security.[1] Recently, much effort has been applied to develop an architecturally sound solution to address these shortcomings (refer, e.g., to [2–5, 9]).

The Host Identity Indirection Infrastructure (*Hi*3) [7, 16] is a novel general-purpose networking architecture, derived from the Internet Indirection Infrastructure (*i*3) [11, 19] and the Host Identity Protocol (HIP) [8, 10, 13, 15]. *Hi*3 benefits from HIP by efficient end-to-end data transfer over IPsec, IPv4/v6 interoperability,

---

[1]This section includes text and figures from an article published in Computer Communications 29(17):3591–3601 by D. Korzun and A. Gurtov, On Scalability Properties of the Hi3 Control Plane, © Elsevier 2006.

D. Korzun and A. Gurtov, *Structured Peer-to-Peer Systems: Fundamentals of Hierarchical Organization, Routing, Scaling, and Security*, DOI 10.1007/978-1-4614-5483-0_12, © Springer Science+Business Media New York 2013

basic mobility, multi-homing, Denial-of-Service (DoS) protection, as well as IETF standardization and strong support of the industry. Integration with $i3$ allows for simultaneous mobility of end points, additional DoS protection, and the initial rendezvous service for $Hi3$ hosts. $Hi3$ can be gradually deployed, has been included into public HIPL and OpenHIP implementations, and tested on PlanetLab [6].

$Hi3$ consists of separate control and data planes. The control plane running over $i3$ is used only for HIP messages; the user data flows via the end-to-end data plane over IPsec. Main performance issues for a control plane infrastructure are scalability, DoS resilience, and robustness to node failures. Despite initial promising measurement results, the $Hi3$ performance at larger scales remains unknown. In this chapter, we develop an analytical model of the first two issues and apply it to quantify the $Hi3$ performance. The third property, robustness, is inherited by $Hi3$ from Chord and has been analyzed elsewhere [18].

Our model can be used as a first approximation of a large-scale Internet control plane before its deployment. In practical terms, the model provides us with a conservative estimate of the $Hi3$ capacity. Despite conservative estimating, our results show that the $Hi3$ control plane can support Internet applications for a large number of mobile hosts.

Our analysis shows that the $Hi3$ control plane scales well; a few hundred infrastructure servers are sufficient to support millions of clients. The internal latency of the infrastructure is satisfactory and does not exceed a few hundred milliseconds in most cases. Additionally, we describe $Hi3$ behavior under a load of Distributed DoS (DDoS) attacks generated by zombies, a set of compromised and exploited PCs.

The rest of the chapter is organized as follows. In Sect. 12.2, background material on HIP and $i3$ is provided. In Sect. 12.3, we describe the $Hi3$ control plane, introduce our modeling approach and derive cost estimates of basic requests to the control plane. Workload scenarios are given in Sect. 12.4. The scenarios illustrate the use of $Hi3$ capacity by mobile, stationary and zombie hosts. In Sect. 12.5, scalability properties of $Hi3$ are analyzed. Based on the workload scenarios, we derive trends for the size and latency of the control plane.

12.2   Background

We begin by giving the necessary background on $i3$, HIP, and the $Hi3$ architecture.

12.2.1   Host Identity Protocol (HIP)

In HIP [13, 14], IP addresses are used to locate and route the packets just as today. Only in the upper parts of the stack the addresses are replaced with the host identifiers. These host identifiers form a new Internet-wide name space, the host identity name space. The identifiers in this name space are public cryptographic keys. With HIP, each host is directly identified with a public key that corresponds to a private key, possessed by the host. Each host generates one or more public/private key pairs to provide identities for itself. A host can prove that it corresponds to the identity by signing data with its private key. All other parties use the host identifier, i.e., the public key, to identify and authenticate the host.

Typically, a host identifier is represented by a 128-bit long identifier, the Host Identity Tag (HIT). A HIT is constructed by applying a cryptographic hash function over the public key. The purpose and function of HITs is similar to $i3$ identifiers used in triggers (see Sect. 12.2.2), but HITs are constructed entirely cryptographically.

The actual HIP protocol [14] consists of a two-round-trip, end-to-end Diffie–Hellman key exchange protocol (called the *HIP base exchange*), a mobility exchange, and some additional messages. The purpose of the HIP base exchange is to create assurance that the peers indeed possess the private key corresponding to their host identifiers. Additionally, the exchange creates a pair of Encapsulated Security Payload (ESP) security associations (SAs), one in each direction.

12.2.2 Internet Indirection Infrastructure (i3)

To ease the deployment of services, Stoica et al. proposed an $i3$ overlay network that offers a rendezvous-based communication abstraction [19]. Instead of explicitly sending a packet to a destination, each packet is associated with a destination identifier; this identifier is then used by the infrastructure to deliver the packet. As an example, a host A may insert a trigger $[ID|IP_A]$ in the $i3$ infrastructure to receive all packets that have the destination identifier ID.

$i3$ provides natural support for mobility. When a host changes its address, the host needs only to update its trigger. When the host changes its address from IP_1 to IP_2, it updates its trigger from $[ID|IP_1]$ to $[ID|IP_2]$. As a result, all packets with the identifier ID are correctly forwarded to the new address.

The next step of $i3$ evolution is the Secure-$i3$ proposal [1]. Its main goal is to provide enhancements against DoS attacks and misuse of the infrastructure. The basic idea is hiding the IP addresses of the end-hosts from other users of the network. In Secure-$i3$, there are two types of triggers, public $[ID|ID']$ and private $[ID'|IP]$. Public triggers are used to announce the existence of a service with identifier ID. Private triggers are used for the actual communication between sender and receiver(s).

12.2.3 Host Identity Indirection Infrastructure (Hi3)

The $Hi3$ sketch [16] by Nikander et al. was based on the observation that a HIP rendezvous server and a single $i3$ server are functionally close to each other.

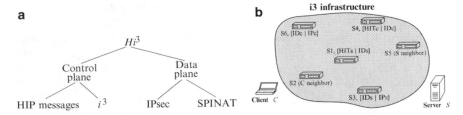

Fig. 12.1 The $Hi3$ architecture. (**a**) Separating data and control in $Hi3$. (**b**) Schematic distribution of HIT-based triggers in $i3$ for $C \leftrightarrow S$ communication

Therefore, the basic idea in $Hi3$ is to allow direct IP end-to-end traffic (the $Hi3$ data plane) while using an indirection infrastructure to route the HIP control packets (the $Hi3$ control plane). The concept of this separation is shown in Fig. 12.1a. $i3$ has an important capability to map HITs from a flat namespace to IP addresses using the underlying Chord DHT, a key property missing from a stand-alone rendezvous server.

The control plane is used to relay HIP messages during the association establishment and when the direct end-to-end connectivity is lost, e.g., after a simultaneous movement of both hosts. The main benefit of using the control plane during association establishment is protection against DoS attacks. This way, the IP addresses of communicating hosts are not revealed until mutual authentication is completed. The data plane provides further protection against DoS if IP addresses become revealed to a group of DDoS attackers after one of them establishes a HIP connection over $Hi3$.

For each host A the pair of triggers $[HIT_A|ID_A]$ and $[ID_A|IP_A]$ can be stored in $i3$, where HIT_A acts as a public $i3$ identifier of the host A and ID_A is a private $i3$ identifier constructed by $i3$. Let C and S be two communicating hosts. Figure 12.1b shows how the corresponding HIT-based public and private triggers are distributed in $i3$ for $C \leftrightarrow S$ communication.

In the data plane, end-to-end traffic is encrypted with ESP, but other encapsulation methods are possible too [7]. The IPsec-aware NAT (SPINAT) provides privacy and protection against DoS of the data plane [20] by dynamic mapping between the actual IP addresses and the NAT addresses using the IPsec Secure Parameter Index (SPI). For the rest of this chapter, we concentrate on the $Hi3$ control plane only.

12.3   Control Plane

In this section, we describe available request types to the $Hi3$ control plane and estimate their computational cost and latency.

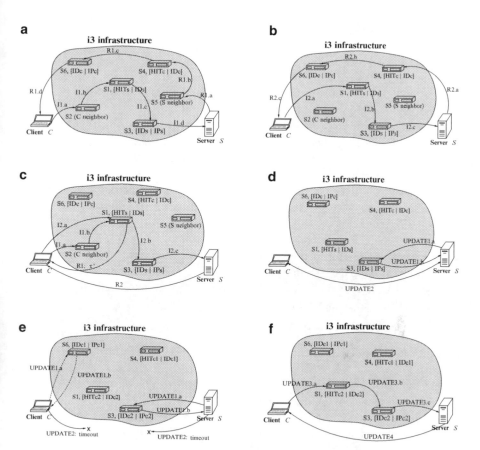

Fig. 12.2 Request types in *Hi*3. (**a**) Pure association setup: I1 and R1 packets. (**b**) Pure association setup: I2 and R2 packets. (**c**) Optimized association setup. (**d**) Location update. (**e**) Failure of the location update. (**f**) Double-jump resolving

12.3.1 Request Types

Consider two hosts that use *Hi*3 for communication. Let one of them be a HIP initiator *C* (e.g., a mobile client) and the other one be a HIP responder *S* (e.g., a stationary Internet server). Figure 12.2 shows main scenarios when *C* ↔ *S* communication involves the *Hi*3 control plane. For details refer to a complete *Hi*3 description [7].

12.3.1.1   Association Setup

The client C starts an $Hi3$ *association setup* with the server S for initiating the communication. This requires $i3$ to contain a server's HIT as a valid public trigger $[HIT_S|ID_S]$ (at node $S1$ in Figs. 12.1b and 12.2a) and a valid private trigger $[ID_S|IP_S]$ (at node $S3$).

The association setup is implemented via the HIP base exchange [14]. The client C uses an IP address of either the $i3$ node that keeps the S's public trigger (node $S1$), or a random $i3$ node ($S2$). The latter case is typical, and the nearest $i3$ server would be typically contacted.

There are two variants of the association setup in $Hi3$, pure and optimized. In pure setup, the client C inserts, perhaps temporarily, its HIT into $i3$, the public trigger $[HIT_C|ID_C]$ (at node $S4$) and the private trigger $[ID_C|IP_C]$ (at node $S6$). These triggers are needed for replying to an I1 packet with an R1 packet during the HIP base exchange. Figure 12.2a, b show the packet flow.

In optimized setup the insertion of a client's HIT into $i3$ is not necessary; replying to I1 with R1 is delegated to $i3$ and the R2 reply packet is sent directly to the client. The packet flow is presented in Fig. 12.2c.

12.3.1.2   Mobility Update

Let us assume that C and S can change their locations and, consequently, their IP addresses during the communication.

Typically, only one host changes its IP address and performs a *location update* at a time. If the change is by the server S, then S updates its private trigger in $i3$ (Fig. 12.2d). The location update also causes the *HIP update exchange* [17] between C and S running over the data plane. One update to $i3$ is sufficient independently of the number of hosts communicating with S. For C having a permanent trigger pair in $i3$ is optional. Thus, if C changes its location, the HIP update exchange runs directly between hosts (only UPDATE1 messages use $i3$ as shown in Fig. 12.2d).

It may happen that both hosts change their locations at the same time, an event known as the *double-jump problem*. We assume that simultaneous mobility of C and S is rare compared to the usual location update. This scenario proceeds as follows. The hosts update their triggers in $i3$ (for C it is optional) as shown in Fig. 12.2e. In parallel, the hosts start a HIP location update on the $Hi3$ data plane. The exchange fails since each host uses the out-of-date IP address to contact the peer. This failure is discovered by a timeout.

At this point the hosts need to use the control plane to recover from the double-jump (Fig. 12.2f). The double jump can be discovered by both hosts, but the client C is responsible for starting the recovery. C sends the first packet of the HIP update exchange (addressed to HIT_S) to S via $i3$. After receiving this packet, S continues the update talking directly to C.

12.3.1.3 Auxiliary Requests

For *Hi*3 association setup and mobility update, the *Hi*3 control plane has to support three auxiliary requests, namely *HIT insertion*, *HIT refreshment*, and *HIT removal*.

Let HIT_A be a HIT of a host A. HIT_A is inserted as a trigger to *i*3 when A is an initiator of a pure association setup. If A is a server, HIT_A needs to stay in *i*3 permanently. In the current *i*3 implementation[2] a trigger needs refreshing every 30 s. If the *i*3 node crashes, the trigger is lost until the host refreshes the trigger (re-insertion). Therefore, HIT removal can be done by sending a message to *i*3 or automatically after the trigger expires.

12.3.2 Transmission and Processing Costs

In this section, we make a preliminary analysis of transmission and processing costs in *Hi*3 according to the ideas given in previous work [7].

12.3.2.1 Upper Bounds for Costs

Let p be a cost parameter, for instance the round-trip time, latency, or the number of hops. The classical worst-case analysis aims in finding an upper bound $\overline{P}$ such that $p \leq \overline{P}$ for all possible p. A more robust metric is a high-probability bound P such that $p \leq P$ with high probability (w.h.p.).[3] Figure 12.3a shows the parameter domain division based on this bound. Most p values concentrate w.h.p. in the interval $[0, P]$; other values appear rarely and form the worst-case domain of the cost.

Given its clear limitations, this approach stills allows to discover some basic trends for conservative cases. It is often applied for cost analysis of concrete DHT systems [12, 18]. The accuracy of trends can be improved by using typical values such as the average or median for some parameters instead of the high-probability bound.

Finding a theoretically approved high-probability bound is quite challenging and meets a lot of difficulties. Formally, one needs to define exactly what a high probability is and give assumptions on the probability distribution. Alternatively, one can apply statistical methods and calculate, based on measurements, a confidence interval for the parameter, e.g., that the bound estimated is valid for 95%. In many cases, however, various practical reasons are enough to provide a reasonable empirical value for the bound. We shall follow this simpler but more practical approach. It is useful as a first approximation of the parameters.

---

[2] Available at http://i3.cs.berkeley.edu.

[3] Throughout this chapter we use the term w.h.p. to mean with probability at least $1 - c/N$ for some constant $0 < c \ll N$, where N is the size of network.

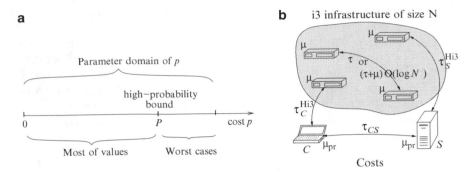

Fig. 12.3 Assumptions for the analysis. (**a**) High-probability bound. (**b**) Costs

12.3.2.2   Costs Inside and Outside of $i3$

Let N be the size of $i3$ (the number of nodes). A request to the $Hi3$ control plane affects two network parts, inside and outside of $i3$ (see Fig. 12.3b).

The outside part is characterized by the one-way trip time τ_A^{Hi3} from a HIP host A to an $i3$ node (or backwards). The parameter τ_A^{Hi3} depends on the host's location, but not on the $i3$ size N. Some requests to the control plane can continue as direct host-to-host communication, e.g., in the optimized association setup an R2 packet is sent by S to C directly (see Fig. 12.2c). The cost of direct IP-based transmission between C and S is estimated as the one-way trip time $\tau_{CS} = \tau_{SC}$.

Communication inside $i3$, i.e. between $i3$ nodes, can be divided into two types. (1) If two nodes use direct IP-based communication, then the one-way trip time is τ. (2) If a node makes a lookup to route a packet to the target node, then the packet visits w.h.p. $O(\log N)$ nodes [18] as a sequence of $i3$ hops towards the destination. Each hop takes time $\tau + \mu$, where μ is the forwarding cost of a packet at an $i3$ node: matching against the $i3$ identifier in the forwarding table, updating the packet header and internal state (if needed), and forwarding to the next hop. Therefore, a lookup takes time $t = (\tau + \mu)O(\log N)$ or, in other words, time $t \le \alpha(\tau + \mu)\log N$ for some positive constant α and for all large N. Considering the w.h.p. case, we assume that $t = \alpha(\tau + \mu)\log N$. The reasonable high-probability bound is $\mu \sim 0.5 \dots 1.0$ ms.

By experimental results of Stoica et al. [18] the average lookup time can be estimated as $\frac{\tau + \mu}{2} \log N$. Taking $\alpha > \frac{1}{2}$ makes the bound more conservative, and $\alpha = 1$ is a high-probability bound.

$Hi3$ inherits cryptographic operations from HIP that load end-hosts and/or $i3$ nodes. The operations include the following:

1. Solving a cryptographic puzzle in the HIP base exchange when C receives an R1 packet and generates an I2 packet with a correct puzzle solution.
2. Checking the puzzle solution and authenticating the client. This is a consequence of the previous step.

A HIP responder (server S) can delegate a part of these operations to $i3$. In the optimized association setup, the $i3$ node $S3$ having the public trigger of the server S (Fig. 12.1b) caches precomputed R1 packets of S. It is able to send R1 packets to initiators in response to I1 packets. Let μ_{pr} be the duration of this operation (Fig. 12.3b). A high-probability bound is $\mu_{pr} \sim 100\dots200$ ms.

All other operations are less time consuming. Delegation of responding with an R1 to $i3$ is comparable with a cost μ of forwarding a packet. According to [1], the cost of a trigger insertion in $i3$ (even with checking trigger constraints) is $20\dots40\,\mu s$ and can also be bounded w.h.p. by μ.

12.3.2.3 Cost of a Request

Requests to the control plane can be classified in two groups: $O(1)$- and $O(\log N)$-*requests* depending on whether Chord routing is required.

$O(1)$-Requests

The requests involve a constant-bounded number of $i3$ nodes regardless of the $i3$ size N. The requests for HITs refreshment, update, and removal can be implemented with direct IP-based communication to the corresponding $i3$ node. However, when a HIT refreshment becomes the HIT re-insertion or when A loses direct connectivity with its peer, $i3$ lookups are needed and the request becomes the $O(\log N)$-type.

$O(1)$-requests load two $i3$ nodes having a public and a private trigger. Exactly one node is needed for an IP address update in $i3$ and in the rare case when the public trigger and the private one are stored at the same node.

$O(\log N)$-Requests

$O(\log N)$-requests use $i3$ lookups and, therefore, affect $O(\log N)$ nodes. The following requests belong to this group: pure and optimized association setup, recovering the double-jump, and HIT (re-)insertion.

Unlike the $O(1)$-requests, the $O(\log N)$-ones are "deep," involving many nodes for a large N. However, $\frac{\log N}{N} \to 0$ when $N \to \infty$ and the relative cost is small in a large-size infrastructure.

12.3.3 Latency Estimates

Each request consists of one or more packets (e.g., the association setup requires four HIP base exchange packets I1, R1, I2, and R2). Denote this number k.

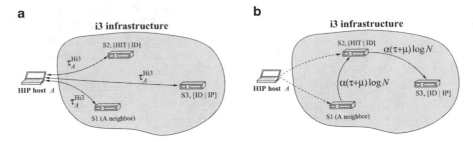

Fig. 12.4 Transmission costs of a packet in $Hi3$. (**a**) Direct communications with $i3$ nodes. (**b**) Lookups in $i3$

Table 12.1 Latency estimates of requests to the $Hi3$ control plane

Request type	k	$T^{\text{Hi3}} = k\tau^{\text{Hi3}}$	$T^{\text{out}} = k\tau^{\text{out}}$
Pure association setup	4	$6\alpha(\tau+\mu)\log N$	$4\tau_C^{\text{Hi3}} + 2\mu_{\text{pr}} + 4\tau_S^{\text{Hi3}}$
Optimized association setup	4	$2\alpha(\tau+\mu)\log N$	$3\tau_C^{\text{Hi3}} + 2\mu_{\text{pr}} + \tau_S^{\text{Hi3}} + \tau_{SC}$
Location update, $A \in \{C,S\}$	2	$\tau+\mu$	$2\tau_A^{\text{Hi3}}$
Double-jump	2	$\alpha(\tau+\mu)\log N$	$\tau_C^{\text{Hi3}} + \tau_S^{\text{Hi3}} + \tau_{SC}$
HIT insertion, $A \in \{C,S\}$	2	$2\alpha(\tau+\mu)\log N$	$2\tau_A^{\text{Hi3}}$
HIT refreshment/removal, $A \in \{C,S\}$	4	$2(\tau+\mu)$	$4\tau_A^{\text{Hi3}}$

($C \leftrightarrow S$ communication)

A packet *enters* $i3$ when it is received by a first-hop $i3$ node and it *leaves* $i3$ when it is forwarded by a last-hop node to the destination end-host. Define τ^{Hi3} and τ^{out} as the time for a packet to be *inside* and *outside* $i3$, respectively. The *request latency* is $L = k(\tau^{\text{Hi3}} + \tau^{\text{out}}) = T^{\text{Hi3}} + T^{\text{out}}$, where τ^{Hi3} and τ^{out} are averaged over all k packets of the request, $T^{\text{Hi3}} = k\tau^{\text{Hi3}}$ and $T^{\text{out}} = k\tau^{\text{out}}$ are *internal* and *external* request latencies.

There are three types of $i3$ nodes that may communicate directly with a HIP host A, see Fig. 12.4a. The node $S1$ is an arbitrary $i3$ node that A contacts. The node $S2$ keeps the public trigger [HIT | ID] inserted by A, or a trigger of an A's peer. The node $S3$ stores the A's private trigger. According to Sect. 12.3.2, the one-way trip time to these nodes is τ_A^{Hi3}.

A request to the $Hi3$ control plane uses $i3$ lookups in the following cases (Fig. 12.4b). (1) An arbitrary $i3$ node ($S1$ in our case) is requested by A to forward a message to the public trigger location. (2) The node $S2$, which stores the public trigger, looks up the private trigger. The cost of such a lookup is $\alpha(\tau+\mu)\log N$. The latency estimates are summarized in Table 12.1 and explained below.

Pure Association Setup

There are $k = 4$ packets. Each of them travels to $i3$ nodes and back. HIP crypto-graphic processing of the cost μ_{pr} is needed in both end-hosts: R1/I2 (initiator) and I2/R2 (responder). Therefore, $T^{out} = 4\tau_C^{Hi3} + 2\mu_{pr} + 4\tau_S^{Hi3}$. The two first packets (I1 and R1) use two $i3$ lookups each. For two last packets (I2 and R2) one lookup is sufficient. In total, there are six lookups and $T^{Hi3} = 6\alpha(\tau + \mu)\log N$.

Optimized Association Setup

The I1/R1 processing is delegated to the $i3$ node where $[HIT_S | IP_S]$ is stored. One lookup for the I1 packet is used to find this node. Both I1 and R1 packets are outside $i3$ for time τ_C^{Hi3}. The I2 packet leaves $i3$ after one lookup to route between the nodes with a public and a private trigger, the external time is $\tau_C^{Hi3} + \tau_S^{Hi3}$. The R2 packet does not use $i3$ and travels directly from S to C in time τ_{SC}. The cost of cryptographic processing is the same as for the pure association setup. There are two lookups, $T^{Hi3} = 2\alpha(\tau + \mu)\log N$ and $T^{out} = 3\tau_C^{Hi3} + 2\mu_{pr} + \tau_S^{Hi3} + \tau_{SC}$.

Location Update (IP Update)

When a host A, either C or S, changes its location, the private trigger update is performed $[ID_A | IP_A] \rightarrow [ID_A | IP_A']$, where IP_A' is a new A's address. Two packets are involved ($k = 2$), an update request by A and the response. The request packet travels directly from A to the $i3$ node storing the private trigger, and the node replies back. The external latency is $T^{out} = 2\tau_A^{Hi3}$. The internal latency $\tau + \mu$ includes the update and forward operations.

Double-Jump

When C has discovered the double-jump by a timeout it sends the first packet of the HIP update exchange to $i3$. As for the location update, $k = 2$. We assume that C remembers the IP address of an $i3$ node storing HIT_S. The first packet crosses $i3$ with one lookup with additional latency outside of $i3$ of $\tau_C^{Hi3} + \tau_S^{Hi3}$. Then, S sends the response packet that travels directly to C in time τ_{SC}.

HIT (Re-)insertion

Host A sends a request packet to an $i3$ node. Two $i3$ lookups are performed: routing to a node to insert the public trigger and routing to a node to insert a private trigger. Hence, $T^{Hi3} = 2\alpha(\tau + \mu)\log N$ and $T^{out} = 2\tau_A^{Hi3}$.

HIT Refreshment and Removal

Two requests are used by a host to refresh or remove the triggers. In the worst case, both requests run sequentially and the latency is doubled. For each request the internal $i3$ latency is $\tau + \mu$ and the external latency is $2\tau_A^{\text{Hi3}}$.

12.4   Workload Model

The preceding section described existing request types to the $Hi3$ control plane. Scalability analysis needs to consider scenarios with many end hosts involved so that $Hi3$ is highly loaded. In this section, we present several such scenarios and derive estimates for the control plane workload.

12.4.1   General Workload Pattern

A request of a given type loads several $i3$ nodes; denote this number r. The request is either $O(1)$- or $O(\log N)$-type and $r = c$ or $r = c \log N$, where c is a constant relative to N different for each request type.

Consider a set of H hosts sending requests of a given type to the control plane. Let λ be the corresponding rate (requests per time unit) that is uniform for all hosts. Then, λH is the total number of requests sent by the hosts per time unit.

Let us define the workload W of an $i3$ node as

$$W = \frac{\lambda H r}{N}, \tag{12.1}$$

Several examples for (12.1) are presented in Table 12.2. A parameter $0 \le P_{\text{us}} < 1$ gives the probability that a location update event is a double-jump.

Total workload W_{tot} is defined as the sum of each request type workload. If the hosts use an optimized association setup (W_{so}), location update (W_{u}), and HIT insertion (W_{i}), then $W_{\text{tot}} = W_{\text{so}} + W_{\text{u}} + W_{\text{i}}$. Some important cases are presented in Sect. 12.4.2.

A node involved in serving the request either forwards a HIP packet (primary requests) or manages a HIT-based trigger (auxiliary requests). According to the assumptions in Sect. 12.3.2, the processing cost is bounded w.h.p. by μ. Hence, r can also be interpreted as the number of μ-long operations per a request.

More detailed analysis could differentiate operations in an $i3$ node using different values for the processing costs. In general $\mu = \mu(N)$ since the local state in a node grows with the size N of a DHT-based infrastructure. Measuring the workload in units of μ is CPU-oriented. It is possible to take into account the link bandwidth; it is a subject of further work.

Table 12.2 Workload estimates for requests to the $Hi3$ control plane

Request type	Rate, λ	#($i3$ nodes), r	Workload, W
Pure association setup	λ_s	$6\alpha \log N$	$W_s = \dfrac{6\alpha\lambda_s H \log N}{N}$
Optimized association setup	λ_{so}	$2\alpha \log N$	$W_{so} = \dfrac{2\alpha\lambda_{so} H \log N}{N}$
Location update	λ_u	1	$W_u = \dfrac{\lambda_u H}{N}$
Double-jump	$\lambda_u P_{us}$	$\alpha \log N$	$W_{us} = \dfrac{2\alpha\lambda_u P_{us} H \log N}{N}$
HIT insertion	λ_i	$2\alpha \log N$	$W_i = \dfrac{2\alpha\lambda_i H \log N}{N}$
HIT refreshment/removal	λ_r	2	$W_r = \dfrac{\lambda_r H}{N}$

The workload is generated by H hosts at the rate of λ per host

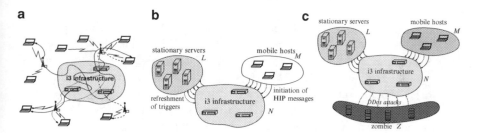

Fig. 12.5 Workload scenarios. (**a**) Mobile peers. (**b**) Mobile hosts and stationary servers. (**c**) Presence of zombie

12.4.2 Workload Scenarios

Let us consider several simple scenarios of $Hi3$ use based on the workload model.

12.4.2.1 Mobile Peers

Let M hosts (mobile peers) use $Hi3$ for communication, see Fig. 12.5a. All hosts generate requests for association setups (pure and optimized) and location updates. This leads to the following estimate of total workload:

$$W_{tot} = W_s + W_{so} + W_u + W_{us} = \alpha\frac{\lambda_{mob}M}{N}\log N + \frac{\lambda_u M}{N}, \qquad (12.2)$$

where $\lambda_{mob} = 6\lambda_s + 2\lambda_{so} + \lambda_u P_{us}$.

For simplicity we do not consider requests of HIT insertion and refreshment. These requests are rare compared to the other requests. Instead, we can use larger values for λ_{mob} and λ_u to avoid underestimating the workload.

12.4.2.2   Mobile Hosts and Stationary Internet Servers

In this scenario, M mobile hosts and L stationary Internet servers (Fig. 12.5b) are considered separately. A mobile host communicates with servers and with other mobile hosts as in the previous scenario. The servers do not initiate communication and do not change their location, but perform HIT insertion and refreshment. Together, all $M + L$ hosts generate the following workload:

$$W_{\text{tot}} = \alpha \frac{\lambda_{\text{mob}}M + 2\lambda_{\text{i}}L}{N} \log N + \frac{\lambda_{\text{u}}M + 2\lambda_{\text{r}}L}{N}. \tag{12.3}$$

The (12.2) is a particular case of (12.3) when $L = 0$.

12.4.2.3   DDoS Attacks from Zombies

We extend the previous scenario to model DDoS attacks to the control plane. A set of Z zombie hosts generates workload to $Hi3$ in a distributed denial-of-service (DDoS) attack as shown in Fig. 12.5c. The goal of the zombie set is to overload $i3$.

Obviously, a zombie prefers $O(\log N)$-requests since these requests add more load to $i3$. Thus, a zombie can (1) initiate the association setup with many other hosts via $Hi3$ and/or (2) insert many HITs to $i3$. We skip the double-jump request as the same load can be achieved easier by (1) and/or (2).

An initiation of the association setup can be implemented by a zombie as follows (1) send an I1 packet to $i3$, (2) receive an R1 packet from $i3$, (3) immediately reply with an I2 packet containing a wrong puzzle solution.

Overall, this generates five (for a pure setup) or two (for an optimized setup) $i3$ lookups. Let λ_{zs}, λ_{zso} and λ_{zi} be request rates for a pure association setup, optimized association setup, and HIT insertion by a zombie, respectively. Together all zombies generate the following injurious workload

$$W_{\text{bad}} = W_{\text{zs}} + W_{\text{zso}} + W_{\text{zi}} = \alpha \frac{\lambda_{\text{z}}Z}{N} \log N, \tag{12.4}$$

where $\lambda_{\text{z}} = 5\lambda_{\text{zs}} + 2\lambda_{\text{zso}} + 2\lambda_{\text{zi}}$.

12.5   Scalability Analysis

In this section, we analyze the scalability of the control plane for workload scenarios described in the previous section. The general idea is to estimate the utilization of the control plane depending on the number of $i3$ nodes N. We obtain an estimate of N that provides a certain utilization level for given workload and internal request processing latency.

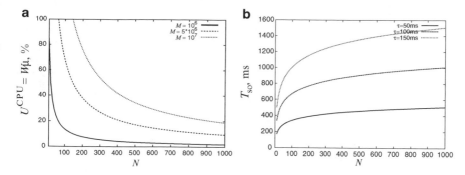

Fig. 12.6 The utilization/latency trade-off for the mobile peers scenario. $\alpha = 1/2$, $\lambda_s = \lambda_{so} = 30\,\text{min}^{-1}$, $\lambda_u = 5\,\text{min}^{-1}$, $P_{us} = 10^{-2}$. (**a**) CPU utilization U^{CPU} for several values of M. (**b**) Internal latency of the optimized association setup

12.5.1 The Utilization/Latency Trade-Off

The utilization of an $i3$ node can be estimated as $U = U^{\text{CPU}} = W\mu > 0$, i.e., what fraction of time a node spends for $Hi3$-related processing. If $U > 1$ then the infrastructure is overloaded by requests. Similarly, the utilization of $i3$ bandwidth is $U' = U^{\text{COM}} = W\tau > 0$, where τ is the one-trip time between two arbitrary $i3$ nodes (see Sect. 12.3.2). Each node involved in a request sends a packet after processing for μ milliseconds. The overload happens when $U' > B$ for a some throughput threshold B.

Intuitively, the greater N is, the less the utilization is. This fact is supported by the workload model, see (12.1). On the other hand, increasing N increases the internal latency T^{Hi3} as shown in Table 12.1. The trade-off between the utilization and latency of the infrastructure is an interesting fact that we consider next. We assume $\mu = 1\,\text{ms}$ and $\alpha = 1/2$ based on the measurements in [6, 18].

Figure 12.6 presents the case when workload is generated in accordance to the mobile peers scenario, (12.2). The plot in Fig. 12.6a shows a rapidly decreasing load with the growth of N (order of $\frac{\log N}{N}$). On the other hand, the internal latency increases slowly with the growth of N (order of $\log N$). Figure 12.6b shows the internal latency for the optimized association setup; the latency of a pure association setup, double jump and HIT insertion is proportional with coefficients 3, 1/2 and 1, respectively. The latency of the location update does not depend on N and equals $\tau + \mu \approx \tau$, i.e., $100\ldots200\,\text{ms}$.

Figure 12.7 shows utilization for M mobile hosts and L stationary Internet servers. These two sets generate good workload W_{good} according to (12.3), as shown in Fig. 12.7a. Injurious utilization is plotted in Fig. 12.7b for different values of $z = \lambda_z Z$.

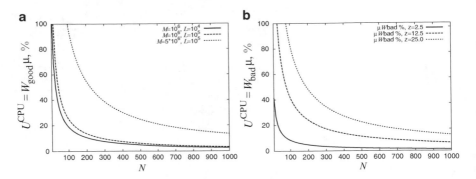

Fig. 12.7 The utilization/latency trade-off in case of M mobile hosts, L servers, and Z zombie hosts. $\alpha = 1/2$, $\lambda_i = 1\,h^{-1}$, $\lambda_r = 30\,s^{-1}$, $\lambda_{mob} = 6\lambda_s + 2\lambda_{so} + \lambda_u P_{us}$, $\lambda_s = \lambda_{so} = 30\,min^{-1}$, $\lambda_u = 5\,min^{-1}$, $P_{us} = 10^{-2}$. (**a**) Utilization by good workload. (**b**) Utilization by injurious workload

12.5.2 Scalability Problems

Rapidly decreasing workload and slowly increasing internal latency are attractive properties of the $Hi3$ control plane inherited from Chord [18]. It enables $Hi3$ to scale well, as shown in detail below.

Let us introduce a simple solution to the utilization/latency trade-off. Despite its simplicity and coarseness, it is useful for evaluating several important scalability problems. We show the estimates only for the CPU utilization; the throughput utilization is a subject of future work.

12.5.2.1 Estimating the Needed $i3$ Size

The problem is to find an interval for N values that keeps utilization reasonable for the workload presented in Sect. 12.4.2. For these values, the internal latency can be estimated using equations in Table 12.1.

Consider the mobile peers scenario with $\tau = 100$ ms. Require the latency of the optimized association setup to be at most 1 s.[4] w.h.p. As shown in Fig. 12.6b, the $i3$ size can reach 1,000 nodes without exceeding the latency bound. As shown in Fig. 12.6a, several hundred $i3$ nodes can serve several million mobile hosts within reasonable utilization bounds.

Consider the workload scenario shown in Fig. 12.7a with L servers added. Increasing L from 10^4 to 10^5 increases utilization less compared to increasing M from 10^6 to $5 \cdot 10^6$. The result confirms that $Hi3$ is more resilient to the number of stationary servers.

---

[4]In this case, a pure association setup and a double-jump would take 3 and 0.5 s, respectively.

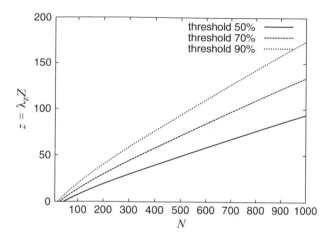

Fig. 12.8 Critical values of the zombie rate×size product acording to (12.5), where W_{good} is defined by (12.3), $M = 10^6$, $L = 10^5$, $\alpha = 1/2$, $\lambda_i = 1\,h^{-1}$, $\lambda_r = 30\,s^{-1}$, $\lambda_{mob} = 6\lambda_s + 2\lambda_{so} + \lambda_u P_{us}$, $\lambda_s = \lambda_{so} = 30\,min^{-1}$, $\lambda_u = 5\,min^{-1}$, $P_{us} = 10^{-2}$

12.5.2.2 Resistance to DDoS Attacks

Below we present analysis for a zombie DDoS attack to the $Hi3$ control plane. Let a threshold $\overline{W}$ for total workload of an $i3$ node be fixed, e.g., $\overline{W}\mu = 50\%, 70\%$, or 90%. We assume that $i3$ can handle the given workload if $W_{good} + W_{bad} \leq \overline{W}$, where W_{good} is good workload generated by $L + M$ legitimate hosts and W_{bad} is injurious workload from zombie DDoS attacks. According to (12.4) the critical value for the rate×size of the zombie set is

$$z = \lambda_z Z = \left(\overline{W} - W_{good}\right) \frac{N}{\alpha \log N} \tag{12.5}$$

The dependency is plotted in Fig. 12.8 for several values of the threshold.

Let us estimate the size of the zombie set based on the critical value z. Assume that links between $i3$ nodes have much higher bandwidth compared with links between zombies and $i3$. In the worst case, a HIP packet has a size of several hundred bytes or approximately $0.5\,kB = 4\,Kbits$. If every zombie host has 1 Mbits/s access to $i3$ then $\lambda_r \leq 0.25$. For $Z = 100$ we have $\lambda_r Z \leq 25$. According to Fig. 12.8, about $120...230$ nodes can resist the attack with utilization at most $90...50\%$. For $1,000$ nodes, the maximal number of zombies is in the range of $400...700$.

For large N, $z = $ rate $\times$ size grows a little slower than linearly. More precisely, $z = z(N) = a\frac{N}{\log N} + b\log N + c$, where a, b, and c are constants [see (12.2) and (12.3)]. Such behavior is due to the Chord protocol where most requests require $O(\log N)$ hops. However, an $i3$ node can cache IP addresses of other nodes reducing the lookup time to $O(1)$. Although this enables $i3$ to be a one-hop network for repeating requests, it also complicates its operation in the presence of churn, when nodes enter and leave the Chord ring. Evaluation of such effects is left for future study.

12.6 Summary

In this chapter, we presented analytical analysis of an Internet control plane created by integration of HIP and $i3$. Our workload model includes a conservative (w.h.p.) case of host requests, as well as the case of heavy load generated by a DDoS attack of zombies. The scalability analysis of the control plane outlines the trade-off in balancing the latency of the control infrastructure vs. decreasing the utilization of $i3$ servers.

Summarizing the results, the $Hi3$ control plane has the following scalability properties.

- The workload and the internal latency scale well: order $\frac{\log N}{N}$ of decreasing and order $\log N$ of increasing, respectively.
- A few $i3$ nodes ($N \sim 10^2$) is sufficient for a large set of HIP hosts ($M + L \sim 10^6 \ldots 10^7$).
- The resilience of $Hi3$ control plane to DDoS attacks behaves as $z(N) = O(\frac{N}{\log N})$, which is close to linear (proportional) behavior for large values of $i3$ size.
- For reasonable values of $i3$ size ($N \sim 10^2 \ldots 10^3$) the internal latency is satisfactory: at most a few seconds for rare requests such as association setup and simultaneous host mobility; for location update the internal latency does not depend on N and is at most a few hundred milliseconds (a substantial part of the total request latency is outside of $i3$).

The model is based on estimates that are high-probability bounds between typical values and the worst case. In practice, the trends of estimated parameters are more optimistic. For instance, when caching of Chord IDs is enabled in $i3$, the control plane is certainly more resilient to DDoS attacks.

At larger scales, performance evaluation using measurements or simulation can be prohibitively expensive or impossible. Even given its clear limitations, the simple model used in this chaper is a suitable tool to study a new large-scale networking architecture before its wide deployment. In future work, we plan to extend our first approximation analysis to include details of the $Hi3$ infrastructure such as caching, and perform more extensive calibration of a model using measurements.

References

1. Adkins, D., Lakshminarayanan, K., Perrig, A., Stoica, I.: Towards a more functional and secure network infrastructure. Tech. Rep. UCB/CSD-03-1242, University of California, Berkeley (2003)
2. Balakrishnan, H., Lakshminarayanan, K., Ratnasamy, S., Shenker, S., Stoica, I., Walfish, M.: A layered naming architecture for the internet. In: Proceedings of ACM SIGCOMM'04, pp. 343–352. ACM, New York (2004)
3. Clark, D., Braden, R., Falk, A., Pingali, V.: FARA: Reorganizing the addressing architecture. ACM SIGCOMM Comput. Commun. Rev. 33(4), 313–321 (2003)
4. Ford, B.: Unmanaged Internet Protocol: taming the edge network management crisis. ACM SIGCOMM Comput. Commun. Rev. 34(1), 93–98 (2004)

5. Francis, P.: IPNL: A NAT-extended internet architecture. In: Proceedings of ACM SIG-COMM'01. ACM, New York (2001)
6. Gurtov, A., Koponen, T.: Hi3 implementation for Linux (2005). Available at http://infrahip.hiit.fi and http://www.openhip.org Available at http://infrahip.hiit.fi/ and http://www.openhip.org/. Accessed Oct. 2012
7. Gurtov, A., Korzun, D., Nikander, P.: Hi3: An efficient and secure networking architecture for mobile hosts. Tech. Rep. TR-2005-2, HIIT (2005)
8. Henderson, T.R., Ahrenholz, J.M., Kim, J.H.: Experience with the Host Identity Protocol for secure host mobility and multihoming. In: Proceedings of the IEEE Wireless Communications and Networking Conference (WCNC'03), 3, pp. 2120–2125, IEEE (2003)
9. Johnson, D.B., Perkins, C., Arkko, J.: Mobility support in IPv6. RFC 3775, IETF (2004) http://tools.ietf.org/html/rfc3775
10. Jokela, P., Nikander, P., Melen, J., Ylitalo, J., Wall, J. Host Identity Protocol: Achieving IPv4 – IPv6 handovers without tunneling. Evolute workshop 2003, University of Surrey, Guildford, UK (2003)
11. Joseph, D., Kannan, J., Kubota, A., Lakshminarayanan, K., Stoica, I., Wehrle, K.: OCALA: an architecture for supporting legacy applications over overlays. In: NSDI'06: Proceedings of the 3rd Conference on 3rd Symposium on Networked Systems Design & Implementation (2006) USENIXAssociation,USA.http://usenix.org/events/nsdi06/tech/full_papers/joseph/joseph.pdf
12. Malkhi, D., Naor, M., Ratajczak, D.: Viceroy: a scalable and dynamic emulation of the butterfly. In: PODC '02: Proceedings of 21st Annual Symposium on Principles of Distributed Computing, pp. 183–192. ACM, New York (2002). doi: http://doi.acm.org/10.1145/571825. 571857
13. Moskowitz, R., Nikander, P.: Host identity protocol architecture: draft-ietf-hip-arch-02.txt (2005, work in progress, expired in August) http://www.ietf.org/rfc/rfc4423.txt
14. Moskowitz, R., Nikander, P., Jokela, P., Henderson, T.R.: Host Identity Protocol: draft-ietf-hip-base-10 (2007, work in progress, expired in May 2008) http://tools.ietf.org/html/rfc5201
15. Nikander, P., Ylitalo, J., Wall, J.: Integrating security, mobility, and multi-homing in a HIP way. In: Proceedings of Network and Distributed Systems Security Symposium (NDSS'03). Internet Society, San Diego (2003)
16. Nikander, P., Arkko, J., Ohlman, B.: Host identity indirection infrastructure (hi3). In: The Second Swedish National Computer Networking Workshop (2004)
17. Nikander, P., Arkko, J., Henderson, T.: End-host mobility and multi-homing with host identity protocol: draft-ietf-hip-mm-05 (2007, work in progress, expired in September 2007) http://tools.ietf.org/html/rfc5206
18. Stoica, I., Morris, R., Karger, D., Kaashoek, M.F., Balakrishnan, H.: Chord: a scalable peer-to-peer lookup service for internet applications. In: Proceedings of ACM SIGCOMM'01, pp. 149–160. ACM, New York (2001)
19. Stoica, I., Adkins, D., Zhuang, S., Shenker, S., Surana, S.: Internet indirection infrastructure. In: Proceedings of ACM nSIGCOMM'02, pp. 73–88. ACM, New York (2002)
20. Tschofenig, H., Shanmugam, M.: Traversing HIP-aware NATs and Firewalls: problem Statement and Requirements: draft-tschofenig-hiprg-hip-natfw-traversal-04 (2006) http://tools.ietf.org/html/draft-tschofenig-hiprg-hip-natfw-traversal-06

Chapter 13
Commercial Applications

Abstract This chapter describes several distributed databases for storing structured data in the Internet. They are known under a common name of NoSQL, to separate from traditional relational database management systems (RDBMSes). NoSQL databases are often built on top of classical DHT functionality with goals of high performance put and get operations for data using a key. Databases such as Cassandra power popular services including Facebook and therefore have to scale up to billions of users.

13.1 OpenDHT

OpenDHT [6] is a publicly available DHT service running on PlanetLab, a worldwide testbed of several hundred servers. Unlike other DHT systems, the user does not have to run a local DHT node to be able to access OpenDHT. There is no registration or accounts required to store and lookup data in OpenDHT; available storage is shared fairly among all users. Applications can access OpenDHT using Sun RPC and XML RPC interfaces. The TCP protocol is used to contact DHT nodes from a client that allows access from behind most NATs and firewalls.

OpenDHT is based on Bamboo DHT servers. While its code is publicly available to set up its own set of DHT servers, it is more convenient for most users to rely on a publicly maintained set of servers instead of its own set. Some advanced HIP extensions, such as proving HIT ownership to DHT, require modification to OpenDHT code to operate. Then, running a private set of modified OpenDHT servers is the only alternative until the proposed changes have been incorporated to the official OpenDHT release.

OpenDHT stores values using a key that can be up to 20 bytes long. A key is typically a fixed-length hash of an actual identifier, such as a DNS host name. Each value has a limited Time-to-Live of a maximum of 604,800 s, which is 1 week.

D. Korzun and A. Gurtov, *Structured Peer-to-Peer Systems: Fundamentals of Hierarchical Organization, Routing, Scaling, and Security*, DOI 10.1007/978-1-4614-5483-0_13, © Springer Science+Business Media New York 2013

Table 13.1 General content
of a put operation to
OpenDHT

Field	Type
Application	String
Client_library	String
Key	Max 20-byte array
Value	Max 1024-byte array
ttl_sec	Four-byte integer (max 604800)

Table 13.2 Content of a get
operation to OpenDHT

Field	Type
Application	String
Client_library	String
Key	Max 20-byte array
Maxvals	Four-byte signed integer (max $2^{31}-1$)
Placemark	Byte array (100 bytes max)

The basic interface provided by OpenDHT implements *put* and *get* operations. In our experiments we are using an XML-RPC interface. XML tags corresponding to field names are used to encapsulate data values in the interface messages.

Table 13.1 shows the content of all OpenDHT put operations. OpenDHT uses application and client_library fields for the logging of requests. The application is given as a string name, and the client_library refers to the particular name and version of the XML-RPC library used in the query. The key supplied by an application selects an OpenDHT server to store the value.

The OpenDHT server can reply to a put operation with three values. "Zero" refers to a successful put, "one" indicates failure because the capacity of the server is exceeded, and "two" indicates a temporal failure and a suggestion to re-try the put operation.

Table 13.2 shows the content of a get operation to OpenDHT server replies with values in an array and a placemark. The placemark can be used in a subsequent query to obtain additional values stored for the key.

To select the closest available server to the client, OpenDHT employs Overlay Anycast Service InfraStructure (OASIS) [3]. By performing a DNS lookup on the host name "opendht.nyuld.net", the OpenDHT client obtains an IP address of the server that has the lowest RTT to the client's host. Alternatively, the client can retrieve a list of available OpenDHT servers from a well-known location and itself select the server to use.

Since OpenDHT is a public open data repository, anyone can place data under any key using the simple put/get interface. This opens a way to drowning attacks, when an adversary places a large number of garbage values under the same key as the client being attacked. To prevent such attacks, OpenDHT supports two mechanisms: immutable puts and signed puts. *Immutable puts* are of the form $k = H(v)$ (the key is a hash of the stored value) and cannot be removed until their lifetime expires. Clearly, immutable puts are not a suitable mechanism to secure HIP name resolution as the key value is fixed for such operations.

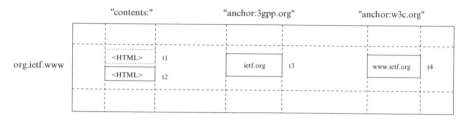

Fig. 13.1 The structure of BigTable

Signed puts include a value and nonce signed by the private key of the client storing the value. When another client retrieves a value stored under the given key, it also provides a hash of the public key of the client that stored the value. Therefore, that client would obtain only values stored with the private key matching the supplied public key. This prevents the client from drowning attacks. Signed puts are a suitable mechanism to secure HIT to IP address mappings for HIP hosts, as the HIT effectively is a hash of the public key of the client that stored the mapping.

13.2 Google's BigTable

BigTable [1] is a distributed storage system developed by Google for its applications such as Google Earth, web index, and Google Analytics, total over 60 products. The system spans thousands of servers capable of storing petabytes of data. BigTable provides high-performance storage for structured data for a wide range of applications with different object size and latency requirements.

Although similar to RDBMS, BigTable does not provide the relational data model. BigTable consists of rows and columns that are treated as uninterpreted strings that applications can store and retrieve using a network API. The data format is follows: $(row : string, column : string, time : int64) \dashrightarrow string$, where time is a timestamp. As an example, Webtable contains a web index with URL names as rows and various web page properties under contents: column and references to this web page under anchor: columns.

Generally row keys are up to 64 kB in size (although typical size is under 100 Bytes) and are stored in lexicographic order. Rows are partitioned to ranges that are stored in a single server and are efficient to access. The table can contain similar data collected at different time and identified by a timestamp. BigTable can be configured to garbage-collect old data columns automatically.

Figure 13.1 shows an example of table storing Web pages. Each row is identified by a reversed URL of the web page. The column labeled "contents" contains the actual HTML code of the web page. The "anchor" columns contains pages that reference the original web page (www.ietf.org). In this example, ietf.org is referenced by two other web pages, 3gpp.org and w3c.org, each having own timestamp ($t3$ and $t4$). The contents cell has two versions of the original web page, with timestamps $t1$ and $t2$.

BigTable design goals were different to those of traditional DHT such as Chord. In particular, variable bandwidth, untrusted nodes, decentralized control and node churn were left out of scope for the BigTable. The data storage model of DHTs is limited to key-value storage, there as BigTable provides richer model including multiple attributes (columns) for a single key. On the other hand, BigTable is simpler than RDBMS systems and does not provide complex queries but does allow the user to specify explicitly which records should be kept in memory or on the disk.

13.3   Amazon's Dynamo

E-commerce services such as provided by amazon.com require data key-value operations at large scales, including hundreds of thousands concurrent sessions. The operations must be fault-tolerant working even in the presence of server and network failures. Traditional relation databases (RDBMS) provide excessive functionality for most e-commerce web applications focusing on data consistency at the expense of speed, hardware and management cost and availability. Therefore, creating a simpler but highly scalable key-value storage service has been a requirement set by Amazon [2].

Dynamo provides a highly-available and scalable data storage service that can be distributed among data centers. It has been in operational use since 2007 and served millions of customers daily running over tens of thousands servers worldwide. Dynamo provides high availability of data somewhat sacrificing consistency in the presence of failures. Compared to complex query interface of relational databases, Dynamo provides simple key-value interface available to applications over the network. In this respect, Dynamo provides similar service as traditional Distributed Hash Tables (DHTs) but presents a next evolution step in scalability and fault tolerance. Each data object up to 1 MB is identified by a unique key and can be accessed by a query operation (no operations span over multiple objects).

The ACID (Atomicity, Consistency, Isolation, Durability) properties of traditional databases are not always adequate for most web e-commerce operations. Strict adherence to these properties affects availability which is not suitable for strict service-level agreements (for example, a response within 300 ms for 99.9 % of its requests up to 500 requests per second) with e-commerce providers. Therefore, less consistency might be acceptable to satisfy 99.9 percentile latency and throughput requirements with cost efficiency and availability. Dynamo provides these as an internal service in Amazon which is assumed to run in trusted environment, without the need to authenticate participating servers.

To achieve high availability, Dynamo permits data updates to proceed in background while permitting the access to potentially inconsistent data objects. Possible conflicts need to be detected and resolved between replicas using a voting procedure, so that eventually all replicas reach a consistent state. Write operations are never rejected, and conflict resolution is pushed towards read operations. The system can provide simple conflict resolution rules such as "last write wins" or alternatively the application can be made aware of replicas differences to perform more intelligent conflict resolutions, such as merging the differences.

Other key design goals of Dynamo include incremental scalability (adding one node at a time), symmetry (all nodes have equal functionality to facilitate system partitioning), decentralization (lack of centralized control and single point of failures), and heterogeneity (addressing nodes with different hardware capabilities).

Dynamo can be viewed as a zero-hop DHT where all keys are reachable from any node without request routing. In this respect, it is different from classic DHTs such as Chord there request traverse several hops before reaching the destination node. Multi-hop routing increases delay variability and therefore Dynamo maintains additional state at each node enabling to reach any data key directly from its responsible node.

The objects are identified by a 128-bit hash of the key. The Dynamo interface is very simple and involves only put() and get() operations. The get(key) operation retrieves the associated object to a key. The put(key, context, object) stores the object under the key with associated metadata in the context field, such as the object version. The partitioning of key space among nodes is similar to Chord: the hashed keys form a ring and each node is allocated a range of keys in the ring. Each node has a predecessor and an immediate successor, the ring can be walked in a clockwise directions. This basic mechanism is extended with a concept of virtual nodes: in fact each node is responsible for multiple ranges on the ring so that each physical node is responsible for several virtual nodes. It brings the advantages of fine-grain load distribution when nodes join or leave the system, as well as load assignment based on capabilities of physical node.

The data are replicated on N nodes including the coordinator node holding the primary replica and its $N - 1$ successors in the ring, as shown in Fig. 13.2. The preference list contains the list of nodes holding the replicas and ensures that at least N physical nodes (in opposite of virtual nodes) have the data replica. The vector clocks are associated to each object and include a list of (node, counter) pairs. Vector clocks are used to synchronized different object branches resulted from updated in portioned state. If the system detects a conflict that cannot be automatically resolved, it returns all replicas to the application for semantically-aware resolution.

To provide consistency, Dynamo uses a quorum-like system to resolve conflicts as shown in Fig. 13.3. R is defined as the minimum number of nodes responding to read request, and W—to the write request. A typical setting is $R + W > N$ to provide quorum, but R and W less than N to reduce the latency. The coordinator node receives the client request and distribute queries to other $N - 1$ nodes combining the results if necessary. A typical value of $N = 3$ for production systems, whereas W can be set to one to guarantee that writes are never rejected if there is at least one active node.

To handle failures, a mechanism of hinted handoff to temporarily re-allocate data from a node to the successor of $N - 1$ nodes holding the data replicas. The nodes in replica set are typically configured from different data centers that ensures data

Fig. 13.2 Key replication in
Dynamo

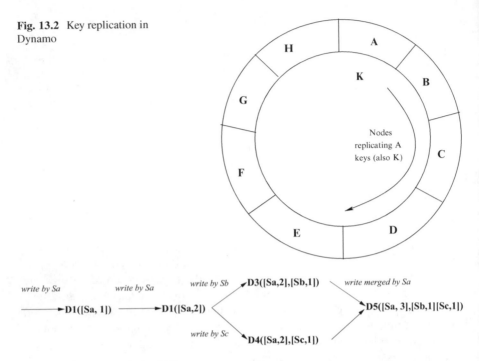

Fig. 13.3 Conflict resolution in Dynamo

availability even in the presence of power outages or network failures. To handle
more permanent failures, replica synchronization mechanism involving Merkle tree
(a tree of hashes) is used. It minimizes the amount of data that needs transferring
between the data centers as it enables quick localization of conflicts.

The node membership is controlled manually by an administrator that can insert
and delete nodes from Dynamo. After joining, the node employs a gossip protocol to
acquire itself a key range from the ring, synchronize its data replicas and learn about
other peers' key ranges. That enable eventually global knowledge about the system
and allows nodes to route requests directly to the node responsible for a given key.

13.4 Facebook's Cassandra

Facebook's Cassandra [4, 5] is a distributed storage system that runs on hundreds of
servers located in multiple data centers. It is highly available service that can tolerate
server and network failures. Cassandra does not provide a full relational data model
of RDBMS, instead focusing on storing large amounts of structured data with high
read and write throughput (Fig. 13.4).

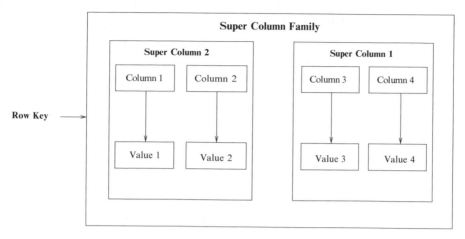

Fig. 13.4 The supertable structure in Cassandra

Cassandra powers the Facebook Inbox search and needs to scale up continuously to account for the growth of the system and number of users. To deal with failures, data needs to be replicated across servers in data centers, yet without sacrificing performance for billions of updates per day. The developers described Cassandra as "a BigTable data model running on an Amazon Dynamo-like infrastructure."

Data is stored in tables, which is a distributed multi-dimensional map indexed by a key. Columns can be organized to Column Families; nested Super columns including another Column Family are supported. The row key is a string of typically 16–36 bytes. The columns are sorted either by time or name.

The API to access a table is quite straightforward:

- Insert(table; key; rowMutation)
- Get(table; key; columnName)
- Delete(table; key; columnName)

where a columnName can be of the form column family:column.

The data values are organized among nodes in a circular fashion, similar to Chord DHT. Each node is responsible for a range of hashed key values. The ring can be traversed in clockwise direction; a node departure affects only its immediate neighbors. For load balancing, a lightly loaded nodes can move on the ring. For reliability, data is replicated at N nodes in one of following ways: Rack Unaware, Rack Aware, and Datacenter Aware. In the first case, data is replicated on $N - 1$ successors on the ring. For other cases, the replication algorithm take into account the physical location of the nodes in racks and datacenters.

Differently from Chord, in Cassandra each node knows about all other nodes and ranges of data they are responsible for. That allows for fast request routing and durability in the presence of node failures.

Fig. 13.5 Consistent hashing
in Voldemort

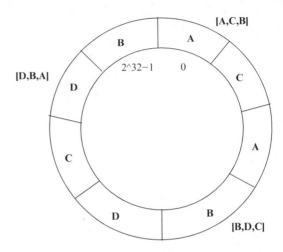

13.5   LinkedIn Voldemort

Project Voldemort is a distributed key-value storage system used by LinkedIn for
high-capacity storage. The data is stored under a key and partition and replicated
among multiple servers. In case of failures, conflicts can be resolved using
versioning of data. Similar to Dynamo, Voldemort is not a relational database. In
summary, developers describe it as a "big, distributed, persistent, fault-tolerant hash
table."

The API of Voldemort is simple and involves only three operations:

- value = store.get(key)
- store.put(key, value)
- store.delete(key)

where keys and values could be a complex object including lists.

Voldemort uses a technique of consistent hashing to distribute the load evenly
among servers. Figure 13.5 illustrates the concept. In case of a server failure, its
keys will be distributed equally to other servers in the cluster. To accomplish this, a
ring node structure of size Q partitions, which is larger than the number of nodes, S.
Each node is assigned Q/S partitions, so that it appears multiple times in a ring. The
keys are held by the primary node as well as its k unique successors in a clockwise
direction.

The simple API of Voldemort has an advantage of predictable performance
comparing to SQL queries of relational databases. The get operation returns all
values associated with a key organized into lists. The data value can be of following
types: int8, int16, int32, int64, float32, float64, string, date, object, bytes, Boolean,
object, array.

Voldemort is available as open-source under Apache license.[1]

---

[1] http://project-voldemort.com/design.php.

13.6   Amazon's S3 and SimpleDB

Amazon's S3 (Simple Storage Service) is a proprietary online storage system offered by Amazon Web Services using public API of Representational State Transfer (REST), Simple Object Access Protocol (SOAP) and BitTorrent. S3 stores over 500 billion objects and is used by popular services such as DropBox to store data. It provides 99.99 % availability of data and 99.999999999 % durability in the presence of network and hardware failures. A single stored object can reach 5 TB and contain up to 2 KB of metadata.

While the design details of S3 are not public, it shares the basic principles with Dynamo. The goals of S3 are somewhat different from Dynamo. S3 focuses on providing simple key-value storage for large data objects, without conflict resolution visible to the user. On the opposite, Dynamo is targeted for highly dynamic services such as web shopping cart management, with plenty of small writes with user-controlled conflict resolution. S3 posses web-service driven interface with authentication and accounting, but conflict resolution is handled internally.

The objects stored in S3 are accessible by normal web clients using the URL scheme such as http://s3.amazonaws.com/bucket/key. Therefore, S3 can host whole static web sites. S3 servers can also act as BitTorrent seeds providing the users opportunity to download the objects using normal BitTorrent clients, that can reduce the bandwidth load (and hence the user bill) on S3 for popular objects. Security is ensured using Access Control Lists for buckets and objects.

SimpleDB is another distributed database from Amazon that provides eventual consistency, a weaker guarantee than RDBMS provide. However, it is highly available and partition-tolerant, operating in the presence of network disconnections. Compared to S3, SimpleDB objects are stored in fast memory while S3 objects are in slow storage. Therefore, having the same amount of data in SimpleDB is more expensive than in S3. S3 has a limited amount of metadata stored with each objects, and in order to find an object based on metadata, the user has to retrieve all metadata from each object in a bucket which can be slow and expensive. SimpleDB provides indexing of key-value pairs for fast querying, and allows updating the metadata for stored objects. In S3, changing the metadata requires deleting and recreating the object.

For applications requiring large objects with rich metadata, a hybrid approach combining S3 and SimpleDB can be used. The large objects are stored in cheaper S3 while the metadata for the object is stored in more expensive but faster SimpleDB together with a pointer to S3 object data.

13.7   BitTorrent

BitTorrent [7] is a peer-to-peer protocols for distributing files in the Internet. It is used by individual users for file sharing, as well as companies including Blizzard forsoftware distribution and updates, and even by Amazon S3 as a content

'announce': 'http://bttracker.debian.org:6969/announce',
'info': 'name': 'Win7.iso',
'piece length': 262144,
'length': 678301696,
'pieces':   '841ae846bc5b6d7bd6e9aa3dd9e551559c82abc1   ...
d14f1631d776008f83772ee170c42411618190a4'

Fig. 13.6 Torrent file for a Debian Linux distribution CD

distribution protocol. By some estimates, BitTorrent traffic accounts for 40–70 %
of all present Internet traffic. The BitTorrent protocol was designed in 2001 by
Bram Cohen.

A set of users distributing a single file is called a swarm. Users having the full
version of the file act as seeders, providing the file to others for download. Users
having only a part of the file are called leechers, they attempt to download the
missing file pieces while providing to other users the pieces they have already. The
pieces are downloaded in random order or rarest-first approach which accelerates
download of a complete file, but complicated the streaming view of media files. The
description of the file including its name, checksum, length and other metadata is
stored in a torrent file which is distributed using web sites.

Figure 13.6 shows an example of .torrent file containing a Linux distribution
CD. The file is separated to pieces of length of about 260 KB (piece size can vary
from 32 KB up to 16 MB). The total length of the file is 647 MB. The list of SHA-1
hashed for all pieces are given in the end of the torrent file.

A server that maintains a list of BitTorrent clients downloading the same file is
called a tracker. The use of tracker accelerates downloads as users can find peers
more easily. In addition, the tracker can maintain a degree of fairness, preventing
free-riding, i.e. just downloading files without providing uploading to other peers.
This is accomplished by monitoring the download/upload ratio and banning the
peers with ratio significantly lower than one.

Since trackers present a single point of failure and can be attacked to disable the
system. A trackerless approach in BitTorrent was introduced that relies on a DHT to
find peers. Two incompatible DHTs are used in the BitTorrent community: the DHT
first implemented by Azureus client and the Mainline DHT used by BitComet and
other clients. Both DHTs are based on Kademlia.

BitTorrent clients can implement various approaches in selecting the peers
and selecting the pieces for uploading. One common strategy is to use tit-for-
tat mechanism, rewarding those peers that have recently uploaded a piece to the
client. However, this approach makes it difficult for new peers to join the swarm, as
those do not yet many pieces to share. To overcome this problem, clients perform
optimistic unchoking, uploading pieces to new peers in hoping to be rewarded in
return. Such approach can be exploited by strategic clients in order to increase their
download speed without providing uploads to other peers. Many BitTorrent clients
employ peer exchange (PEX) to learn about other peers in the swarm.

Recent torrent clients also implement the technique of web seeding. It allows to
download pieces from a web server using HTTP protocol. This way, the web server
can act as a permanent seeder while reducing own traffic load thanks to peer-to-

peer interactions among clients. The web server can decide to stop seeding when the swarm becomes popular enough and there is sufficient number of other seeds. Finally, latest Burnbit software even allows creating a torrent from arbitrary URL using web seeding.

13.8   Summary

In this chapter, we introduced several distributed storage systems utilizing P2P concepts. OpenDHT was an attempt to setup and maintain a public DHT store and lookup service. Although its maintenance stopped, it has served as a one of the first large real-world DHT deployments used by real Internet applications. Its interface is well defined and can serve also in other systems, for example for resolving host identifiers to locators.

Other class of systems is represented by industrial grade systems including Google BigTable, Amazon Dynamo and S3, Facebook's Cassandra, LinkedIn's Voldemort. These are operated in many servers across different datacenters, and therefore are not P2P applications in the traditional sense. However, many of them are using the organization structure developed for classical DHTs. These systems aim for high performance in availability under expense of consistency, which make them different from traditional relational database systems.

Finally, we presented BitTorrent, one of most popular P2P applications on Internet. Its traffic has a major impact on operator's networks. From the start BitTorrent used centralized tracker to connect file sharing peers together, later also supporting decentralized DHT-based model. Nowadays, BitTorrent is used by mainstream software companies to distribute updates to their products.

Reference

1. Chang, F., Dean, J., Ghemawat, S., Hsieh, W.C., Wallach, D.A., Burrows, M., Chandra, T., Fikes, A., Gruber, R.E.: Bigtable: A distributed storage system for structured data. ACM Trans. Comput. Syst. 26, 4:1–4:26 (2008)
2. DeCandia, G., Hastorun, D., Jampani, M., Kakulapati, G., Lakshman, A., Pilchin, A., Sivasubramanian, S., Vosshall, P., Vogels, W.: Dynamo: Amazon's highly available key-value store. In: Proceedings of twenty-first ACM SIGOPS symposium on Operating systems principles, SOSP '07, pp. 205–220. ACM, New York (2007)
3. Freedman, M.J., Laskhminarayanan, K., Mazières, D.: OASIS: Anycast for any service. In: Proceedings of the 3rd Symposium on Networked Systems Design and Implementation, San Jose, CA (2006)
4. Hewitt, E.: Cassandra: The Definitive Guide. O'Reilly Media, USA (2010)
5. Lakshman, A., Malik, P.: Cassandra: structured storage system on a P2P network. In: Proceedings of the 28th ACM symposium on Principles of distributed computing, PODC '09. ACM, New York (2009)

6. Rhea, S., Godfrey, B., Karp, B., Kubiatowicz, J., Ratnasamy, S., Shenker, S., Stoica, I., Yu, H.: OpenDHT: A public DHT service and its uses. In: Proceedings of ACM SIGCOMM'05. ACM, New York (2005)
7. Sherman, A., Nieh, J., Stein, C.: FairTorrent: bringing fairness to peer-to-peer systems. In: CoNEXT '09: Proceedings of the 5th International Conference on Emerging Networking Experiments and Technologies, pp. 133–144. ACM, Boston (2009). doi: http://doi.acm.org/10.1145/1658939.1658955

Summary of Part IV

In this part we covered several architectures utilizing P2P concepts, as well as commercial systems using P2P algorithms. Secure lookup routing and privacy-preserving host rendezvous services are examples of advanced P2P architectures. Identifier-locator split appears as one of new fundamental trends of Future Internet, where P2P name resolution service will play a critical role. We also covered several popular content distribution and commercial key-value storage services that are of decentralized nature and use DHT algorithms internally.

Naturally, in a few chapters it is difficult to describe all versatile P2P applications in use in the Internet. For example, also such popular voice and video communication applications as Skype utilize P2P approach. Services such as Joost help users to enjoy streaming video and IP TV using P2P approach. A music listening service Spotify is also based on P2P. With ossification of Internet architecture on the IP layer, new overlay P2P applications play increasingly important role in implementing services such as peercasting for multicast delivery. Online gaming and virtual worlds is another example where players can benefit from lower latency of direct user-to-user communication.

Advanced P2P concepts such as cooperative web cashing and search, distributed scientific computation, online academic teaching all promise a bright future for P2P architectures, backup up by multi-million R & D investments in these area. In the presence of oppressive governments, citizens will have to increasingly rely on P2P technologies such as Tor to circumvent surveillance and censorship. For example, a web of trust for Pretty Good Privacy was established for P2P user email key certification and verification. In future, it is possible that centralized social networks such as Facebook are replaced by open-source, decentralized P2P services that give the users control over who and when can access their social data.

Index

D. Korzun and A. Gurtov, *Structured Peer-to-Peer Systems: Fundamentals of Hierarchical*   359
Organization, Routing, Scaling, and Security, DOI 10.1007/978-1-4614-5483-0,
© Springer Science+Business Media New York 2013